PROCEEDINGS OF SYMPOSIA
IN PURE MATHEMATICS
Volume XIX

COMBINATORICS

AMERICAN MATHEMATICAL SOCIETY
Providence, Rhode Island
1971

Proceedings of the Symposium in Pure Mathematics
of the American Mathematical Society

Held at the University of California
Los Angeles, California
March 21–22, 1968

Prepared by the American Mathematical Society
under National Science Foundation Grant GP-8436

Edited by
Theodore S. Motzkin

AMS 1970 *Subject Classifications*
Primary 05Axx, 05Bxx, 05Cxx, 10-XX, 15-XX, 50-XX
Secondary 04A20, 05A05, 05A17, 05A20, 05B05, 05B15, 05B20, 05B25, 05B30, 05C15, 05C99, 06A05, 10A45, 10C05, 14-XX, 20Bxx, 20Fxx, 50A20, 55C05, 55J05, 94A20

International Standard Book Number 0-8218-1419-2
Library of Congress Catalog Number 74-153879

Printed in the United States of America

Leo Moser (1921–1970) was active and productive in various aspects of combinatorics and of its applications to number theory. He was in close contact with those with whom he had common interests: we will remember his sparkling wit, the universality of his anecdotes, and his stimulating presence. This volume, much of whose content he had enjoyed and appreciated, and which contains the reconstruction of a contribution by him, is dedicated to his memory.

CONTENTS

Preface

Combinatorics is the theory of finite sets. This is a wide, amorphous, primordial subject matter which in principle includes subareas where more structure and more specific structure is assumed; but whether historically they have grown before or next to combinatorics, or were part of a more general theory encompassing similar infinite structures, or were indeed originally part of combinatorics, many such areas are now considered as separate fields. Number theory and the theories of partitions and finite groups are examples of the first; the theories of finite fields and finite geometries of the second; graph theory and the theory of designs are on their way to be examples of the third kind.

Because of the recent symposia on graph theory, and the existing collections on applied combinatorics (combinatorial geometry, probability theory and extra-mathematical applications), it was decided to emphasize at the present symposium the theory of simple general or homogeneous structures. Of the twenty-four talks, eight treat general structures, nine treat designs (homogeneous structures), six treat applications of the first two topics to sets of integers, algebra and complex analysis, and one is a survey article mainly on general structures and partly on sets of integers. (Asymptotic results occur in seven of the thirteen papers on general structures or applications thereof; computers were used in three papers.) Thus the scope (structures, systems, applications) is close to that of Series A (as of 1971) of the Journal of Combinatorial Theory (Series B will be on graph theory).

Specifically, the first class (general structures) includes the papers of Kleitman, Motzkin, Swift on families (sets of sets), of Crapo/Rota and Klee on simplicial geometries and pregeometries (matroids) or pregeometries, of Tutte on planar graphs, Rothschild on Ramsey theorems for graphs, and Rado on transfinite Ramsey theorems.

The second class (homogeneous structures) includes the articles of Hurwitz on 0-1 matrices, Parker on latin squares, Ray-Chaudhuri/Wilson on Kirkman designs, Hall on designs and groups, Mendelsohn on graphs, semigroups, and groups, Hales on trees and Abelian groups, Bose and Hanani on designs related to finite geometries, and Graham/Rothschild on Ramsey theorems in finite geometry.

Investigations close to the first class are found in the papers of Gordon and Cheema/Motzkin on partitions, Moser and Straus on sets of integers; to the second class in the articles of Taussky on algebraic identities connected with Hadamard matrices, and Atkin/Swinnerton-Dyer on modular forms for discrete groups.

Erdös' survey article deals in sections I, III, IV with families, and in sections II, V, VI with applications including geometry and sets of integers.

I hope that this volume, with its numerous and varied open questions and new methods and results, extending from the solution of a century-old problem on designs to algebro-geometric and number and function theoretic studies, will adequately reflect work done and in progress and contribute to growth and change in combinatorics.

Acknowledgments

On behalf of the contributors and participants, the editor wishes to express gratitude and appreciation to the American Mathematical Society for constant and multifaceted cooperation, to the National Science Foundation for financial support, to the University of California, Los Angeles, for the use of its facilities and to William Clowes & Sons, Ltd. for the excellent transformation into book form.

Theodore S. Motzkin

MODULAR FORMS ON NONCONGRUENCE SUBGROUPS

A. O. L. ATKIN AND H. P. F. SWINNERTON-DYER

1.1. **Introduction.** This paper is an interim account of work still in progress. It started with an intention to discover something about modular forms on non-congruence subgroups of the classical modular group. We were successively led into function theory, group theory, computing, numerical analysis, and combinatorics. This last gives rise to our rather slender justification for including the paper in this volume, though we should say that at the time when one of us was asked to speak at the Symposium, the combinatorial aspect seemed relatively greater than it does now.

The main outcome of our work is the discovery of objects associated with forms on modular subgroups which, in the congruence case, are just the eigenvalues of the Hecke operators T_p. By "discovery" we imply total confidence in their existence, without at present any foreseeable prospect of proof. We are not yet even in a position to make formal conjectures, since there remain some marginal uncertainties as to what properties these objects enjoy. We hope however that the reader who gets as far as §5 will share our certainty of their existence.

Apart from the interest of this discovery, we think that there is some general interest in the techniques by which we arrived at it. From the beginning, we approached the problem with the possibility of using computers in mind. In the event, much of what we have done would have been impossible without their aid, and even those who deny the validity of a proof requiring computed facts cannot cavil at conjectures arrived at by inspection of them. We consider that there must remain large areas of mathematics, and of number theory in particular, where computer-aided investigations should produce significant results. For this reason, we have attempted to make this account generally intelligible, in some places preferring simplicity and explicit example to stating the most general theorem possible.

We thank the Directors of the Atlas Computer Laboratory and the Cambridge University Mathematical Laboratory for making the necessary machine time available, and Mr. Stephen Muir of the Atlas Computer Laboratory for his assistance with some of the programming.

1.2. *The modular group.* The classical modular group Γ consists of all linear fractional transformations $V\tau=(a\tau+b)/(c\tau+d)$, where a, b, c, and d are rational integers and $ad-bc=1$. It is generated by the transformations S and T where

$$S\tau = \tau+1, \qquad T\tau = -1/\tau,$$

and is the free product of the cyclic groups $\{T\}$ and $\{P\}$ of orders 2 and 3 respectively, where

$$P = TS, \qquad P\tau = -1/(\tau+1), \qquad T^2 = P^3 = I.$$

It is a discontinuous group acting on H, the upper half-plane $\operatorname{Im} \tau>0$, and has a fundamental domain F given by $|\tau|>1$, $-\frac{1}{2}\leq \operatorname{Re} \tau<\frac{1}{2}$ and $|\tau|=1$, $-\frac{1}{2}\leq \operatorname{Re} \tau\leq 0$ shown (without the boundary details) as region 1 in Figure 1. The only fixed points in F of elliptic transformations of Γ are $\tau=i$ (fixed by T) and $\tau=\rho=e^{2\pi i/3}$ (fixed by P and P^{-1}); the local uniformizing variables are $(\tau-i)^2$ and $(\tau-\rho)^3$ respectively. In addition ∞ is fixed by the parabolic transformation S, and the local variable is $x=e^{2\pi i\tau}$. The vertical sides of the boundary of F are mapped into each other by S and S^{-1}, and the curved side into itself by T.

We take the Hauptmodul $j=j(\tau)$ of Γ as $j=x^{-1}+744+196884x+\cdots$ $(x=e^{2\pi i\tau})$. Then j has a simple pole at ∞, $j=0$ at ρ, and $j=1728$ at i.

1.3. *Subgroups of the modular group.* Suppose that G is a subgroup of Γ of finite index μ. Then G has a connected fundamental domain D consisting of μ copies of F, the transforms of F by a set of coset representatives for G in Γ. If the elements of G conjugate in Γ to T and P form respectively e_2 and e_3 conjugacy classes in G, then the boundary of F will have e_2 and e_3 inequivalent fixed point vertices of orders 2 and 3 respectively. Suppose further that every element of G conjugate in Γ to a (nonzero) power of S is conjugate in G to some power of one of

$$S^{\mu_1}, g_2S^{\mu_2}g_2^{-1}, \ldots, g_hS^{\mu_h}g_h^{-1}, \qquad \text{(in } G\text{)}$$

where $g_1=I, g_2, \ldots, g_h$ are in Γ, and no $g_ig_j^{-1}$ is in G. Then the boundary of F will have h inequivalent parabolic fixed point vertices (called *cusps*) at ∞ and at $g_2\infty, \ldots, g_h\infty$ which are rational points on the real axis. We then have

$$\mu = \mu_1+\mu_2+\cdots+\mu_h$$

and

$$g = 1+\mu/12-h/2-e_2/4-e_3/3,$$

where g is the genus of the Riemann surface H/G. Following Wohlfahrt [**10**] we define the *level* l of G to be the least common multiple of $\mu_1, \mu_2, \ldots, \mu_h$. We also introduce here the notations (G) for the set of conjugates of G in Γ, and $(G)^\infty$ for the set of μ_1 conjugates of G by S^i $(i=0$ to $\mu_1-1)$, which we call "∞-conjugates". Finally we write G^N for the intersection of the conjugates of G in Γ, so that G^N is the maximal normal subgroup of Γ contained in G.

1.4. *Congruence subgroups.* Let $\bar{\Gamma}$ be the group of matrices $\left(\begin{smallmatrix} a & b \\ c & d \end{smallmatrix}\right)$ with a, b, c, and d rational integers and $ad-bc=1$. Then $\Gamma\cong\bar{\Gamma}/\{\pm I\}$ and any subgroup G of Γ induces a subgroup $\bar{G}$ of $\bar{\Gamma}$ and conversely. If $\bar{G}$ consists of all matrices of $\bar{\Gamma}$ which are congruent to a fixed set M of matrices modulo l, and l is the least integer for which this is true, we say that G is a congruence subgroup of level l. It can be shown that, in this case, l is also the level as defined in §1.3. The subgroup $\Gamma(l)$ for which

M consists of $\pm I$ modulo l is called the principal congruence subgroup of level l, and is normal in Γ; a subgroup of Γ containing $\Gamma(l)$ has level l or a divisor of l.

The classical development of modular theory, and its relation to number theory, rests very largely on deriving modular forms from theta-functions, and on the definition of the Hecke operators. Both these aspects lead naturally to congruence subgroups. However such subgroups are not typical, and it can for instance be shown quite easily that if $C(\mu)$ and $N(\mu)$ are the numbers of congruence and noncongruence subgroups with index $\leq \mu$, then $C(\mu)/N(\mu) \to 0$ as $\mu \to \infty$.

1.5. *The transformation* $\tau \to -\bar{\tau}$. If we write $W\tau = -\bar{\tau}$, then Γ and W generate a group Γ^+ in which Γ has index 2. We have

$$W^2 = I, \qquad WTW = T, \qquad WPW = TP^{-1}T, \qquad WSW = S^{-1}.$$

Thus P and P^{-1} are conjugate in Γ^+, though not in Γ. If G is a subgroup of Γ consisting of transformations $V\tau = (a\tau + b)/(c\tau + d)$ then WGW consists of transformations $WVW\tau = (a\tau - b)/(-c\tau + d)$. The set of conjugates (WGW) may or may not be the same as (G); even if it is there may be no conjugate G^* of G in Γ such that $WG^*W = G^*$. Finally WG^NW may differ from G^N, though they have the same factor group Γ/G^N. We make these negative statements to avoid possible improper assumptions derived from familiarity with well-known congruence subgroups.

It can be seen that Γ^+ is equally defined by adjoining WT to Γ, and that conjugating Γ by WT corresponds to the more simple outer automorphism $T \to T$, $P \to P^{-1}$. However W is more natural for function-theoretic purposes (see §2.4).

2.1. **Modular forms.** Suppose that G is a subgroup of finite index in Γ. We say that $f(\tau)$ is a *modular form* on G if

(i) $f(\tau)$ is *analytic*[1] *in* H *and at the cusps of the fundamental domain of* G,

(ii) $f(V\tau) = (c\tau + d)^{2w} f(\tau)$ *for all* V *in* G, *where* $V\tau = (a\tau + b)/(c\tau + d)$ *and* w *is a fixed integer.*

We are for simplicity ignoring possible multiplier systems. We call w the *weight* of the form (so that $-2w$ is the "dimension", but we wish to avoid confusion with the dimensions of vector spaces). If $w = 0$ we have a modular function on G, while $w = 1$ corresponds to a differential. We also say that $f(\tau)$ is *entire* if it has no poles inside H, that $f(\tau)$ is *regular* if it is entire and also regular at the cusps of G, and that $f(\tau)$ is a *cuspform* if it is entire and zero at the cusps of G. Nontrivial regular forms and cuspforms exist only for $w > 0$. In particular the cuspforms for a given G and w form a finite-dimensional vector space which we denote by $(G, w)_0$. This can be made into a metric space by defining a suitable scalar product, but lacking anything analogous to the Hecke operators we have not made any use of this fact for noncongruence subgroups.

2.2. The function-theoretic basis of our technique is Theorem 1 below. While this is implicit in Klein-Fricke [**4**, especially pp. 611 et ff.], the formal details seem hard to disentangle, especially with regard to the sufficiency, so that we give a proof.

[1] We use "analytic" throughout the paper to mean single-valued and regular except for poles.

THEOREM 1. *A necessary and sufficient condition that $f(\tau)$ be a modular function on a subgroup of finite index in Γ is that $f(\tau)$ should be an algebraic function of $j(\tau)$ and that its only branch points should be branch points of order 2 at which $j=1728$, branch points of order 3 at which $j=0$, and branch points at which j is infinite.*

We first prove the necessity. Suppose that $f(\tau)$ is a modular function on a subgroup G of index μ in Γ, and that G is the largest subgroup on which $f(\tau)$ is invariant. The functions $f(\tau)$ and $j(\tau)$ induce on the Riemann surface H/G analytic functions, also denoted by f and j. Since H/G is compact and j has valence μ on it, there is an irreducible algebraic equation between f and j which is of degree precisely μ in j. Thus f is a μ-valued algebraic function of j.

Now let $Q\colon (f,j)=(f_0,j_0)$ be any branch point of f as a function of j at which the value j_0 of j is finite. Without loss of generality we may assume that the value f_0 of f is finite. Let $m>1$ be the order of Q as a branch point, so that $(j-j_0)^{1/m}$ is a local uniformizing variable at Q. But $j-j_0=c_0(\tau-\tau_0)^\theta+O((\tau-\tau_0)^{\theta+1})$ where $\theta=2$ when $j_0=1728$, $\theta=3$ when $j_0=0$, and $\theta=1$ otherwise; and f and j are single-valued functions of τ. This completes the proof of necessity.

For sufficiency, the algebraic conditions on $f(\tau)$ in terms of $j(\tau)$ are enough to prove that $f(\tau)$ is analytic in H. Let R be the Riemann surface for f as a function of j. Then there are natural analytic maps

$$H \xrightarrow{\alpha} R \xrightarrow{\beta} H/\Gamma,$$

and the total map $H \xrightarrow{\beta\alpha} H/\Gamma$ is locally conformal except where $j=0$ or $j=1728$, so that $H \xrightarrow{\alpha} R$ is locally conformal except perhaps at some of these points, and similarly for $R \xrightarrow{\beta} H/\Gamma$. Now let Q be any point on H at which $j\neq 0$, 1728. Then Q is fixed only by the identity in Γ. The point set $\alpha^{-1}(\alpha Q)$ is a subset of $(\beta\alpha)^{-1}(\beta\alpha Q)$ which is just the set of points equivalent to Q under Γ, so that we may define G as the subset of Γ consisting of those elements of Γ which map Q to a point of $\alpha^{-1}(\alpha Q)$. Note that $f(\tau)$ cannot be invariant by an element of $\Gamma-G$, since such an element changes the value of $f(\tau)$ at Q. We wish to prove that $f(\tau)$ is invariant under G, and that G is a subgroup of finite index in Γ.

The set G depends on Q. Let $N=N(Q)$ be a neighborhood of Q in H so small that the map $\beta\alpha\colon N\to\beta\alpha N$ is conformal. It follows that each connected component of $(\beta\alpha)^{-1}(\beta\alpha N)$ is conformal with $\beta\alpha N$ and hence with N; indeed this merely restates the fact that Γ is discontinuous. But clearly $\beta^{-1}(\beta\alpha N)$ has finitely many connected components each conformal with $\beta\alpha N$, and one of these is just αN. Thus each connected component of $\alpha^{-1}(\alpha N)$ is conformal with N, and the set of these components is just the set of transforms of N by G. It follows that $G(P)=G(Q)$ for all P in N. Now let H^* be the subset of H obtained by deleting those points at which $j=0$ or 1728. We have just shown that the set of points P for which $G(P)$ is any assigned $G_0=G(Q)$ is open; and by considering all G_0 this enables us to write H^* as a disjoint union of open sets. Since H^* is connected, all but one of these must be null; thus G does not depend on the choice of Q. Using continuity to deal with the points where $j=0$ or 1728, we deduce that G is precisely the set of those elements of the modular group which leave $f(\tau)$ invariant; thus in particular G must be a group. Considering it as a subgroup of Γ it must be of finite index; for any element of Γ takes $f(\tau)$ into one of its finitely many conjugates as an algebraic function of $j(\tau)$, and elements of Γ which have the same effect on $f(\tau)$ must lie in the same left coset with respect to G. This completes the proof of the theorem.

2.3. *Subgroups of genus zero.* We now proceed to apply Theorem 1 in some detail to the case when the subgroup G is such that H/G has genus zero. We have in fact considered cases where the genus is 1 or 2, and the reader will see that the principles can be applied quite generally. However the possibilities of Weierstrass points, and the details of specifying the values of a function of high valence on G, make it difficult to state a general theorem of any practical use.

Suppose that we have (for general g) a subgroup G with

$$\mu = \mu_1+\mu_2+\cdots+\mu_h,$$
$$g = 1+\mu/12-h/2-e_2/4-e_3/3,$$

as in §1.3. If $h>1$, let $\mu_2,\ldots,\mu_h$ be in some order ν_1 taken α_1 times, ..., ν_s taken α_s times, so that $\nu_1>\nu_2>\cdots>\nu_s$, $s\geq 1$, $\alpha_i\geq 1$ $(i=1$ to $s)$ and

$$\mu-\mu_1 = \alpha_1\nu_1+\alpha_2\nu_2+\cdots+\alpha_s\nu_s,$$
$$h-1 = \alpha_1+\alpha_2+\cdots+\alpha_s.$$

By a *specification* we shall understand a set of integers $g\geq 0$, $\mu\geq 1$, $h\geq 1$, $e_2\geq 0$, $e_3\geq 0$, $\mu_1\geq 1$, and, if $h>1$, a set (α_i,ν_i) with $\alpha_i\geq 1$, $\nu_1>\nu_2>\cdots>\nu_s$, for $i=1$ to s, satisfying the above equations.

Reverting now to $g=0$, let a specification be given, and let G be a subgroup of Γ with this specification. The local uniformizing variable at the cusp ∞ of G is $\xi=e^{2\pi i\tau/\mu_1}$, and a Hauptmodul $\zeta=\zeta(\tau)$ of G is uniquely determined by the condition that the expansion of ζ at ∞ is

$$\zeta = \xi^{-1}+0+O(\xi)$$

and that ζ is elsewhere regular in the fundamental domain D of G. Writing $2f_2=\mu-e_2$, $3f_3=\mu-e_3$ (with integral f_2, f_3), we define the following polynomials with undetermined coefficients; to avoid an otherwise intolerable notation we use c to denote each of these coefficients, which are of course in general distinct.

$$A_i = \zeta^{\alpha_i}+c\zeta^{\alpha_i-1}+\cdots+c \qquad (i = 1 \text{ to } s),$$
$$F_3 = \zeta^{f_3}+c\zeta^{f_3-1}+\cdots+c,$$
$$E_3 = \zeta^{e_3}+c\zeta^{e_3-1}+\cdots+c,$$
$$F_2 = \zeta^{f_2}+c\zeta^{f_2-1}+\cdots+c,$$
$$E_2 = \zeta^{e_2}+c\zeta^{e_2-1}+\cdots+c.$$

If any of e_2, f_2, e_3, or f_3 is zero, then the corresponding polynomial is taken as unity, and no undetermined coefficients arise. We now define "the j-equations" of the specification as

$$j\cdot A_1^{\nu_1}\cdot A_2^{\nu_2}\cdots A_s^{\nu_s} = F_3^3\cdot E_3,$$
$$(j-1728)\cdot A_1^{\nu_1}\cdot A_2^{\nu_2}\cdots A_s^{\nu_s} = F_2^2\cdot E_2,$$

together with the condition that $A_1 \cdot A_2 \cdots A_s \cdot F_3 \cdot E_3 \cdot F_2 \cdot E_2$ has no repeated zero as a polynomial in ζ. Eliminating j and equating the coefficients of ζ^i ($i=0$ to $\mu-1$) leads to μ simultaneous nonlinear equations between the $\sum \alpha_i+e_3+f_3+e_2+f_2=\mu+1$ unknowns c; there is also a linear relation between the c derivable from the zero constant term in the expansion of ζ of ∞. These $\mu+1$ equations for the c, together with the nonequalities equivalent to the condition in the j-equations, we call "the c-equations". Solutions of the c-equations are solutions of equations E algebraic over the rational field Q less solutions of various equations $(E+E^*)$ algebraic over Q. Hence, if the number of solutions of the c-equations is finite, the values of the c must lie in an algebraic number field.

The group G uniquely determines the values of the c (by the necessity clause in Theorem 1, and since the canonical relation between j and ζ is unique). Conversely, any solution of the c-equations uniquely determines the group G. For the sufficiency clause in Theorem 1 shows that the solution determines an algebraic equation between j and ζ, of which any solution for ζ is a Hauptmodul of some subgroup G with the specification. Further, in the expansion

$$\zeta = \xi^{-1}+0+a(1)\cdot\xi+a(2)\cdot\xi^2+\cdots$$

the Fourier coefficients $a(1), a(2), \ldots$ can be successively and uniquely determined from the j-equations, so that ζ and hence G is unique.

Now let $\omega=e^{2\pi i/\mu_1}$ be a primitive μ_1th root of unity. For $1\le r<\mu_1$ let G_r be the conjugate group S^rGS^{-r}. Then $\zeta(S^{-r}\tau)$ is a Hauptmodul for G_r, and is also a solution of the j-equations for G, while $\omega^{-r}\zeta(S^{-r}\tau)$ is clearly the canonical Hauptmodul for G_r. Thus we may obtain the values of the c for G_r by multiplying the values of the c for G by appropriate powers of ω^r. These considerations lead to the following theorem.

Theorem 2. *Suppose that for a given specification there exist N different sets of ∞-conjugate groups $(G_i)^\infty$, $i=1$ to N. Then to each set there corresponds a solution of the j-equations in which the coefficients of ζ are of the form $k_i^a\cdot b$, where $k_i^{\mu_1}$ and b lie in an algebraic number field $\mathscr{A}_i$, a is integral, and $k_i^{\mu_1}$ depends only on i. The degree of $\mathscr{A}_i$ over Q is at most N. Within the set the different j-equations for the ∞-conjugates are given by replacing k_i by $\omega^r k_i$ ($r=1$ to μ_1).*

It is also easy to see that the Fourier coefficients $a(n)$ in $\zeta=\xi^{-1}+\sum_{n=1}^{\infty} a(n)\xi^n$ are of the form $a(n)=k_i^{n+1}\cdot b(n)$ with $b(n)$ in $\mathscr{A}_i$.

2.4. *Feasible computation.* At this point one can consider just writing down j-equations and solving them by hand, which is not difficult for subgroups of small index (several examples are given in Klein-Fricke [4, pp. 636 et seq.]). In these cases our canonical normalization of the additive constant in ζ is not always the most convenient, and if any of e_2, f_2, e_3, f_3, or an α_i, is 1, one can make $\zeta=0$ at the appropriate point and specify only $\zeta=\xi^{-1}+O(1)$ at ∞, without destroying the correspondence between subgroups and j-equations. Also the number of unknowns c can be cut in half at once by observing that $dj/d\zeta$ has factors $F_3^2\cdot F_2$. We give two examples.

(i) $g=0, \mu=\mu_1=5, h=1, e_2=1, e_3=2.$

$$j = \zeta^3(\zeta^2+5k\zeta+40k^2),$$
$$j-1728 = (\zeta^2+4k\zeta+24k^2)^2(\zeta-3k) \qquad (k^5 = 1).$$

This gives a set of 5 congruence subgroups of level 5.

(ii) $g=0, \mu=9, h=3, e_2=1, e_3=0, \mu_1=7, \nu_1=1, \alpha_1=2.$

$$j(\zeta^2+13k\zeta/4+8k^2) = (\zeta^3+4k\zeta^2+10k^2\zeta+6k^3)^3,$$
$$(j-1728)(\zeta^2+13k\zeta/4+8k^2) = \zeta(\zeta^4+6k\zeta^3+21k^2\zeta^2+35k^3\zeta+63k^4/2)^2,$$

where $k^7=64$. This gives a set of 7 noncongruence subgroups of level 7, which we denote by Γ_{711} and to which we shall return later. In both cases, the direct calculation shows that no further subgroups with these specifications exist; in the notation of Theorem 2 we have $N=1$ and $\mathscr{A}_1=Q$. We have pushed these hand methods up to a case where $N=5$ (Examples 8 and 9 in Table 1, §4.3), but they are inadequate for serious investigation, and the use of computers must now be considered.

The basis of our technique is the principle "a rational number can be guessed from a good approximation". In fact, the continued fraction expansion of the approximation should yield some large partial quotient, which one then takes to be ∞. Even quadratic irrationals may be difficult to recognize, while cubic and higher irrationals are impossible. Thus any attempt to solve the j-equations for some set $(G_i)^\infty$ must involve solving them also for the corresponding sets in the conjugate fields, and we therefore prefer the degree of $\mathscr{A}_i$ to be reasonably small. The question then arises whether we can, for a specification with large N, foresee the presence of some $\mathscr{A}_i$ with small degree. Now suppose that some subset E of the $(G_i)^\infty$, say for $i=1$ to $N_0<N$, is such that the groups involved have overgroups (other than Γ) or subgroups with certain specifications, which the $(G_i)^\infty$ for $i=N_0+1$ to N do not. Then the c in the c-equations for G in E will satisfy additional algebraic relations which are not satisfied by the G outside E, and so the degrees of $\mathscr{A}_i$ for $i=1$ to N_0 will be at most N_0. In practice the only useful *subgroup* is the maximal subgroup of G normal in Γ. We consider in §3 an effective technique involving permutations for determining all the G with a given specification, and obtaining information about their overgroups and maximal normal subgroups. In the meantime we suppose that some small set $(G_i)^\infty$, $i=1$ to N_0, is given, such that if $c=k_i^a\cdot b$ is any fixed coefficient in the j-equations, then the values of b as i varies satisfy an algebraic equation of degree N_0 over Q (as also do the values of $k_i^{\mu_1}$). It remains possible, of course, that these equations are reducible.

Let a subgroup G be given in such detail that we know not only its specification but a suitable fundamental domain D together with the precise way in which the sides on its boundary correspond in pairs. The assertion that $\zeta(\tau)=\xi^{-1}+0+O(\xi)$ at ∞, is elsewhere regular in D, and has the same value at any two corresponding points on the boundary of D, is theoretically sufficient to determine ζ uniquely. We attempt to approximate this as follows. For each cusp, write down a *finite* expansion in integral powers of the appropriate local uniformizing variable with undetermined coefficients. On each transform l_r of the side of F containing i we select a finite number of equally spaced points. If l_r is inside D, we equate the two expansions from the two associated cusps at these points. If l_r is on the boundary of D we equate the expansions from its associated cusp at these points to the expansions at corresponding points on its corresponding side l_s (associated with the same cusp). It should be observed that the expansions converge rapidly at the relevant points. In this way we can obtain any number of simultaneous *linear* equations for our unknown coefficients, which may be solved by least squares or linear programming. From the viewpoint of numerical analysis, these equations are of course very ill-conditioned. The power series converge so rapidly that one must be careful not to take too many terms, and the equality conditions at adjacent points in a subdivision of the sides are nearly equivalent. However, by judicious choice of the number of terms in the power series and the number of subdivision points, for which we can give no universal prescription, we have been able to determine the first 8 or so coefficients of ζ in powers of ξ with 7 significant figures in many cases, using a linear programming package in the Hartran library at the Atlas Computer Laboratory. For the purpose of computing values of ζ, the lesser accuracy of later coefficients is not important, and one can obtain the values of ζ at the points in D where $j=0$ or $j=1728$ with 7 significant figures, which effectively determines the coefficients c in the j-equations to this accuracy.

For a given ∞-conjugate set $(G_i)^\infty$, it is only necessary to solve the c-equations for one member G_i. The group WG_iW clearly has the same specification as G_i and satisfies the same specification conditions with regard to overgroups and subgroups. Thus $(WG_iW)^\infty$ is some $(G_p)^\infty$ with $1 \leq p \leq N_0$. It may happen that $p=i$. In this case, we have $WG_iW = S^\alpha G_i S^{-\alpha}$ for some α, and hence

$$WS^qG_iS^{-q}W = S^{-q}WG_iWS^q = S^{\alpha-2q} \cdot S^qG_iS^{-q} \cdot S^{2q-\alpha}.$$

If we can choose q with $\alpha - 2q \equiv 0 \pmod{\mu_1}$, we say that the set $(G_i)^\infty$ is *real* and replace G_i by $S^qG_iS^{-q}$ to give $WG_iW = G_i$. If not (when α is odd and μ_1 even), we say that $(G_i)^\infty$ is *semireal*. If $p \neq i$ we say that $(G_i)^\infty$ is *complex*.

If $\zeta(\tau) = \xi^{-1} + 0 + \sum_{n=1}^\infty a(n)\xi^n$ is on G_i, then $\zeta^*(\tau) = \xi^{-1} + 0 + \sum_{n=1}^\infty \overline{a(n)}\, \xi^n$ is on WG_iW, since $\tau \to -\bar{\tau}$ corresponds to $\xi \to \bar{\xi}$, and $\overline{\zeta(-\bar{\tau})} = \zeta^*(\tau)$. Thus if $(G_i)^\infty$ is real, and $G_i = WG_iW$, we have $a(n) = \overline{a(n)}$ for all n, i.e., $a(n)$ is real. If $(G_i)^\infty$ is semireal, we can choose G_i so that $WG_iW = S^{-1}G_iS$, which involves $\overline{a(n)} = \omega^{-n-1} \cdot a(n)$ with $\omega = e^{2\pi i/\mu_1}$, and thus that $a(n) \cdot \omega^{-(n+1)/2}$ is real. In both these cases the field $\mathscr{A}_i$ is real, and $a(n) = k_i^{n+1} b(n)$ with $K = k_i^{\mu_1}$ and $b(n)$ in $\mathscr{A}_i$; in the real case, $K > 0$ and for one G_i in $(G_i)^\infty$ the $a(n)$ are real, while in the semireal case $K < 0$ and no G_i in $(G_i)^\infty$ has all its coefficients real. In the complex case there are two sets of $(G_i)^\infty$ with two conjugate complex fields $\mathscr{A}_i$.

For computational purposes, the real and semireal cases are much easier than the complex case, where we have to use complex coefficients in the power series, and so need twice as many unknowns. The semireal case can be made effectively real by conjugation by $\tau \to \tau + \frac{1}{2}$ (not in Γ). In practice one may, for a given set $(G_i)^\infty$, choose k_i so that the first nonzero coefficient $a(n)$ in the expansion of ζ is precisely $(k_i)^{n+1}$; with this normalization one's remaining computations are solely over the fields $\mathscr{A}_i$. Finally, once the values of the c in the j-equations have been guessed, their verification is a perfectly rigorous matter involving exact computation. This may be feasible by hand even when obtaining the values is not, as in the group of genus 1 which we give below.

(iii) $g=1$, $\mu=\mu_1=9$, $h=1$, $e_2=1$, $e_3=0$. Let $k^9=27/256$,

$$\begin{aligned} Y^2 &= X^3+(225/4)k^2X^2+960k^4X+4096k^6,\\ f_1(X) &= X^3+54k^2X^2+933k^4X+5041k^6,\\ 2f_2(X) &= 9kX^4+642k^3X^3+16425k^5X^2+174303k^7X+612480k^9,\\ f_3(X) &= X^3+48k^2X^2+672k^4X+2176k^6,\\ f_4(X) &= X^4+72k^2X^3+1872k^4X^2+20544k^6X+81792k^8. \end{aligned}$$

Then

$$\begin{aligned} j &= Yf_1(X)-f_2(X),\\ Y^2f_1^2(X)-f_2^2(X) &= f_3^3(X),\\ Y^2f_1^2(X)-(f_2(X)+1728)^2 &= Xf_4^2(X). \end{aligned}$$

This gives a set of 9 noncongruence groups of level 9, which we denote by Γ_9. We may add that, in the first instance, we attempted to solve the corresponding c-equations by a direct Monte Carlo attack. Unfortunately there is at least one infinite family of solutions and one isolated solution for the c-equations ignoring the nonequality conditions, which proved to be fatal; we obtained (rather surprisingly) this incorrect isolated solution from the computer. However, the possibility of using the c-equations should not be ignored if the computation of power series is insufficient. Accuracy may be limited there by ill-conditioning, but an initial approximation to the values of the c is given. This may now be improved by use of the c-equations themselves, for there is no reason to suppose that they are ill-conditioned near a genuine solution.

Finally, if one is concerned solely to obtain reasonable certainty as to the fields $\mathscr{A}_i$, without formal proof via the j-equations, one may use the power series for modular forms of positive weight without poles. This gives better approximate solutions than the use of functions.

3.1. **Subgroups and permutations.** We confine ourselves in this subsection to stating the necessary results, since the theory is developed in detail by Ashworth[2] [1] and Millington[2] [6], [7]. We consider permutations on μ letters named as the integers 1 to μ, where 1 is specially distinguished. We say that a pair (t, p) of permutations is *legitimate* if $t^2=p^3=I$ and the group Σ generated by t and p is transitive. If σ is any element of S_μ, we write $(t, p)\sim(\sigma t\sigma^{-1}, \sigma p\sigma^{-1})$, and if σ is any element of S_μ fixing 1, we write $(t, p)\sim_1(\sigma t\sigma^{-1}, \sigma p\sigma^{-1})$. Then we have

THEOREM 3. *There is a one-to-one correspondence between subgroups of index μ in the modular group and equivalence classes of legitimate pairs of permutations (t, p)*

[2] These authors are identical.

under the equivalence relation $\sim_1$. *If G is a subgroup and* (t, p) *a representative of the corresponding equivalence class, then*

(i) e_2 *and* e_3 *are the number of letters fixed by t and p respectively,*

(ii) $s = tp$ *has h cycles of lengths* $\mu_1, \ldots, \mu_h$, *and* μ_1 *is the length of the cycle containing* 1,

(iii) Σ *is isomorphic to the factor group* Γ/G^N,

(iv) *G is maximal if and only if* Σ *is primitive.*

In fact there is a homomorphism $\theta: \Gamma \to \Sigma$ defined by $\theta(T) = t$, $\theta(P) = p$, and we may consider 1 as representing G, and 2 to μ its left cosets in Γ. The subgroup of Σ fixing 1 is just $\theta(G)$, and the identity of Σ is $\theta(G^N)$. The permutations t and p give the left action of T and P on the left cosets of G in Γ. We may accordingly number the μ copies of F which constitute the fundamental domain D of Γ, with F itself as 1; if this is done then t and s define precisely the pairing of the sides of D, while p ensures the consistency of the 3 cycles.

We obtain the conjugates of G in Γ by conjugating t and p by elements σ_i ($i = 2$ to μ) of S_μ where σ_i takes i into 1. Thus there is a one-to-one correspondence between sets of conjugate subgroups of Γ, and equivalence classes of legitimate pairs (t, p) under the relation $\sim$. The results (i) to (iv) of Theorem 3 still apply, except that μ_1 and 1 are no longer distinguished.

Finally if (t, p) is a representative pair for G, then $(t, tp^{-1}t)$ is a representative pair for WGW. In practice it is simpler to get a pair for WGW by leaving t fixed, and replacing s by s^{-1}, p then following as ts.

3.2. We now give some examples to illustrate the use of these permutations.

$g = 0$, $\mu = \mu_1 = 5$, $h = 1$, $e_2 = 1$, $e_3 = 2$.

We may take $D = F + SF + S^2F + S^3F + S^4F$, numbered 1 to 5, so that $s = (12345)$. Since $e_2 = 1$, there is only one fixed letter in t, and given any group with this specification, its 5 conjugates will have respectively each of the 5 different letters fixed; thus one member of the set will have

$$t = (1)(2, x-1)(x5),$$
$$p = (12x)(\quad)(\quad),$$

which implies $x = 4$ and

$$\text{(i)} \qquad t = (1)(23)(45), \quad p = (124)(3)(5).$$

The conjugates are formed most simply by conjugating by powers of s, giving

$$t = (2)(34)(51),$$
$$p = (235)(4)(1), \quad \text{etc.}$$

There is thus only one conjugate set (in fact the first example in §2.4) with this specification, which must be its own transform by W. In fact the actual group given by (i) is self-conjugate by W, since $s^{-1} = (15432)$, $t = (1)(23)(45)$, and conjugating this by (25)(34) which fixes 1 we arrive at (i) again. The case of one cusp (called *cycloidal*) is in general particularly simple; we may always choose $s = (12 \cdots \mu)$, and conjugation by W corresponds to $n \to 2-n$ for n in t, and conjugation by S^{α} to $n \to n+\alpha$, where $+$ and $-$ are modulo μ. Also of course $(G)=(G)^{\infty}$ for cycloidal subgroups. In the remaining examples we give fewer details.

$g=0, \mu=\mu_1=6, h=1, e_2=0, e_3=3.$

There is one conjugate set, of which a member G is given by $s=(123456)$, $t=(12)(34)(56)$, $p=(135)(2)(4)(6)$. The group $\{t, p\}$ is here imprimitive, fixing the equivalence relation $1'\colon 1\sim 3\sim 5$, $2'\colon 2\sim 4\sim 6$, and so G is a subgroup of the group Γ_2 of index 2 in Γ given by $s'=(1'2')$, $t'=(1'2')$, $p'=(1')(2')$. Adding 2 to the numbers in s, t and p leaves G unaltered, so there are only two conjugates and Γ_2 is the normalizer of G in Γ; these conjugates G and SGS^{-1} are also conjugates by W, so that we have a semireal case (§2.4). If ζ_2 and ζ_6 are suitable Hauptmoduln of Γ_2 and G, we have

$$j = \zeta_2^2+1728 = \zeta_6^3(\zeta_6^3-2k^3),$$
$$j-1728 = \zeta_2^2 = (\zeta_6^3-k^3)^2,$$

where $(k^3)^2=-1728$, $\zeta_6^3=\zeta_2+k^3$. In this case only k^3 and not k itself occurs in the Fourier coefficients of ζ_6; with $\zeta=e^{2\pi i\tau/6}$ we have $\zeta_6=\xi^{-1}(1+\sum_{n=1}^{\infty} k^{3n}\cdot b(3n)\cdot \xi^{3n})$ with $b(3n)$ integral. These are congruence subgroups.[3]

$g=0, \mu=\mu_1=8, h=1, e_2=e_3=2.$

We take $s=(12345678)$ and there are four conjugate sets given by

$$\begin{aligned}(G_1)\colon\quad t &= (1)(26)(34)(5)(78),\\ (G_2)\colon\quad t &= (1)(23)(48)(5)(67),\\ (G_3)\colon\quad t &= (15)(2)(34)(67)(8),\\ (G_4)\colon\quad t &= (1)(2)(38)(45)(67).\end{aligned}$$

Of these (G_1) and (G_2) are complex (conjugate by W), (G_3) is real, and (G_4) semireal. On this information only, we might have four conjugate quartic fields, two real and two complex conjugate. However, (G_1) and (G_2) are imprimitive, corresponding to an overgroup of index 4 in Γ, and so we obtain the fields $\mathscr{A}_1, \mathscr{A}_2=Q(\sqrt{-2})$ and $\mathscr{A}_3, \mathscr{A}_4=Q(\sqrt{2})$. For $\mathscr{A}_3, \mathscr{A}_4$ we may take k as given by $k^8=-3(1\mp 2\sqrt{2})^2(1\mp\sqrt{2})$ which is positive for $\mathscr{A}_3$ and negative for $\mathscr{A}_4$. One can also distinguish the groups by their maximal normal subgroups; (G_1) and (G_2) are congruence groups with $\Gamma/G_i^N\cong \mathrm{LF}(2, 8)$[4] while (G_3) and (G_4) are noncongruence groups with $\Gamma/G_i^N\cong \mathrm{PGL}(2, 7)$.

$g=0, \mu=9, h=3, e_2=1, e_3=0, \mu_1=7, \alpha_1=2, \nu_1=1.$

We find

$$\begin{aligned} s &= (1234567)(x)(y)\\ t &= (1)(24)(3x)(57)(6y)\\ p &= (125)(3x4)(6y7)\end{aligned}$$

[3] There is no immediate indication from t and p as to whether a subgroup is congruence; one must either just recognize it, or find the order of Γ/G^N.

[4] We write LF(2, 8) if matrix entries in the ring $Z/8Z$ are implied; LF(2, 2^3) for entries in the field of 2^3 elements.

and its 7 conjugates by powers of s as the only solutions. The subgroup G corresponding to the permutations given has $WGW=G$; its fundamental domain is shown in Figure 1, and the symmetry about the imaginary axis is evident. We show also the pairing of the sides, and the corresponding generators of G; these generators are (since $g=0$) elliptic and parabolic, namely S^7, $S^{-3}TS=S^{-2}T\cdot S\cdot TS^2$, T, $STS^{-3}=S^2T\cdot S\cdot TS^{-2}$, and we have the canonical relation

$$S^7\cdot S^{-3}TS\cdot T\cdot STS^{-3} = I.$$

The copies 1 to 7, x, and y, of F are respectively F, SF, S^2F, S^3F, $S^{-3}F$, $S^{-2}F$, $S^{-1}F$, S^2TF, and $S^{-2}TF$. It is quite easy to derive all this information from the permutations. Every point in H is equivalent under G to some copy g_iF ($i=1$ to 9), and if $V\tau \in g_iF$ then $g_i^{-1}V\tau \in F$, so that the copies g_iF correspond to the left coset decomposition

$$\Gamma = \sum_{i=1}^{9} g_i^{-1}G.$$

If we consider the conjugate curved sides of 2 and 4 in the diagram, for which $g_i=S$ and S^3, then the left action of T given by t implies that $T\cdot S^{-1}G=S^{-3}G$, i.e., STS^{-3} is in G, and of course STS^{-3} fixes 1 as expected. The same generator is given by the continuation of these sides as sides of x in the form $S\cdot(S^2T)^{-1}\cdot G=(S^2T)^{-1}\cdot G$ or $S^2TSTS^{-2}=STS^{-3}$ in G.

We have already considered the ∞-conjugate set $(G)^\infty$ as Γ_{711} in §2.4. The field $\mathscr{A}$ is Q, and $k^7=64$. The two conjugates of G in Γ with a $\mu_1=1$ specification give two conjugate complex sets $(G_1)^\infty$ and $(G_2)^\infty$ each consisting of a single group; here $k=1$ and $\mathscr{A}_i=Q(\sqrt{-7})$. The fact that conjugate groups in this case involve the fields generated by $(64)^{1/7}$ and by $\sqrt{-7}$, which are not even conjugate over the rationals, reflects the way in which we have normalized the Hauptmodul ζ. This was chosen to be analytically rather than algebraically simple, and moreover it involves picking out one of the cusps in preference to the others. Presumably there is a best field of definition from an algebraic viewpoint, and a corresponding algebraic normalization of the Hauptmodul; but we have not yet considered this problem.

3.3. We have written a computer program which systematically enumerates the cycloidal subgroups of Γ of given index μ, on the basis of the representation in Theorem 3; and we have used this program to list all such subgroups for $\mu\leq 18$. The program essentially searches through a tree of possibilities, and the only difficulty is in keeping the tree as small as possible. At any stage the store contains a certain number of partial descriptions of legitimate pairs (t, p); one of these is being worked on and the others are in a 'last-in, first-out' list, to be worked on when the present one is finished. A pair (t, p) is described by listing, insofar as they have yet been determined, the values of αt and αp for $\alpha=1, 2, \ldots, \mu$. Since we are concerned with cycloidal subgroups, $\alpha(tp)=\alpha+1$; for convenience we identify $\mu+1$ with 1 here. Thus we can make the following deduction:

(i) if $\alpha p=\beta$ then $(\beta-1)t=\alpha$,

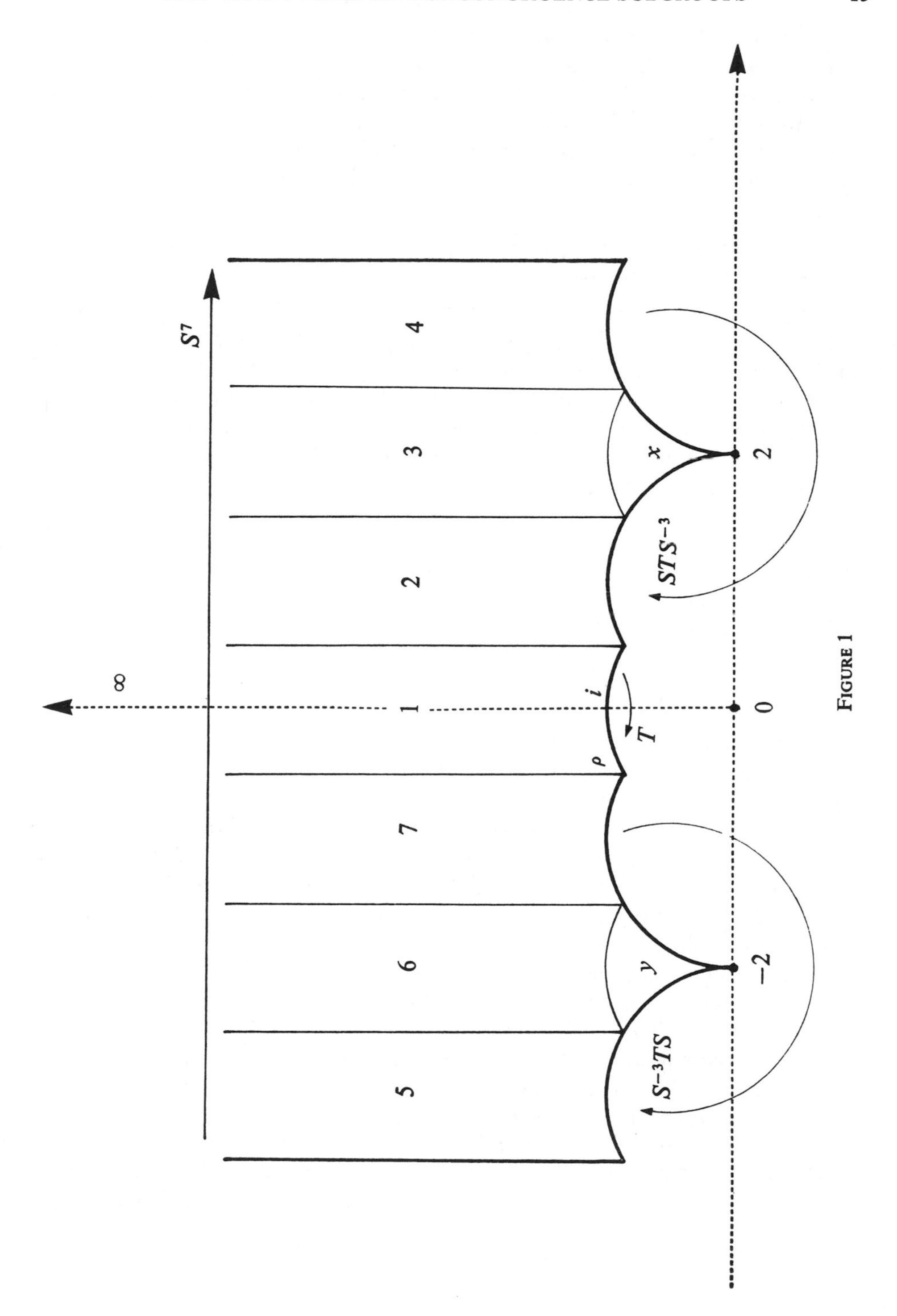

FIGURE 1

(ii) if $\alpha t=\beta$ then $\beta p=\alpha+1$ and $\beta t=\alpha$,

(iii) if $\alpha p=\beta$ and $\beta p=\gamma$ then $\gamma p=\alpha$.

A partial description of (t, p) is an incomplete list of values of αt and αp. Suppose we have such a partial description; the first thing to do is to add as many logical deductions based on the rules above as possible. Suppose that this is done and the description is still incomplete. We then select an unknown αp or αt according to the following rule:

> If there is a known $\beta p=\gamma$ such that γp is unknown, select that γp; otherwise select any unknown αt.

For this unknown αp or αt there are finitely many possible values, to wit those of $1, 2, \ldots, \mu$ which are not already assigned as images under p or t respectively. Each of these possibilities can be inserted in our old partial description to give a new and more extensive one; one of these new descriptions is worked on immediately and the others are put in the push-down list to be worked on in due course. Each partial description is eventually either completed or proved impossible in this way. The reason for making the list a push-down one is to keep it as small as possible; in this way we always ensure that the next partial description to be worked on is that one in the list which is most nearly complete. The rule which selects the αp or αt next to be chosen is designed to ensure that as many further logical deductions as possible can be made before we have to make a further choice.

In the form described above, the process finds each member of a set of conjugate subgroups. To find only one member of each such set, we have to introduce a partial ordering into the set of legitimate pairs (t, p)—say by looking at the value of $1t$, and then at that of $2t$ in the case of a tie, and so on. A partial description of a pair (t, p) can then be rejected as soon as it is clear that the pair would necessarily be dominated by one of its conjugates. This step cuts the machine time needed, for a given value of μ, by a factor of about $\frac{1}{2}\mu$.

It should be noted that for each index μ this program determines all the conjugate sets of cycloidal subgroups without regard to e_2, e_3, and g, which are worked out at the end. The ideas of Conway described in the next section enable us to derive all cycloidal subgroups from those with $e_2=e_3=0$ and a possibly smaller value of μ. The program can easily be modified to deal only with the case $e_2=e_3=0$; this leads to a further increase of speed and (which is even more important) reduction of output. In fact there is just one such set of conjugate cycloidal subgroups with $\mu=6$, and just nine such sets with $\mu=18$. The modifications needed to enumerate non-cycloidal subgroups are also trivial, though it is no longer so easy to choose one out of each set of conjugates.

3.4. *Combinatorial aspects and Conway diagrams.* In the course of this work we observed the following curious fact, which has no apparent relation to modular theory. Let g and h be fixed, and let $N(a, b, c)$ be the number of conjugate sets of modular subgroups with $\mu=a$, $e_2=b$, $e_3=c$ (and of course $g=1+\mu/12-h/2-e_2/4-e_3/3$, so that $a-3b-4c$ is constant). Then

$$N(a, b, c) = N(a+b-c, c, b).$$

To prove this, consider any legitimate pair (t, p) on μ letters with $e_2>0$. Then if x is any letter fixed by t, we may define a legitimate pair (t', p') on $\mu+1$ letters, where $p'=p+(y)$, $t'=t-(x)+(x, y)$, and y is a new letter. Clearly $\{t', p'\}$ is transitive if $\{t, p\}$ is. This process π increases e_3 and μ by 1, and decreases e_2 by 1, and leaves unaltered the number of cycles of $s=tp$ (so that g and h are unaltered). Similarly if $e_3>0$ we may reverse this process. In general the correspondence between all the groups when $e_2, e_3 \to e_2 \pm n, e_3 \mp n$, is very intricate. However, there *is* a one-to-one correspondence between legitimate pairs (up to conjugacy in their respective S_μ's), if we start with $e_2=b$, $e_3=c$ and apply π to *all* the b letters fixed by t, and π^{-1} to *all* the c letters fixed by p, to produce (t', p') with $e_2=c$, $e_3=b$. This is the result above. Note that it is not possible to keep a given letter distinguished, so that the result applies only to conjugate sets and not to the number of individual groups. We communicated this result to Dr. J. H. Conway, who found the above construction independently of ourselves, and in addition found an ingenious geometrical representation which we call a Conway diagram. A Conway diagram for a legitimate pair (t, p) consists of μ points $1, 2, \ldots, \mu$, some of which may be colored red, blue or green, together with red, blue and green lines joining them. A pair of equations $\alpha t=\beta$ and $\beta t=\alpha$ is represented by an unoriented red line joining α and β; if $\alpha t=\alpha$ the point α is simply colored red. Thus each point either is colored red or is at the end of just one red line. A triplet of equations $\alpha p=\beta$, $\beta p=\gamma$, $\gamma p=\alpha$ is represented by an oriented blue triangle $\alpha\beta\gamma$; if $\alpha p=\alpha$ then the point α is simply colored blue. Cycles under $s=tp$ are similarly represented in green. The identity $s=tp$ is equivalent to the statement that for each α the point αp immediately follows αt on some s-cycle; this translates into the existence of enough triangles suitably oriented and with one side of each color. There is a one-one correspondence between legitimate pairs (t, p) and legitimate Conway diagrams; and the various possible constructions can be most intelligibly interpreted in terms of collapsing or expanding parts of a Conway diagram so as to produce another Conway diagram.

Conway also found a further construction as follows. Suppose that t contains a fixed letter (x). Then p has a cycle (xyz) if $\mu>1$ (by transitivity), and t has cycles $(ya)(zb)$. For $\mu>3$, we cannot have either $a=z$, $b=y$ or $a=y$, $b=z$ (by transitivity); thus either a, b, y, z are all distinct or, say, $a=y$, $b\neq z$. In both these cases we remove (xyz) from p; while in t we either replace $(x)(ya)(zb)$ by (ab), or replace $(x)(y)(zb)$ by (b). In any event we obtain (t', p') with $\mu'=\mu-3$, $e_2'=e_2-1$, and e_3, h, and g are unaltered. We may conversely always insert a new 3-cycle in p, though in so many different ways that no one-to-one correspondence seems available by this construction.

One can give further constructions to remove cycles of lengths 1, 2, or 3 in s, with exceptions for small μ. The value of these constructions is twofold. In the first place they can be used computationally to derive all subgroups of one specification from knowledge of those with a simpler specification. Secondly, Millington [7] has shown the existence of a modular subgroup with any given consistent g, μ, h, e_2, and e_3, and also given examples of specifications (i.e., with the additional condition $\mu=\mu_1+\cdots+\mu_h$) where no subgroup exists. These constructions simplify the proof of her theorem, and may throw some light on what partitions $\mu=\mu_1+\cdots+\mu_h$ are impossible. For existence purposes the lack of one-to-one correspondence with the constructions is no object.

3.5. For any legitimate pair (t, p) it is desirable to be able to determine the structure of the group Σ generated by t and p, since this is isomorphic to the factor group Γ/G^N. More generally, let Σ be any group of permutations on the μ symbols $1, 2, \ldots, \mu$; and for $\alpha=0, 1, 2, \ldots, \mu$ let Σ_α be the subgroup of Σ consisting of those permutations which leave $1, 2, \ldots, \alpha$ each fixed. We suppose that Σ is described by giving a set of generators for it. We have written a program which finds for each α the index of Σ_α in $\Sigma_{\alpha-1}$. While this information is not always sufficient to identify Σ uniquely, it is so in the cases with which we have been concerned. In most cases Σ is the alternating group A_μ or the symmetric group S_μ; the interest lies in the exceptions. Several instances of $\mathrm{LF}(2, p^n)$ and $\mathrm{PGL}(2, p^n)$ appear, and it might be fruitful to start from these groups and determine all the appropriate specifications that can give rise to them. (This is almost, but not exactly, equivalent to finding all the ways of generating these groups by X and Y satisfying $X^2=Y^3=I$.) Similarly, we have determined the only four essentially distinct pairs X, Y of this sort which generate the Mathieu group M_{12}; M_{11} cannot be generated in this way. Leech **[5]** has used these to obtain elegant presentations of M_{12}.

The program which finds the index of Σ_α in $\Sigma_{\alpha-1}$ is based on the following scheme for the description of a permutation group Σ.

LEMMA. *Let $\mathscr{S}_1, \ldots, \mathscr{S}_\mu$ be sets of permutations on the μ symbols $1, 2, \ldots, \mu$ with the properties*

(i) *each $\mathscr{S}_\alpha$ contains the identity permutation I;*

(ii) *each element of $\mathscr{S}_\alpha$ leaves the symbols $1, 2, \ldots, \alpha-1$ fixed;*

(iii) *two distinct elements of the same $\mathscr{S}_\alpha$ have different effects on α.*

Let $\mathscr{S}$ be the set of those permutations which have a representation in the canonical form

$$S = S_\mu S_{\mu-1} \cdots S_2 S_1$$

where each S_α is in $\mathscr{S}_\alpha$. Then each S in $\mathscr{S}$ has just one representation in canonical form, and S leaves $1, 2, \ldots, \alpha$ fixed if and only if $S_1=\cdots=S_\alpha=I$. Moreover, a necessary and sufficient condition that the elements of $\mathscr{S}$ form a group is that for each pair S', S'' in $\mathscr{S}_\alpha$ the products $S'S''$ and $S''S'$ should be in $\mathscr{S}$.

Conversely, any permutation group can be described in this way.

The proof of the direct part of this lemma is a straightforward but notationally tedious induction on μ, and we omit it. The proof of the converse is essentially contained in the description of the program below. With the definition of the Σ_α above, the index of Σ_α in $\Sigma_{\alpha-1}$ is just the cardinality of $\mathscr{S}_\alpha$; to find these indices, it is therefore enough to describe Σ in this way.

The program builds up the sets $\mathscr{S}_\alpha$, using for this purpose an auxiliary list $\mathscr{L}$ of permutations which are known to belong to Σ but have not yet been put into canonical form. Initially each $\mathscr{S}_\alpha$ contains only the identity permutation, and $\mathscr{L}$

consists of the given generators of Σ. The basic cycle of the program, which continues until $\mathscr{L}$ is exhausted, is as follows. Take an element L of $\mathscr{L}$, and delete it from $\mathscr{L}$. If L is the identity then the cycle is complete; otherwise let α be the first symbol which is not fixed under L, and suppose that $\alpha L=\beta$. There are now two possibilities:

(i) There is an element $S_{\alpha\beta}$ of $\mathscr{S}_\alpha$ such that $\alpha S_{\alpha\beta}=\beta$. In this case we form $LS_{\alpha\beta}^{-1}$ and repeat the process with $LS_{\alpha\beta}^{-1}$ (which fixes $1, 2, \ldots, \alpha$ inclusive) instead of L.

(ii) There is no such element $S_{\alpha\beta}$. In this case we adjoin L to $\mathscr{S}_\alpha$ and insert into the list $\mathscr{L}$ all the permutations LS and SL, where S runs through all the elements of $\bigcup \mathscr{S}_\gamma$ including L itself; this completes this cycle.

In practice, in the latter case we need only insert in $\mathscr{L}$ those LS for which S does fix $1, 2, \ldots, \alpha-1$ and those SL for which S does not fix all of $1, 2, \ldots, \alpha$. Moreover we can save space but not time by only forming products when they are taken out of $\mathscr{L}$ for examination, not when they are put in. When $\mathscr{L}$ is exhausted the sets $\mathscr{S}_\alpha$ are complete.

If Σ is in fact the alternating or symmetric group—as it usually is—we can deduce this long before the program is complete, by making use of known theorems about permutation groups. For example, if $\mathscr{S}_\alpha$ has its maximum cardinality $\mu+1-\alpha$ for each of $\alpha=1, 2, \ldots, \alpha_0$ then Σ is at least α_0-ply transitive and hence is already very nearly identified, at least for $\alpha_0 \geq 4$. Again, if Σ contains $(1\ 2 \cdots \mu)$, as it does in the cycloidal case, and if $\mathscr{S}_{\mu-1}$ contains $(\mu-1\ \mu)$, which is the only element other than the identity which it can contain, then Σ is the full symmetric group. There is a similar test for the alternating group, involving $\mathscr{S}_{\mu-2}$. These shortcuts are valuable because the speed of the program depends on the size of Σ, and one therefore needs to curtail the calculations in precisely those cases when Σ is large.

4.1. **Fields of Fourier coefficients.** Suppose that G is any subgroup of Γ with a cusp of width μ_1 at ∞, and local variable $\xi=e^{2\pi i\tau/\mu_1}$. Let

$$F_r(\tau) = \sum_{n=M}^{\infty} a_r(n)\xi^n \qquad (M = M(r) > -\infty).$$

Then we say that $\mathscr{A}$ is "a field of G" if there exists a (countably infinite) basis of *entire* forms $(F_r(\tau))$ with $F_r(\tau)$ on G and $a_r(n)$ in $\mathscr{A}$ such that *every* entire form of integral weight on G is a finite linear combination of a finite subset of $(F_r(\tau))$. The smallest $\mathscr{A}$ which is a field of G is called *the* field of G. The field of G appears to be the field defined by the solutions of the c-equations (§2.3) for *all* conjugates of G; if the genus is zero we can give a rather tedious and unilluminating proof of this. It is clear that some restriction has to be made in order to give a sensible definition; the restriction to entire forms is in accordance with experience on congruence subgroups.

If G is a congruence subgroup of level l, then it also appears that the field of G is contained in the field of the lth roots of unity. Indeed the field of $\Gamma(l)$ itself appears to be Q, and while (for example) our $\zeta_6(\tau)$ in §3.2 of index and level 6 has $b(3n)$ in $Q(\sqrt{-3})$, we may write $\zeta_6(\tau)=\alpha(\tau)+\sqrt{-3}\,\beta(\tau)$, where $\alpha(\tau)$ and $\beta(\tau)$ are on a subgroup of index 12 and the same level 6, and the Fourier coefficients of $\alpha(\tau)$

and $\beta(\tau)$ are rational integers. Finally it appears that we may choose the $F_r(\tau)$ for $\Gamma(l)$ not merely with rational coefficients, but with *integral* coefficients. These are interesting questions, which we raise in positive form since we know no counter-examples, but we do not wish any formal conjectures to be understood.

4.2. *Fields of noncongruence groups.* The situation for noncongruence groups appears to be different in three ways, which we discuss in order of complexity.

4.2.1. There certainly exist infinitely many noncongruence groups whose field is Q. For instance, if $\zeta(\tau)=\eta(\tau)/\eta(13\tau)$, where

$$\eta(\tau) = e^{\pi i\tau/12}\prod_{r=1}^{\infty}(1-e^{2\pi i r\tau}),$$

then $\zeta(\tau)$ is the Hauptmodul of a congruence group with $g=0$, $\mu=28$, $h=2$, $e_2=e_3=4$, $\mu_1=2$, $\mu_2=26$; $\zeta=\infty$ at the cusp ∞ and $\zeta=0$ at the cusp 0. For any $m>1$, $\zeta^{1/m}(\tau)$ is the Hauptmodul of a noncongruence group with $g=0$, $\mu=28m$, $h=2$, $e_2=e_3=4m$, $\mu_1=2m$, $\mu_2=26m$; and the Fourier coefficients of $\zeta^{1/m}(\tau)$ are clearly rational. However, they have unbounded denominators, whereas those of $\zeta(\tau)$ are all integral, and this distinction is in all the cases we know a precise criterion for distinguishing congruence and noncongruence groups.

4.2.2. Suppose next that the field $\mathscr{A}_i$ of Theorem 2 is Q, and $\mu_1>1$, so that $K=k^{\mu_1}\in Q$, and the field of G contains that of the μ_1th roots of K. It remains possible that the field of the maximal normal subgroup G^N is Q itself. Even if this is so there is still the distinction that G involves, not the μ_1th roots of unity (as is apparently so in the congruence case), but the μ_1th roots of (various) integers $K\neq 1$, in our examples. The factors of K may be -1, 2, 3, and those of the index of G in Γ.

4.2.3. Finally the field $\mathscr{A}_i$ of Theorem 2 may be greater than Q, in a way which is not retrievable (as for ζ_6 in §4.1) in terms of G^N. This is a phenomenon which has no congruence analogy, for there is only one *congruence* subgroup $\Gamma(l)$ with factor group $\Gamma/\Gamma(l)\cong \mathrm{LF}(2, l)$. But, for example, our two groups numbered 9 in Table 1 with $\Gamma/G^N\cong A_{12}$ correspond to two *different* G^N which are conjugate by W.

4.3. We give in Table 1 below a list of specifications for noncongruence groups, mainly cycloidal, and the fields $\mathscr{A}_i$ corresponding to their ∞-conjugate sets. We give also the factor groups Γ/G^N, when these have been used to distinguish them from other sets with the same specification. An asterisk after the minimum polynomial of a field denotes that we have not formally verified it from the j-equations, but been content with the approximate computation of forms. A particular point of interest is the primes at which these fields ramify. In every case except one (number 9), these are primes dividing the level (which is also the index in most cases). Again it is not clear yet exactly what conjecture to make, but one would be surprised if such a field ramified at primes other than 2, 3, or those dividing the level or index. In particular, it is not too easy to construct directly fields which ramify only at specified primes, and these cycloidal groups may provide a feasible technique for doing this.

We do not give the values of $k_i^{\mu_1}$ in Table 1, nor the many examples when $i=1$

Table 1

	μ	g	h	e_2	e_3	μ_1	$\mu_i\,(i>1)$	Γ/G^N	number of ∞-conjugate sets (real, semireal, complex)	fields $\mathscr{A}_i$	ramify at	level
1.	7	0	2	1	1	6	1	—	$2=0+0+2$	$\sqrt{-3}$	3	6
2.	8	0	1	2	2	8	—	PGL(2, 7)	$2=1+1+0$	$\sqrt{2}$	2	8
3.	9	0	1	1	3	9	—	—	$2=0+0+2$	$\sqrt{-3}$	3	9
4.	9	0	2	3	0	8	1	—	$2=0+0+2$	$\sqrt{-2}$	2	8
5.	9	0	3	1	0	1	$1+7$	—	$2=0+0+2$	$\sqrt{-7}$	7	7
6.	10	0	1	4	1	10	—	—	$5=0+1+4$	λ^5-4	2, 5	10
7.	12	1	1	2	0	12	—	PGL(2, 11)	$2=2+0+0$	$\sqrt{3}$*	2, 3	12
8.	12	0	2	4	0	11	1	M_{12}	$2=0+0+2$	$\sqrt{-11}$	11	11
9.	12	0	2	4	0	11	1	A_{12}	$3=1+0+2$	$\lambda^3-11\lambda-22$	2, 11	11
10.	13	0	1	5	1	13	—	LF(3, 3)	$4=0+0+4$	$\lambda^4+13\lambda+39$*	13	13
11.	14	1	1	0	2	14	—	PGL(2, 13)	$3=1+2+0$	$\lambda^3+\lambda^2-2\lambda-1$*	7	14
12.	17	1	1	1	2	17	—	LF(2, 2^4)	$2=2+0+0$	$\sqrt{17}$*	17	17
13.	18	2	1	0	0	18	—	PGL(2, 17)	$3=2+1+0$	$\lambda^3-3\lambda+1$*	3	18

and $\mathscr{A}_i = Q$. Finally all the groups involved are in fact maximal in Γ. It will be seen that all the fields are subfields of cyclotomic fields except in cases 6 and 9.

5.1. **Hecke operators on congruence subgroups.** We state first the main results involving Hecke operators on congruence subgroups in a form relevant to our later analogy in §5.2 for noncongruence subgroups. We define the congruence subgroups $\Gamma_0(l)$, $\Gamma_0^0(l)$ by $c\equiv 0 \pmod{l}$, $b\equiv c\equiv 0 \pmod{l}$ in $V\tau=(a\tau+b)/(c\tau+d)$. The Hecke operators can be defined for forms on $\Gamma(l)$, but this gives essentially no more number-theoretic information than confining them to $\Gamma_0^0(l)$. Since $f(\tau)$ is on $\Gamma_0^0(l)$ if and only if $f(l\tau)$ is on $\Gamma_0(l^2)$, we may further confine our account to $\Gamma_0(l)$. For

$$F(\tau) \in \{\Gamma_0(l), w\}_0 = C_0$$

and

$$F(\tau) = \sum_{n=1}^{\infty} a(n)x^n \qquad (x = e^{2\pi i\tau})$$

we define, for prime p with $(p, l)=1$, the operator T_p by

$$T_p\cdot F(\tau) = p^{-1}\sum_{\lambda=0}^{p-1} F\left(\frac{\tau+\lambda}{p}\right)+p^{2w-1}F(p\tau) = \sum_{n=1}^{\infty} \{a(np)+p^{2w-1}a(n/p)\}x^n,$$

where $a(n/p)=0$ if n/p is not an integer. The operators T_p and $T_{p'}$ commute, and T_p maps C_0 into itself. In addition there is a scalar product

$$(F, G) = \iint_D F(\tau)\overline{G(\tau)}\sigma^{2w-2}\, d\rho\, d\sigma \qquad (\tau = \rho+i\sigma)$$

where D is a fundamental domain of $\Gamma_0(l)$, and T_p is hermitian with respect to (F, G), i.e., $(T_pF, G)=(F, T_pG)$. We may therefore conclude that if d is the dimension of the vector space C_0, there exists a basis $F_r(\tau)$, $r=1$ to d, for C_0 which are simultaneous eigenforms (we avoid "eigenfunction" since this suggests in our context a form of weight 0) of all the T_p, with real eigenvalues $\lambda_r(p)$, so that

$$T_p\cdot F_r(\tau) = \lambda_r(p)\cdot F_r(\tau) \qquad (r = 1 \text{ to } d).$$

Now if $G(\tau)$ is on $\Gamma_0(l')$, where l' is a proper divisor of l and $l=ml'$, then $G(d'\tau)$ is on $\Gamma_0(l)$ for all $d'|m$. The properties of all such forms $G(d'\tau)$ essentially arise on the groups $\Gamma_0(l')$, and they generate a subspace C_0^- of C_0 fixed under each T_p. If now C_0^* of dimension d^* is the orthogonal complement of C_0^- in C_0, then the eigenforms $F_r(\tau)$ in C_0^* we call the "newforms" on C_0, and for these one can show that the coefficient of x is nonzero. We may then normalize so that the newforms are

$$F_r(\tau) = x+\sum_{n=2}^{\infty} a_r(n)x^n;$$

the eigenvalue $\lambda_r(p)$ is clearly $a_r(p)$, so that we have for all prime $p\nmid l$ and all n the multiplicative relation

$$a_r(np)-a_r(p)a_r(n)+p^{2w-1}a_r(n/p) = 0 \qquad (r = 1 \text{ to } d^*).$$

We also have in fact for prime $q|l$ the relation

$$a_r(nq)-a_r(q)a_r(n) = 0,$$

where $a_r(q)=0$ if $q^2|l$ and $a_r(q)=\pm q^{w-1}$ otherwise. All this is discussed fully by Atkin and Lehner [2].

The effect of these results is that the number-theoretic properties of the Fourier coefficients of cuspforms on $\Gamma_0(l)$ are completely determined from a knowledge of the eigenvalues $a_r(p)$ for $p\nmid l$. If, as seems likely, C_0 has some basis with integral Fourier coefficients, then the $a_r(p)$ are algebraic of degree at most d^*. In every case known to us they are algebraic *integers*. The "Ramanujan conjecture" that $|a_r(p)|<2p^{w-1/2}$ has only been proved for $w=1$ (Igusa [3]), but the evidence for its truth is overwhelming. In addition there is good evidence that, unless there is a formula for $a_r(p)$ as a sum of complex divisors of p in an imaginary quadratic field, the distribution of θ with $2\cos\theta=a_r(p)/p^{w-1/2}$ and $0<\theta<\pi$ is $\sin^2\theta\cdot d\theta$, as suggested by Tate. Finally when $a_r(p)$ is integral there may exist congruence properties modulo various small primes, for which a sophisticated explanation and generalization has recently been suggested by Serre [8].

5.2. *Fourier coefficients on noncongruence subgroups.* We now suppose that G is a noncongruence subgroup such that $(G)^\infty$ gives rise to the rational field $\mathscr{A}=Q$ and $k^{\mu_1}=K$ is rational, in the notation of Theorem 2. This is as yet the only case where we have any evidence, and we have only used the groups Γ_{711} and Γ_9 of §2.4. The vector space $C_0=\{G, w\}_0$ will have a basis of forms $F(\tau)$ with

$$F(\tau) = \sum_{n=1}^{\infty} a(n)\xi^n = \sum_{n=1}^{\infty} k^{n-1}b(n)\xi^n$$

and the $b(n)$ rational. We say that p is a "good" prime if it does not occur in the denominator of any $b(n)$ nor in the numerator or denominator of K. In general, the primes 2 and 3, and those dividing the index or level, may be bad (contrast the congruence case where only primes dividing the level are distinguished). For a good prime, we consider expressions of the form

$$T(A,p,n) = T(n) = a(np)-A(p)\cdot a(n)+p^{2w-1}a(n/p)$$

for an integral $A(p)$. If we define $\pi_p(c)=\pi(c)$ for integral c by $p^{\pi(c)}|c$, $p^{\pi(c)+1}\nmid c$, and for rational c/d by $\pi(c/d)=\pi(c)-\pi(d)$, so that $\pi(c)$ is just the p-adic valuation of c, we may extend π to $T(n)$ by defining $\pi(T(n))=\pi(k^{1-n}T(n))$. For

$$k^{1-n}T(n) = m^n b(np)-A(p)\cdot b(n) \quad \text{if } p\nmid n,$$

and

$$k^{1-pn}T(pn) = m^{pn}b(np^2)-A(p)\cdot b(np)+p^{2w-1}m^{-n}b(n),$$

where $m=k^{p-1}$. Now $K=k^{\mu_1}$ is in Q, so that $K^{p-1}\equiv 1 \pmod p$ since $\pi(K)=0$ if p is good. We may then, since $\pi(\mu_1)=0$, interpret m p-adically as the unique solution of $m^{\mu_1}=K^{p-1}$ with $m\equiv 1 \bmod p$, and the other terms in $k^{1-n}T(n)$ are in Q.

We now state our empirical discoveries for $\{\Gamma_9, 1\}_0$ and $\{\Gamma_{711}, 2\}_0$, where in each case the vector space of cuspforms has dimension 1, and we write the unique form canonically as

$$F(\tau) = \xi + \sum_{n=2}^{\infty} a(n)\xi^n.$$

For each good prime p there exists an integer $A(p)$ such that for all n we have

$$\pi(T(n)) \geq (2w-1)(\alpha+1) \quad \textit{if } \pi(n) = \alpha.$$

Further we have $|A(p)| < 2p^{w-1/2}$ and the distribution of θ with $2\cos\theta = A(p)/p^{w-1/2}$ is $\sin^2\theta\cdot d\theta$. Finally $A(p)$ has congruence properties modulo some small primes of a form analogous to those of Hecke eigenvalues on congruence subgroups.

We have verified this subject to the constraints $p<100$, $n<1000$, $n\cdot p^{(2w-1)(\alpha+1)} < 6.7\times 10^{10}$. The computation involved is considerable, for one must, for each p, work out the coefficients $b(n)$ modulo a high power of p from the j-equations and known forms on Γ. It is not feasible to find $b(n)$ exactly once for all and then reduce (mod p^β); for instance on $\{\Gamma_9, 1\}_0$ the expansion for $F(\tau)$ is

$$\xi - k\xi^2 - 3k^2\xi^3 - k^3\xi^4 - 5k^4\xi^5 \cdots + 2152{,}80386{,}03144{,}71068{,}67277{,}46400\cdot 3^{-40}\cdot k^{81}\xi^{82} + \cdots$$

with $k^9 = 27/256$. For the distribution of θ we ran a χ^2-test.

We now turn to $\{\Gamma_9, 2\}_0$ and $\{\Gamma_{711}, 3\}_0$, where in each case the vector space of cuspforms has dimension 2, with a basis

$$F(\tau) = \xi + \sum_{n=2}^{\infty} k^{n-1}b(n)\cdot\xi^n,$$

$$G(\tau) = \sum_{n=2}^{\infty} k^{n-1}c(n)\cdot\xi^n,$$

and $b(n)$, $c(n)$ in Q. In this case we find the following. For each good prime p, there exists a quadratic field $Q(\sqrt{P})$, integers $A_1(p)$ and $A_2(p)$ in $Q(\sqrt{P})$, and λ_1 and λ_2 lying in the extension of the p-adic field by $\sqrt{P}$ (when P is a quadratic nonresidue of p), such that if

$$F_r(\tau) = \xi + \sum_{n=2}^{\infty} k^{n-1}\{b(n)+\lambda_r c(n)\}\xi^n = \xi + \sum_{n=2}^{\infty} k^{n-1}a_r(n)\xi^n,$$

$$T_r(n) = a_r(np) - A_r(p)a_r(n) + p^{2w-1}a_r(n/p),$$

then for $r=1, 2$ we have

$$\pi(T_r(n)) \geq (2w-1)(\alpha+1) \qquad \text{if } \pi(n) = \alpha.$$

Our computations here have so far been restricted to $p \leq 19$. However, $|A_r(p)| < 2p^{w-1/2}$ in all cases, and we also find that all the fields $Q(\sqrt{P})$ are real. We give below the left-hand sides of the quadratic equations satisfied by $A_r(p)$.

p	$\{\Gamma_9, 2\}_0$	field $Q(\sqrt{P})$	$\{\Gamma_{711}, 3\}_0$	field $Q(\sqrt{P})$
3	bad	—	$\mu^2+20\mu-168$	$\sqrt{67}$
5	$\mu^2+3\mu-216$	$\sqrt{97}$	$\mu^2+60\mu-756$	$\sqrt{46}$
7	$\mu^2-25\mu-422$	$\sqrt{257}$	bad	—
11	μ^2-2889	$\sqrt{321}$	$\mu^2+24\mu-106992$	$\sqrt{186}$
13	$\mu^2+26\mu-1856$	Q	$\mu^2+84\mu-1152676$	$\sqrt{5890}$
17	$\mu^2+21\mu-7128$	$\sqrt{3217}$	$\mu^2+552\mu-3244500$	$\sqrt{10249}$
19	$\mu^2-55\mu-1694$	Q	$\mu^2-2628\mu-864136$	$\sqrt{647683}$

There are some striking differences between these results and those in the congruence case, where we would have λ_1 and λ_2 the *same* for *all* p, the eigenvalues $A_1(p)$ and $A_2(p)$ all lying in the *same* real quadratic field (that of λ_1 and λ_2), and of course $T_1(n)$ and $T_2(n)$ would be identically zero and not merely p-adically small. We should emphasize that λ_1 and λ_2 *are* apparently p-adic, i.e., their field is the extension of the p-adic field by $\sqrt{P}$ and not $Q(\sqrt{P})$ itself, whereas by contrast $A_1(p)$ and $A_2(p)$ are in $Q(\sqrt{P})$.

It is hard to resist the conclusion that for any subgroup G of Γ where the field $\mathscr{A}$ of $(G)^\infty$ is Q, and any d-dimensional vector space of cuspforms $(G, w)_0$ on G, there exists for each good p an equation of degree d over Q, whose roots $A_r(p)$ give rise to p-adic properties of the Fourier coefficients as above. For congruence subgroups, this certainly happens. There, the Hecke operators act first on the forms and after simultaneous diagonalization we reach the $A_r(p)$ for a fixed p as a by-product. One might hope, however, that there is some entirely different approach (which we cannot visualize at all at present) which could start with a fixed p in relation to the given $(G, w)_0$ and derive the basic equation of degree d which we conjecture, whose roots are the values $A_r(p)$. In particular, this might throw some light on those conjectures for Hecke eigenvalues which remain unsolved, such as $|A_r(p)|<2p^{w-1/2}$ and the distribution of θ, since these appear to remain valid for our noncongruence objects which are not accessible via Hecke operators.

5.3. *Differentials.* For $w=1$ the space $\{G, 1\}_0$ of $F(\tau)$ corresponds to differentials $F(\tau)\,d\tau$ of the first kind on H/G. In this case our conjecture is that, after the p-adic diagonalization, we have

$$a(np)-A(p)\cdot a(n)+p\cdot a(n/p) \equiv 0 \pmod{p^{\alpha+1}} \quad \text{if } n \equiv 0 \pmod{p^\alpha}.$$

If we write $a(n)=k'a'(n)$, where $(k')^{p-1}=m'\equiv 1 \pmod p$, this becomes

$$(m')^n a'(np)-A(p)\cdot a'(n) \equiv 0 \pmod p \quad \text{if } (n, p) = 1$$

and

$$(m')^{pn}a'(np^2)-A(p)\cdot a'(np)+p\cdot(m')^{-n}a'(n) \equiv 0 \pmod{p^{\alpha+2}} \quad \text{if } n \equiv 0 \pmod{p^\alpha}.$$

Now $(m')^n\equiv 1 \pmod{p^{\alpha+1}}$ if $n\equiv 0 \pmod{p^\alpha}$, so that the powers of m' in the above equations can be removed, showing that the congruences are *independent of* k, and invariant under any substitution $\xi \to k'\xi$ (if $\pi(k')=0$). An accidental insertion of a wrong data card in the computer led us to the much more general theorem below.

THEOREM 4. *Let $p \neq 2$ or 3, and let $y^2 = x^3 - Bx - C$ be an elliptic curve over the field of p elements. Let $x = \xi^{-2} + \sum_{n=-1}^{\infty} c(n)\xi^n$ be any expansion, with $y = \xi^{-3} + \cdots$, and write*

$$F = -\xi\, dx/2y\, d\xi = \xi + \sum_{n=1}^{\infty} a(n)\xi^n.$$

Here B, C, c(n), and a(n), are p-adic integers. Then

$$a(np) - A\cdot a(n) + p\cdot a(n/p) \equiv 0 \pmod{p^{\alpha+1}} \quad \text{if } n \equiv 0 \pmod{p^{\alpha}},$$

where

$$A = -\sum_{\lambda=0}^{p-1} \left(\frac{\lambda^3 - B\lambda - C}{p}\right).$$

We proved this by first showing that the property was invariant under any substitution $\xi \to \xi + M\xi^N$ (from a succession of which any p-adic expansion can be derived), and then choosing $x = \xi^{-2}$. This proves our conjectures for forms of weight 1 on subgroups of genus 1, when $\{G, 1\}_0$ has dimension 1. We do not reproduce this proof since we learned subsequently from Professor Cartier that he has proved a much more general result using the methods of p-adic Lie groups. His result in particular implies most of what we conjecture about forms of weight 1 when the field $\mathscr{A}$ for $(G)^\infty$ is Q.

It is worth observing that if $\mathscr{A}$ is Q and H/G has genus 1, then the elliptic curve involved is defined over Q. According to a conjecture of Weil [9], the constants $A = A(p)$ of our Theorem 4 must be the actual Fourier coefficients of x^p in a new form on $(\Gamma_0(N), 1)_0$, where N is the "conductor" of the curve.

5.4. *Further investigations.* We summarize here the further work which our present results make obviously desirable. The most desirable consequence would be that someone should find a theory; failing this, more computation is needed. Since the time involved is not trivial in either human or machine terms, we venture to suggest that anyone interested in pursuing these computations should correspond with one of us to avoid duplication of effort.

If we have the set-up of Theorem 2, with $(G_i)^\infty$, $\mathscr{A}_i$, and k_i ($i = 1$ to N), then our results can be summed up by the statement: if $\mathscr{A}_i = Q$, then for each good prime p and each d-dimensional space $\{G_i, w\}_0$ there exists an equation of degree d over Q, whole roots A_r occur in p-adic expressions $a_r(np) - A_r \cdot a_r(n) + p^{2w-1}a_r(n/p)$. If $d > 1$ and one of the A_r is zero modulo p, there remains a slight uncertainty as to the nature of the relations satisfied by these p-adic expressions, and computation directed to this end might be worthwhile. We do not feel any need for further confirmation of the existence of the A_r.

It would seem certain that for two sets $(G_i)^\infty$ and $(G_j)^\infty$ which are conjugate in Γ, the A_r must be the same. The Fourier coefficients of $(G_j)^\infty$ merely correspond to the expansions at another cusp of $(G_i)^\infty$. This we propose to investigate with our Γ_{711}.

One would expect that the $\pm\sqrt{-7}$ involved will appear as two separate parts of the p-adic relations.

The case where $\mathscr{A}_i$ is greater than Q is the problem of most immediate interest. One hopes for some A_r defined by an equation of degree d over $\mathscr{A}_i$, but whether one will need to consider the ideal factorization of (p) in $\mathscr{A}_i$ is not at all clear.

It would also be of interest to have more examples of the actual fields $\mathscr{A}_i$ as in §4.3. Finally there is the special problem of finding subgroups of genus 1, especially those for which $\mathscr{A}_i = Q$. One could make a cheap conjecture that every elliptic curve defined over an algebraic number field is realizable as the curve of H/G for some subgroup G of Γ, but there is no evidence for this; however, the problem is an interesting one.

The possible connexions with algebraic geometry of our objects A_r, even when these are eigenvalues of Hecke operators in the congruence case for $w > 1$, remain wholly obscure. Professor Weil tells us that this is an area where he would not dare to conjecture, so that we are not inclined to rush in.

References

1. M. H. Ashworth, *Congruence and identical properties of modular forms*, D. Phil. Thesis, Oxford, 80 pages, 1968.

2. A. O. L. Atkin and J. Lehner, *Hecke operators on* $\Gamma_0(m)$, Math. Ann. **185** (1970), 134–160.

3. Jun-ichi Igusa, *Kroneckerian model of fields of elliptic modular functions*, Amer. J. Math. **81** (1959), 561–577. MR **21** #7214.

4. F. Klein and R. Fricke, *Vorlesungen über die Theorie der elliptischen Modulfunktionen*, Band 1, Stuttgart, 1890, reprinted New York, 1966.

5. J. Leech, *A presentation of the Mathieu group* M_{12}, Canad. Math. Bull. **12** (1969), 41–43.

6. M. H. Millington, *On cycloidal subgroups of the modular group*, Proc. London Math. Soc. (3) **19** (1969), 164–176.

7. ———, *Subgroups of the classical modular group*, J. London Math. Soc. (2) **1** (1969), 351–357.

8. J.-P. Serre, *Une interprétation des congruences relatives à la fonction* τ *de Ramanujan*, Séminaire Delange-Pisot-Poitou, Paris, **9** (1967/8) n° 14, 17 pp.

9. A. Weil, *Über die Bestimmung Dirichletscher Reihen durch Funktionalgleichungen*, Math. Ann. **168** (1967), 149–156. MR **34** #7473.

10. K. Wohlfahrt, *An extension of F. Klein's level concept*, Illinois J. Math. **8** (1964), 529–535. MR **29** #4805.

The Atlas Computer Laboratory, Chilton, Didcot
Trinity College, Cambridge

SELFCONJUGATE TETRAHEDRA WITH RESPECT TO THE HERMITIAN VARIETY $x_0^3+x_1^3+x_2^3+x_3^3 = 0$ IN PG(3, 2^2) AND A REPRESENTATION OF PG(3, 3)[1]

R. C. BOSE

1. **Introduction.** If h is any element of a Galois field GF(q^2), where q is a prime or a prime power, then $\bar{h}=h^q$ is defined to be *conjugate* to h. Since $h^{q^2}=h$, h is conjugate to $\bar{h}$. A square matrix (h_{ij}), $i, j=0, 1, \ldots, N$, with elements from GF(q^2) is said to be *Hermitian* if $h_{ij}=\bar{h}_{ji}$ for all i, j. Let PG(N, q^2) denote the finite projective space of N dimensions over the Galois field GF(q^2). The set of all points in PG(N, q^2) whose row vectors $\mathbf{x}^T=(x_0, x_1, \ldots, x_N)$ satisfy the equation $\mathbf{x}^T H\mathbf{x}^{(q)}=0$ are said to form a *Hermitian variety* V_{N-1}, if H is Hermitian and $\mathbf{x}^{(q)}$ is the column vector whose transpose is $(x_0^q, x_1^q, \ldots, x_N^q)$. The variety is said to be *nondegenerate* if H has the rank $N+1$. By a transformation of coordinates the equation of a nondegenerate Hermitian variety can be expressed in the canonical form

$$x_0^{q+1}+x_1^{q+1}+\cdots+x_N^{q+1} = 0. \tag{1.1}$$

The properties of the Hermitian curve $x_0^{q+1}+x_1^{q+1}+x_2^{q+1}=0$ were studied by the author in some detail in [**1**], and the general Hermitian variety (1.1) was studied by I. M. Chakravarti and the author in [**2**]. We shall assume the results of these two papers to which the reader may refer for detailed proofs.

If C and D are two points of PG(N, q^2) with coordinates $\mathbf{c}^T=(c_0, c_1, \ldots, c_N)$ and $\mathbf{d}^T=(d_0, d_1, \ldots, d_N)$ then C and D are defined to be *conjugate* to each other if $\mathbf{c}^T\mathbf{d}^{(q)}=c_0d_0^q+c_1d_1^q+\cdots+c_Nd_N^q=0$, or equivalently $\mathbf{d}^T\mathbf{c}^{(q)}=d_0c_0^q+d_1c_1^q+\cdots+d_Nc_N^q=0$. The set of all points conjugate to C lie on a hyperplane Σ with equation $c_0^qx_0+c_1^qx_1+\cdots+c_N^qx_N=0$, which is defined to be the *polar hyperplane* of C. The point C is called the *pole* of Σ. The polar hyperplane of a point on V_{N-1} is defined to be the *tangent hyperplane* to V_{N-1} at C.

In this paper we shall confine our attention to the Hermitian variety V_2 with equation $x_0^3+x_1^3+x_2^3+x_3^3=0$ in PG(3, 2^2). Each point C then has a polar plane. A

[1] This research was supported by the National Science Foundation Grant No. GP-5790 and the Air Force Office of Scientific Research Grant No. AFOSR-68-1415.

tetrahedron with vertices C_1, C_2, C_3, C_4 is said to be *selfconjugate* with respect to V_2 if C_i is conjugate to C_j for $i \neq j$, $i, j = 1, 2, 3, 4$. Then the plane determined by any three vertices is the polar plane of the remaining vertex.

It was shown in [2] that the number of points on V_2 is $(2^3+1)(2^2+1)=45$. Since there are 85 points in PG$(3, 2^2)$, there are 40 points not on V_2. These points will be called *external points*. The points on V_2 will be called *internal points*. If C is any point on V_2, the tangent plane at C meets V_2 in three lines passing through C. Thus through each point of V_2 there pass three lines which are completely contained in V_2. These lines are called *generators* of V_2. There are 27 generators. Through C there pass 5 lines lying in the tangent plane at C, three of which are generators. The other two lines meet V_2 only at C. These are defined to be *tangents* to V_2. Hence there are 90 tangents to V_2, lying two by two on the 45 tangent planes. A line which is not a tangent or a generator intersects V_2 in exactly 3 points and is called a *secant*. A plane which is not a tangent plane meets V_2 in a nondegenerate Hermitian cubic with $2^3+1=9$ points. A nontangent plane will be called an *external plane* or *secant plane*. Each of the 40 external planes has thus 9 internal points and 12 external points. We shall show that the twelve external points on any external plane can be divided into four disjoint triplets, such that any two points which belong to the same triplet are conjugate to each other. Thus the 12 external points on any secant plane Π form four selfconjugate triangles. Let P be the pole of Π, then P is an external point. Each external point P is in this way a vertex of four selfconjugate tetrahedra, the other 12 vertices of which are the external points on the polar plane of P. Clearly there are 40 selfconjugate tetrahedra.

Let the 40 external points be called *pseudopoints*. If Π is a secant plane then the set of 13 points consisting of the 12 external points on Π and the pole of Π will be defined to be a *pseudoplane*. Thus there are 40 pseudoplanes. The set of 4 external points on any tangent to V_2 is defined to be a *pseudoline of the first kind*. Thus there are 90 pseudolines of the first kind. The set of 4 external points which are the vertices of a selfconjugate tetrahedron is defined to be a *pseudoline of the second kind*. Thus there are 40 pseudolines of the second kind. Now we note that the number of pseudopoints, pseudolines and pseudoplanes is the same as the number of points, lines and planes in PG(3, 3), viz., 40, 130 and 40 respectively. Also the number of points on a pseudoplane or a pseudoline agrees with the number of points on a plane or a line of PG(3, 3). This leads one to surmise that the system of pseudopoints, pseudolines and pseudoplanes is isomorphic with PG(3, 3). The main result of the present paper is to prove that this is indeed true.

2. **Association schemes.** The concept of association schemes was first introduced by Shimamoto and the author in [4]. Here we need consider only *two-class association schemes*. Such a scheme is defined as a scheme of relations between v objects such that

(i) any two objects are either first associates or second associates, but not both;

(ii) each object has n_i ith associates ($i=1, 2$);

(iii) if two objects are ith associates, then the number of objects common to the

jth associates of the first and kth associates of the second is p^i_{jk} and is independent of the pair of treatments with which we start, and $p^i_{jk}=p^i_{kj}$.

Clatworthy and the author [3] showed that it is unnecessary to assume the constancy of all the p^i_{jk}'s. If we assume that n_1, n_2, p^1_{11}, p^2_{11} are constant, then the constancy of p^1_{12}, p^1_{21}, p^1_{22}, p^2_{12}, p^2_{21} and p^2_{22} follows as well as $p^1_{12}=p^1_{21}$, $p^2_{12}=p^2_{21}$. Also it appears from their proof that

$$p^1_{12} = n_1-p^1_{11}-1 = p^1_{21}, \qquad p^1_{22} = n_2-n_1+p^1_{11}+1, \tag{2.1}$$

$$p^2_{12} = n_1-p^2_{11} = p^2_{21}, \qquad p^2_{22} = n_2-n_1+p^2_{11}-1. \tag{2.2}$$

Since a generator to V_2 does not contain any external point, the line joining two external points is either a tangent or a secant. We shall define two external points to be first associates if the line joining them is a tangent and second associates if the line joining them is a secant.

3. **The association scheme of the 8 external points on a tangent plane to V_2.** Let Σ be a tangent plane to V_2 at the point C. Then through C there pass three generators g_1, g_2 and g_3 of V_2. The other two lines through C in Σ are tangents to V_2. Let these be t_1 and t_2. Since each line in PG(3, 2^2) has 5 points t_i contains 4 external points B_{i1}, B_{i2}, B_{i3}, B_{i4}. Any two external points lying on the same tangent are first associates according to the definition given in the previous paragraph. Now consider the two external points B_{1j} and B_{2k} the first of which lies on t_1 and the second on t_2 (j, $k=1, 2, 3, 4$). The line $B_{1j}B_{2k}$ meets g_1, g_2, g_3 in points C_1, C_2, C_3 lying on V_2 and distinct from C. Thus $B_{1j}B_{2k}$ is a secant. Hence B_{1j} and B_{2k} are second associates. Thus the 8 external points on Σ can be divided into two groups of 4 points on the same tangent belonging to the same group. Two points belonging to the same group are first associates and two points belonging to different groups are second associates. It is then easy to check that the 8 points form a two class association scheme for which $n_1=3$, $n_2=4$, $p^1_{11}=2$, $p^2_{11}=0$. The remaining parameters can be deduced from (2.1) and (2.2). We thus obtain

LEMMA 1. *The eight external points on a tangent plane to V_2 form a two-class association scheme with parameters*

$$n_1 = 3, \qquad n_2 = 4, \qquad (p^1_{jk}) = \begin{pmatrix} 2 & 0 \\ 0 & 4 \end{pmatrix}, \qquad (p^2_{jk}) = \begin{pmatrix} 0 & 3 \\ 3 & 0 \end{pmatrix}. \tag{3.1}$$

4. LEMMA 2. *A secant to V_2 has exactly* 2 *external points which are conjugate to each other.*

Let l be a secant to V_2. Then l intersects V_2 in three points. Since l contains exactly five points, there are just two external points on l. Let these be P_1 and P_2. If the polar plane of P_1 does not pass through P_2, then it will meet l in a point C on V_2. Hence the polar plane of C must pass through P_1. Since the polar plane Σ of C is the tangent plane to V_2 at C, it follows that the line l which is identical with the line CP_1 is a tangent line, which is a contradiction. This shows that the polar of P_1 passes through P_2.

5. **The association scheme of the 12 external points in a secant plane to V_2.** Let Π be a secant plane to V_2. Then it meets V_2 in a nondegenerate Hermitian cubic V_1 which has $2^3+1=9$ points (cf. [1], [2]). Since there are $4^2+4+1=21$ points on Π, the number of external points on Π is 12. We shall show that these 12 points form a two-class association scheme in the sense of §2.

Given any external point P_1 on Π, the polar plane of P_1 meets it in a line l, which is the polar line of P_1 with respect to V_1. Let l meet V_1 in C_1, C_2, C_3. Then P_1C_1, P_1C_2, P_1C_3 are tangents to V_1 and therefore to V_2. P_1C_i contains 3 other points besides P_1 and C_i. It is clear that P_1 has 9 first associates among the external points on Π. Again there pass two secants through P_1 lying on Π. From Lemma 1, each of these contains exactly one external point. Let l_2 and l_3 be the two secants through P_1 and let P_2 and P_3 be the external points (other than P_1) contained in l_2 and l_3 respectively. Thus P_1 has two second associates. We have thus shown that $n_1=9$, $n_2=2$.

From Lemma 2, P_2 and P_3 are conjugate to P_1. Thus P_2P_3 is the polar line to P_1. Since P_1 is an external point P_2P_3 cannot be a tangent line. Hence P_2P_3 is a secant line. From Lemma 2, P_2 and P_3 must be second associates and conjugate to each other. Thus P_1, P_2, P_3 is a triangle selfconjugate with respect to V_1 and therefore with respect to V_2.

Each external point on Π is the vertex of a uniquely determined selfconjugate triangle. Hence we have

LEMMA 3. *If Π is a secant plane to V_2, then the* 12 *external points on Π can be partitioned into* 4 *triplets, such that the points in the same triplet are mutually second associates and form the vertices of a triangle selfconjugate with respect to V_2.*

Let Δ_1, Δ_2, Δ_3, Δ_4 be the four selfconjugate triangles on Π referred to in the above lemma. If Δ_1 is the triangle $P_1P_2P_3$, then for symmetry of notation we may change the designation of these points to P_{11}, P_{12}, P_{13} denoting the vertices of Δ_i by P_{i1}, P_{i2}, P_{i3} $(i=1, 2, 3, 4)$. We have already shown that $n_1=2$, $n_2=9$. The two second associates of P_{ij} are the other two vertices of the triangle Δ_i. The nine first associates of P_{ij} are then the nine vertices of the three triangles other than Δ_i. Since the line joining any two second associates must be a secant line, the 12 sides of the triangles Δ_1, Δ_2, Δ_3, Δ_4 are the 12 secants in Π. The remaining nine lines of Π are the tangents to V_1 at the nine points of V_1. The three vertices of Δ_1 can be joined to the three vertices of Δ_2 in nine ways giving rise to the lines $P_{1j}P_{2k}$ $(j, k=1, 2, 3)$. These must be the nine tangents to V_1. Hence each of the nine tangents to V_1 passes through one vertex of each of the triangles Δ_1, Δ_2, Δ_3, Δ_4. We thus have

LEMMA 4. *In Lemma* 3, *the line joining any two points belonging to different triplets must be a tangent to V_2 and therefore to the section V_1 of V_2 by Π. Each of the nine tangents to V_1 passes through exactly one point of each triplet.*

If two of the 12 external points on Π are first associates, then they belong to different triangles, and their common first associates are the six vertices of the two remaining triangles. Hence $p^1_{11}=6$.

Again if two of the external points on Π are second associates they belong to the same triangle and their common first associates are the nine vertices of the remaining three triangles. Hence $p_{11}^2=9$. The remaining parameters of the association scheme can now be calculated from (2.1) and (2.2). Hence we have

LEMMA 5. *The* 12 *external points on a secant plane to* V_2 *form a two-class association scheme with parameters*

$$n_1 = 9, \qquad n_2 = 2, \qquad (p_{jk}^1) = \begin{pmatrix} 6 & 2 \\ 2 & 0 \end{pmatrix}, \qquad (p_{jk}^2) = \begin{pmatrix} 9 & 0 \\ 0 & 1 \end{pmatrix}.$$

6. **The association scheme of the 40 points external to V_2.** We have already seen that there are 40 points external to V_2. Let us consider the association scheme formed by these points. Let P be any external point and let Π be polar plane to P. Then Π intersects V_2 in the cubic V_1 with nine points C_i ($i=1, 2, \ldots, 9$). The line PC_i is tangent to V_2 at C_i. Also if P_{ij} ($i=1, 2, 3, 4; j=1, 2, 3$) are the external points in Π then the lines PP_{ij} are the secants to V_2 through P. The first associates of P are the external points on the tangents through P (other than P). Each tangent has three such points. Hence P has 27 first associates, i.e., $n_1=27$. The second associates are the external points on the secants through P. Each secant has exactly one such point. Hence P has 12 second associates, i.e., $n_2=12$. We have incidently proved that

LEMMA 6. *The second associates of any external point P are the* 12 *external points on the polar plane* Π *of* P.

Lemmas 3, 4, and 5 thus provide us information about the geometric relations between the 12 second associates of any external point P.

Let P_1 and P_2 be any two external points which are first associates. We want to determine the number p_{11}^1 of points which are first associates of both P_1 and P_2. The line P_1P_2 must be a tangent to V_2 at some point C. Let Σ be the tangent plane to V_2 at C. Then from Lemma 1, P_1 and P_2 have two common first associates lying in Σ, which in the proof of this lemma have been shown to be the other two external points on the tangent P_1P_2. Let these be P_3 and P_4. Besides Σ there pass four other planes of PG(3, 2^2) through the line P_1P_2. These are secant planes, and from Lemma 5 each of these contains 6 points which are first associates of both P_1 and P_2. Two of these six points are P_3 and P_4. Hence each secant plane contains four first associates of P_1 and P_2 (other than P_3 and P_4). Hence $p_{11}^1=2+4\cdot 4=18$.

Again suppose P_1 and P_2 are second associates. The line joining P_1 and P_2 is a secant line l. Let the polar planes of P_1 and P_2 intersect in the line m. Then m is a secant line intersecting V_2 in three points which may be denoted by D_1, D_2, D_3. Besides these m has two external points Q_1 and Q_2. Then the planes of PG(3, 2^2) through l are the three planes $P_1P_2D_i$ ($i=1, 2, 3$) which are tangent planes, and the two planes $P_1P_2Q_j$ ($j=1, 2$) which are secant planes. The line l does not contain any external points other than P_1 and P_2. From Lemma 1, none of the planes $P_1P_2D_i$ ($i=1, 2, 3$) contains a common first associate of P_1 and P_2. Again from Lemma 5 each of the two planes $P_1P_2Q_j$ ($j=1, 2$) contains nine points which are first associates of both P_1 and P_2. Hence $p_{11}^2=2\cdot 9=18$. The other parameters are now given by (2.1) and (2.2). Hence we have

LEMMA 7. *The* 40 *points of* PG(3, 2^2) *external to* V_2 *form a two-class association scheme with parameters*

$$n_1 = 27, \qquad n_2 = 12, \qquad (p_{jk}^1) = \begin{pmatrix} 18 & 8 \\ 8 & 4 \end{pmatrix}, \qquad (p_{jk}^2) = \begin{pmatrix} 18 & 9 \\ 9 & 2 \end{pmatrix}.$$

7. **Pseudopoints, pseudolines and pseudoplanes.** Let P be any point external to V_2 and let Π be the polar plane of P. We have already studied in §5 the association scheme of the 12 external points in Π. Let these points be denoted as before by P_{ij} $(i=1, 2, 3, 4; j=1, 2, 3)$ where two points belong to the same selfconjugate triangle if they have common subscript i.

Consider the set of 13 points constituted by P and the 12 points P_{ij}. This set will be defined to be the *pseudoplane corresponding to* P, and will be denoted by $\Pi(P)$. If external points are called pseudopoints then each pseudoplane contains 13 pseudopoints, and there are 40 pseudoplanes.

Let t be a tangent to V_2 at any point C, then the set of four external points or pseudopoints on t will be defined to be a pseudoline $l(t)$ of the first kind. Since there are 90 tangents to V_2 there are 90 pseudolines of the first kind each containing four pseudopoints,

Four points which are pairwise conjugate with respect to V_2 may be said to form a selfconjugate tetrahedron with respect to V_2. The polar plane of each vertex of a selfconjugate tetrahedron is the plane determined by the other three vertices. In the notation already developed if P is any external point, then there are four selfconjugate tetrahedra with P as one vertex, viz., $P, P_{i1}, P_{i2}, P_{i3}$ $(i=1, 2, 3, 4)$ where P_{ij} are the external points of Π, the polar plane of P. Starting from each external point we thus get four selfconjugate tetrahedra. However each such tetrahedron can be obtained from any of its four vertices. Hence the number of selfconjugate tetrahedra is 40. A set of four external points or pseudopoints which form the vertices of a selfconjugate tetrahedron will be defined to be a pseudoline of the second kind. Thus there are 40 pseudolines of the second kind each containing 4 pseudopoints. Altogether there are $90+40=130$ pseudolines. We state here the following obvious lemma

LEMMA 8. *The four external points constituting a pseudoline of the first kind are pairwise first associates. The four external points constituting a pseudoline of the second kind are pairwise second associates.*

We note that the number of pseudopoints, pseudolines and pseudoplanes, viz., 40, 130 and 40 respectively, agrees with the number of points, lines and planes in PG(3, 3). Also the number of pseudopoints in a pseudoplane or a pseudoline agrees with the number of points in a plane or a line of PG(3, 3). This leads to the surmise that the system of pseudopoints, pseudolines and pseudoplanes is isomorphic to PG(3, 3). In what follows we shall prove that this is indeed true if incidence is defined in the natural way by the relation of containing or being contained in.

8. The configuration of pseudopoints or pseudolines incident with a given pseudoplane.

THEOREM 1. *Two distinct pseudopoints are incident with one and only one pseudoline.*

Consider the two distinct pseudopoints P_1 and P_2. We have to consider two separate cases.

Case I. Let the external points P_1 and P_2 be first associates. The line $t=P_1P_2$ is a tangent to V_2 at some point C. Then P_1 and P_2 are contained in the unique pseudoline $l(t)$ of the first kind. From Lemma 8 there is no pseudoline of the second kind containing both P_1 and P_2.

Case II. Suppose the external points P_1 and P_2 are second associates. The line $l=P_1P_2$ is a secant. The polar planes of P_1 and P_2 intersect in a secant m, containing the two external points P_3 and P_4 which are conjugate to each other by Lemma 2. Hence $P_1P_2P_3P_4$ is a selfconjugate tetrahedron. Thus P_1 and P_2 are contained in a unique pseudoline of the second kind. From Lemma 8 there is no pseudoline of the first kind containing both P_1 and P_2.

THEOREM 2. *If the pseudopoints P_1 and P_2 are incident with a pseudoplane, then all four points of the pseudoline containing P_1 and P_2 are incident with this pseudoplane.*

Let $\Pi(P)$ be the pseudoplane with which both P_1 and P_2 are incident. We have to consider two cases.

Case I. Let P_1 be P and let P_2 be one of the external points P_{ij} in the secant plane Π. Then from Lemma 3, P_{ij} is one vertex of a selfconjugate triangle in Π. The three vertices of this triangle together with P then constitute the unique pseudoline determined by P_1 and P_2 and they all lie in $P(\Pi)$.

Case II. Let P_1 and P_2 be two external points in the secant plane Π. If P_1 and P_2 are second associates they are vertices of a selfconjugate triangle (cf. Lemma 3). The three vertices of this triangle together with P constitute the unique pseudoline determined by P_1 and P_2 and all these lie in $P(\Pi)$. If P_1 and P_2 are first associates then they are vertices of two different selfconjugate triangles of Π. Then from Lemma 4 the line $t=P_1P_2$ is a tangent to V_2, and contains exactly one vertex of each of the four selfconjugate triangles in Π. Thus P_1 and P_2 are contained in the unique pseudoline $l(t)$, all of whose points belong to $\Pi(P)$.

THEOREM 3. *The system of pseudopoints and pseudolines incident with a pseudoplane is isomorphic with the system of points and lines of* PG(2, 3).

Consider the set of 13 pseudopoints of the pseudoplane $\Pi(P)$ where P is an external point. Any two of these pseudopoints determine a unique pseudoline contained in $\Pi(P)$. There are 78 pairs formed by the pseudopoints, since a particular pseudoline is determined by six different pairs, the number of pseudolines in $\Pi(P)$ is thirteen. Let V_1 be the section of V_2 by Π. Let t_i $(i=1, 2, \ldots, 9)$ denote the nine tangents to V_1, i.e., the nine tangents of V_2 lying in Π. Then $\Pi(P)$ contains nine pseudolines of the first kind, viz., $l(t_i)$ $(i=1, 2, \ldots, 9)$. Again $\Pi(P)$ contains four pseudolines of the second kind, viz., $m_i=(P, P_{i1}, P_{i2}, P_{i3})$ $i=1, 2, 3, 4$; where the points P_{ij} are defined as before. It is easily checked that each pseudopoint is contained in four of these 13 lines. Thus P is contained in m_1, m_2, m_3, m_4. Again P_{i1} is contained in m_i and the three pseudolines corresponding to the tangents to V_1 through P_{i1}, viz., the pseudolines determined by $P_{i1}P_{j1}, P_{i1}P_{j2}, P_{i1}P_{j3}$ where $j\neq i$.

Again any two pseudolines in $\Pi(P)$ intersect in exactly one point. If they are both of the second kind they intersect in P. If they are both of the first kind, say $l(t_i)$ and $l(t_j)$, $i\neq j$, then they intersect in the external point where the tangent t_i and t_j to V_1 meet. Again consider the pseudolines m_i and $l(t_j)$. Now by Lemma 4, t_j passes through just one of the points P_{i1}, P_{i2}, P_{i3}. Hence m_i and $l(t_j)$ intersect in exactly one pseudopoint.

Finally it is easy to find four pseudopoints of $P(\Pi)$ no three of which are contained in the same pseudoline. Since all the axioms of a finite projective plane are satisfied by the pseudopoints and lines of $\Pi(P)$, the required result follows as it is well known that there is essentially one finite projective plane of order 3, viz., PG(2, 3).

9. **The pseudoplane determined by three distinct pseudopoints not incident with the same pseudoline.** Let P_1, P_2, P_3 be three external points. If P_2 and P_3 are ith associates, P_3 and P_1 are jth associates, and P_1 and P_2 are kth associates we say that the triangle $P_1P_2P_3$ is of the type (i, j, k). We shall first prove the following lemma.

LEMMA 9. *If P_1, P_2, P_3 are three external points not contained in the same pseudoline, then the triangle $P_1P_2P_3$ cannot be type* (2, 2, 2).

Suppose the triangle $P_1P_2P_3$ is of type (2, 2, 2). Then P_1, P_2, P_3 are pairwise second associates, and the lines P_2P_3, P_3P_1, P_1P_2 are secant lines. Let Π be the plane of PG(3, 2^2) determined by P_1, P_2, P_3. The plane Π cannot be a tangent plane since by Lemma 1, two external points in a tangent plane cannot have a common second associate lying in the plane, as $p^2_{22}=0$ for the association scheme of the external points in a tangent plane. Thus Π is a secant plane. Let P be the pole of Π. Then P, P_1, P_2, P_3 are the vertices of a selfconjugate tetrahedron, and constitute a pseudoline of the second kind which is a contradiction.

THEOREM 4. *Three distinct pseudopoints not incident with the same pseudoline are incident with a unique pseudoplane.*

Let P_1, P_2, P_3 be three distinct external points. We have to consider three distinct cases according to the nature of the triangle $P_1P_2P_3$. From the previous lemma it cannot be of type (2, 2, 2).

Case I. Let $P_1P_2P_3$ be of type (1, 1, 2), i.e., P_1 and P_2 are second associates and P_3 is a first associate of both P_1 and P_2. The plane Π determined by P_1, P_2, P_3 must be a secant plane, since for the association scheme of external points in a tangent plane $p^2_{11}=0$ by Lemma 1. Let P be the pole of Π. Then the pseudopoints P_1, P_2, P_3 are contained in the pseudoplane $\Pi(P)$. The only other way for P_1, P_2, P_3 to be contained in a pseudoplane would be for one of the points, say P_i, to be the pole of a secant plane Π containing the other two points P_j and P_k, where (i, j, k) is some permutation of $(1, 2, 3)$. Then P_i is second associate of both P_j and P_k and the triangle $P_1P_2P_3$ is not of the type $(1, 1, 2)$.

Case II. Let $P_1P_2P_3$ be of type $(1, 2, 2)$, i.e., P_2 and P_3 are first associates and P_1 is a second associate of both. Let Σ be the plane determined by P_1, P_2, P_3. Then Σ must be a tangent plane for by Lemma 5, for the association scheme of external points in a secant plane $p_{12}^2=0$. P_2P_3 is a tangent to V_2 at some point C, and P_1 lies on the second tangent to V_2 through C. By Lemma 1, P_1 is conjugate to both P_2 and P_3. Let Π_1 be the polar plane of P_1. Then Π_1 passes through both P_2 and P_3. Then P_1, P_2, P_3 are contained in the pseudoplane $\Pi_1(P_1)$, P_2 and P_3 being vertices of different selfconjugate triangles in Π_1 (cf. proof of Lemma 5). Then the only other way P_1, P_2, P_3 could be contained in a pseudoplane is for P_1, P_3 to be in the polar plane Π_2 of P_2, or for P_1, P_2 to be in the polar plane of Π_3 of P_3. In any case P_2 and P_3 would be second associates from Lemma 6 which is a contradiction.

Case III. Let $P_1P_2P_3$ be of the type $(1, 1, 1)$. Then P_1, P_2, P_3 are pairwise first associates. Let Π be the plane determined by P_1, P_2, P_3. Then Π cannot be a tangent plane, for if Π is a tangent plane to V_2 at C, then we have shown during the proof of Lemma 1 that any two first associates must lie on the same tangent. If P_1, P_2, P_3 lie on the tangent t at C they are contained in the pseudoline $l(t)$ contrary to the hypothesis. Thus Π is a secant plane. Let P be the pole of Π. Then P_1, P_2, P_3 are contained in the pseudoplane $\Pi(P)$. They must be vertices of different self-conjugate triangles in Π. The only other way in which P_1, P_2, P_3 could be contained in a pseudoplane would be for one of them, say P_1, to be the pole of a secant plane Π_1 containing the other two points P_2 and P_3. In this case P_2 and P_3 would be second associates of P_1 from Lemma 6 which is a contradiction.

If the triangle $P_1P_2P_3$ is of any other type, except $(2, 2, 2)$ which is impossible from Lemma 9, we can reduce the situation to one of the Cases I, II, and III by suitably renaming the points.

COROLLARY 1. *A pseudoline $l(t)$ and a pseudopoint not contained in $l(t)$ are incident with a unique pseudoplane.*

Let P_1 be an external point not belonging to $l(t)$. Let P_2 and P_3 be any two pseudopoints of $l(t)$. Then from the above theorem P_1, P_2, P_3 are contained in a unique pseudoplane. By Theorem 2, the other points of $l(t)$ are also contained in this pseudoplane.

COROLLARY 2. *Two distinct pseudolines $l(t_1)$ and $l(t_2)$ which have a common pseudopoint are contained in a unique pseudoplane.*

COROLLARY 3. *Two distinct pseudolines $l(t_1)$ and $l(t_2)$ which have no common pseudopoint are not contained in any common pseudoplane.*

10. **Intersection of pseudolines and pseudoplanes.**

THEOREM 5. *The intersection of two distinct pseudoplanes is a pseudoline.*

Consider two distinct pseudoplanes $\Pi_1(P_1)$ and $\Pi_2(P_2)$, where Π_1 and Π_2 are the polar planes of the external points P_1 and P_2.

Case I. Let P_1 and P_2 be second associates. Then P_1P_2 is a secant. Π_1 and Π_2 intersect in a secant m containing two external points which are conjugate to each other by Lemma 2. Thus $P_1P_2P_3P_4$ is a selfconjugate tetrahedron. Thus $P_1(\Pi_1)$ and $P_2(\Pi_2)$ intersect in a pseudoline of the second kind consisting of the pseudopoints P_1, P_2, P_3, P_4. $\Pi_1(P_1)$ and $\Pi_2(P_2)$ cannot have any other point in common, otherwise they would coincide by Corollary 1 to Theorem 4.

Case II. Let P_1 and P_2 be first associates. The line P_1P_2 is a tangent to V_2 at some point C. The tangent plane Σ to V_2 at C contains another tangent t to V_2 passing through C and lying in Σ. It was shown in the proof of Lemma 1 that any external point Q on t is a second associate of both P_1 and P_2. Hence QP_1 and QP_2 are secants and by Lemma 2 Q is conjugate to both P_1 and P_2. Thus Π_1 and Π_2 pass through any external point on t. The pseudoline $l(t)$ of the second kind is therefore contained in $\Pi_1(P_1)$ and $\Pi_2(P_2)$. As in Case I the pseudoplanes $\Pi_1(P_1)$ and $\Pi_2(P_2)$ cannot have any other common pseudopoint otherwise they would coincide.

THEOREM 6. *A pseudoplane and a pseudoline not contained in the pseudoplane intersect in exactly one pseudopoint.*

Let $\Pi(P)$ be a pseudoplane, where P is an external point and Π is the polar plane of P. Consider a pseudoline not contained in P. We have to consider two cases.

Case I. Let $l(t)$ be a pseudoline of the second kind where t is tangent to V_2 at some point C. Let Σ be the tangent plane to V_2 at C. Suppose that $l(t)$ is not contained in $\Pi(P)$. Then Π does not pass through t. Hence Π intersects t in a single point of PG$(3, 2^2)$ which is either C or an external point of t. We thus have to consider two subcases. (a) Let Π intersect t in C. The line of intersection of Π and Σ must be the second tangent t^* to V_2 at C (i.e., the tangent other than t), since Π being a secant plane cannot meet V_2 in a generator. Now t^* and t are conjugate lines, i.e., the polar plane of any point on t passes through t^* and vice versa. Hence the pole of Π, i.e., P lies on t. Thus P is a pseudopoint common to $\Pi(P)$ and $l(t)$. There cannot be any other common pseudopoint otherwise $l(t)$ would be contained in $\Pi(P)$. (b) Let Π intersect t in an external point Q. Then the pseudopoint Q is common to $\Pi(P)$ and $l(t)$. As before there cannot be any other common point.

Case II. Let m be a pseudoline of the second kind consisting of the selfconjugate tetrahedron with vertices P_1, P_2, P_3, P_4, which are external points. Since m is not contained in $\Pi(P)$, P cannot coincide with one of the vertices P_1, P_2, P_3, P_4 and in consequence Π cannot coincide with one of the faces of the tetrahedron. Since there are only two external points on any edge P_iP_j $(i, j=1, 2, 3, 4, i\neq j)$ P cannot lie on any edge of the tetrahedron. Since each plane has 12 external points, so each face of the tetrahedron contains nine external points other than the vertices of the tetrahedron. This shows that the four faces contain altogether 36 distinct external points other than the vertices. These together with the four vertices make up all the 40 external points. This shows that P must be contained in one of the faces, say in the face $P_2P_3P_4$. Then the polar plane Π of P must pass through P_1. Hence P_1 is a pseudopoint common to the pseudoline $m=(P_1, P_2, P_3, P_4)$ and the pseudoplane $\Pi(P)$. As before there cannot be any other common point.

COROLLARY 1. *Three distinct pseudoplanes* $P_1(\Pi_1)$, $P_2(\Pi_2)$ *and* $P_3(\Pi_3)$ *which have no common pseudoline meet in exactly one pseudopoint.*

From Theorem 5, $P_1(\Pi_1)$ and $P_2(\Pi_2)$ intersect in a pseudoline $l(t)$. By hypothesis this line is not contained in $P_3(\Pi_3)$. The required result follows from Theorem 6.

THEOREM 7. *The system of pseudopoints, pseudolines and pseudoplanes forms a geometry isomorphic to* PG(3, 3).

This follows from Theorems 1–6 and their corollaries.

REFERENCES

1. R. C. Bose, *On the application of finite projective geometry for deriving a certain series of balanced Kirkman arrangements*, Golden Jubilee Commemorative Volume (1958–59), part II, Calcutta Math. Soc., Calcutta, 1963, pp. 341–354. MR **27** #4769.

2. R. C. Bose and I. M. Chakravarti, *Hermitian varieties in a finite projective space* PG(N, q^2), Canad. J. Math. **18** (1966), 1161–1182. MR **34** #668.

3. R. C. Bose and W. H. Clatworthy, *Some classes of partially balanced designs*, Ann. of Math. Statist. **26** (1965), 212–232.

4. R. C. Bose and T. Shimamoto, *Classification and analysis of partially balanced incomplete block designs with two associative classes*, J. Amer. Statist. Assoc. **47** (1952), 151–184. MR **14**, 67.

UNIVERSITY OF NORTH CAROLINA

MULTIPARTITIONS AND MULTIPERMUTATIONS

M. S. CHEEMA AND T. S. MOTZKIN

1.1. **Introduction.** In the following we study multipartition numbers. We consider ordered s-tuples $(n_1, \ldots, n_s)$ whose components $n_j, j=1, \ldots, s$, are nonnegative integers. A *multipartition* will mean a representation of an s-tuple as a sum of nonzero s-tuples; the latter are called *parts*. Sums which differ only in the order of parts are regarded as one and the same multipartition. For $s=1$ the multipartitions become the well-known partitions of nonnegative integers, on whose numbers an extensive literature exists starting with Euler **[11]** and culminating in such results as those of Ramanujan and Hardy **[18]**.

1.2. The asymptotic theory of bipartition numbers was developed by Auluck **[1]** and Wright **[33]**. Wright remarks that an asymptotic formula, comparable in accuracy to that of Hardy, Ramanujan and Rademacher for $s=1$, probably does not exist for $s>1$. He gave without proof the terms of highest and next highest order of an asymptotic formula for log $U(n_1, \ldots, n_s)$ (for the definition of u, u', $U, U', p, p', P, P', q, q', Q, Q'$, see §2), when all $n_i \to \infty$ in such a way that n_i/n_j is bounded for all i, j. The highest term had been derived by Meinardus **[22]**. Wright **[34]** obtains asymptotic expressions for $U(n_1, n_2)$, for $n_1^{1/2+\varepsilon_1} < n_2 < n_1^{2-\varepsilon_2}$, this is a substantial relaxation of restrictions on $n_1 n_2$.

1.3. Rieger **[24]** obtains an asymptotic formula and a bound for the number of partitions of n into exactly r parts; Glaisher **[13]** and Sylvester **[27]** had obtained these results, as mentioned in Wright **[36]**. We obtain similar asymptotic formulas and bounds for multipartition numbers. Auluck **[1]** has a formula for $u(n_1, n_2)$ when n_2 is fixed and n_1 large and Nanda **[23]** has shown that this formula remains valid for n_2 large provided $n_2 = O(n_1^{1/4})$. Robertson **[25]** extends Nanda's method to asymptotic formulas for $u(n_1, n_2, \ldots, n_s)$, $u'(n_1, n_2, \ldots, n_s)$, when one particular n_i tends to infinity more rapidly than the fourth power of every other n_i. We show that this is also true if only vectors with at most j nonzero components are allowed as parts of $(n_1, n_2, \ldots, n_s)$. We also show that asymptotic results of Robertson **[25]** for $U(n_1, n_2, \ldots, n_s)$, $U'(n_1, n_2, \ldots, n_s)$, when one particular n_i tends to infinity more slowly than the cube root of every other n_i, follow from the asymptotic behavior of $P_r(n_1, n_2, \ldots, n_s)$. Robertson **[26]** also extends Wright's method to derive a formula for $u(n_1, \ldots, n_s)$ where $n_1 n_2 \cdots n_s < \bar{n}^{s+1-\varepsilon}$, $\bar{n} = \min n_i$, $s > 1$.

1.4. We show that the partition numbers satisfy recurrence relations and establish a connection between multipartitions and certain ordered sets of permutations, (multipermutations) similar to certain results on lattice functions by MacMahon [21] for plane and solid partitions. Some properties of $q_r(n_1, n_2, \ldots, n_s)$ and $\lambda_r(m_1, m_2, \ldots, m_s)$ (defined in §2) will be proved, e.g. it will be shown that $\lambda_r(m_1, \ldots, m_s)$ is bounded for fixed m_i as $r \to \infty$. Formulas for $q_r(n_1, n_2, \ldots, n_s)$, $r \leq 5$, follow from the relations connecting multipartitions and multicompositions, i.e. partitions in which the order of the parts also matters.

1.5. Formulas are obtained for $\lambda_r(m_1, m_2)$, the number of permutations having characteristic numbers m_1, m_2 defined in 4.1 for $m_2 \leq 3$. Two methods for the evaluation of $q_r(n_1, \ldots, n_s)$, $\lambda_r(m_1, \ldots, m_s)$ on high speed computing machines are described, one based on recurrence relations among $q_r(n_1, \ldots, n_s)$ and the other involving multipermutations.

The recursive technique for evaluation of $q_r(n_1, \ldots, n_s)$, $\lambda_r(m_1, \ldots, m_s)$ is suitable for high speed computing when $s \leq 3$. The method based on permutations for evaluating $\lambda_r(m_1, m_2)$ is described in detail. For tables of values of $q_r(n_1, n_2)$, $\lambda_r(m_1, m_2)$, $q_r(n_1, n_2, n_3)$, $\lambda_r(m_1, m_2, m_3)$, see [9].

Carlitz [3] has shown that $\pi(n, m)$ the number of partitions of (n, m) such that $\min(n_j, m_j) \geq \max(n_{j+1}, m_{j+1})$ $(j = 1, 2, \ldots)$ is equal to the number of partitions into parts $(n, n-1)$, $(n-1, n)$, $(2n, 2n)$ $(n = 1, 2, \ldots)$. He also notes from the Jacobi identity that $\alpha(n, m)$ the number of partitions of (n, m) into distinct parts of the type $(n, n-1)$, $(n-1, n)$ is equal to $p(n-(n-m)(n-m+1)/2)$; conversely a combinatorial proof of this result gives another proof of the Jacobi identity, see [7].

1.6. There are many interesting unsolved problems in the theory of multipartitions. Multipartition numbers $u(n_1, \ldots, n_s)$ defined in §2, do not satisfy congruences similar to those satisfied by $p(n)$, the number of partitions of an integer n. To show that $u(n_1, n_2, \ldots, n_s) \equiv a \bmod m$ has an infinite number of solutions for all m, a and given s (even in the case $s = 1$) is an open problem.

Gordon [15] proved that $p(n)$ is equal to the number of two row plane partitions of n distinct along rows. Other problems are to find the forms for the generating functions of self-conjugate plane partitions, solid and higher dimensional partitions. Szekeres [28] proved that $P_r(n)$ attains a unique maximum for fixed n and r varying. The tables of $q_r(n_1, \ldots, n_s)$ display this unique maximum property of $P_r(n_1, \ldots, n_s)$ for fixed n_i and r varying.

Lastly it would be of interest to find the number of partitions of $(n_1, \ldots, n_s)$ into nonparallel vectors; in two dimensions this becomes the problem of finding the number of monotone convex polyhedral paths joining two points of a plane grid.

2.1. **Definitions and interconnections.** All numbers to be considered (except variables in power series) are nonnegative integers.

Let $(n_1, n_2, \ldots, n_s)$ be an ordered s-vector. Let $q_r(n_1, \ldots, n_s)$, q'_r, Q_r, Q'_r be the number of partitions of $(n_1, n_2, \ldots, n_s)$ into *at most* r parts, distinct parts, parts with positive components, distinct parts with positive components, and p_r, p'_r,

P_r, P'_r the number of partitions into *exactly* r parts (four categories as above). By u, u', U, U' we denote the respective numbers of partitions of $(n_1, n_2, \ldots, n_s)$ with no restrictions on the number of components. We make the convention that $u(0, \ldots, 0) = u'(0, \ldots, 0) = U'(0, \ldots, 0) = 1$. Note that $U(n_1, \ldots, n_s) = U'(n_1, \ldots, n_s) = 0$ if at least one, but not all $n_i = 0$. It is obvious that $q_r(n_1, n_2, \ldots, n_s)$ is also the number of solutions of the vector equation,

$$(2.1.1) \qquad (n_1, n_2, \ldots, n_s) = \sum_{i=1}^{r} (a_i, b_i, \ldots, k_i)$$

in which the order of the vectors on the right does not matter and zero vectors are permitted. The numbers q'_r, Q_r, Q'_r can be expressed in a similar way.

The generating function of q_r is given by

$$(2.1.2) \qquad \prod (1 - x_1^{k_1} x_2^{k_2} \cdots x_s^{k_s} Z)^{-1} = \sum_{r=0}^{\infty} \phi_r(x_1, x_2, \ldots, x_s) Z^r$$

where the product on the left extends over all $(k_1, \ldots, k_s)$; we have $\phi_0(x_1, \ldots, x_s) = 1$, and

$$(2.1.3) \qquad \phi_r(x_1, x_2, \ldots, x_s) = \sum q_r(n_1, \ldots, n_s) x_1^{n_1} x_2^{n_2} \ldots x_s^{n_s},$$

for $r \geq 1$ (the sum is over all $(n_1, n_2, \ldots, n_s)$). This is clear if we expand the product as a formal power series. The series converges if $|x_i| < 1$, $|Z| < 1$.

If the product in 2.1.2 extends over all nonnegative k_i except $(0, 0, \ldots, 0)$, it defines the generating function for $p_r(n_1, \ldots, n_s)$; we then write $\prod'$ instead of $\prod$.

If we write

$$(2.1.4) \qquad \prod (1 + x_1^{k_1} x_2^{k_2} \cdots x_s^{k_s} Z) = \sum_{r=0}^{\infty} \psi_r(x_1, x_2, \ldots, x_s) Z^r,$$

the product being taken over all nonnegative k_i except $(0, 0, \ldots, 0)$, then $\psi_0(x_1, x_2, \ldots, x_s) = 1$, and

$$(2.1.5) \qquad \begin{gathered} \psi_r(x_1, x_2, \ldots, x_s) = \sum p'_r(n_1, n_2, \ldots, n_s) x_1^{n_1} x_2^{n_2} \cdots x_s^{n_s}, \\ p'_r(0, 0, \ldots, 0) = 1, \text{ for } r \geq 1. \end{gathered}$$

If the products 2.1.2, 2.1.4 extend over all $k_i \geq 1$, they define the generating functions of Q_r and P'_r; we write $\prod''$ instead of $\prod$ and $\prod'$.

Next we note that

$$(2.1.6) \qquad \prod{}' (1 - x_1^{k_1} x_2^{k_2} \ldots x_s^{k_s})^{-1} = \sum u(n_1, n_2, \ldots, n_s) x_1^{n_1} x_2^{n_2} \ldots x_s^{n_s},$$

is the generating function for $u(n_1, n_2, \ldots, n_s)$. Writing $\prod''$ we obtain the generating function for $U(n_1, n_2, \ldots, n_s)$. Similarly

$$(2.1.7) \qquad \prod{}' (1 + x_1^{k_1} x_2^{k_2} \cdots x_s^{k_s}) = \sum u'(n_1, n_2, \ldots, n_s) x_1^{n_1} x_2^{n_2} \cdots x_s^{n_s}$$

is the generating function for $u'(n_1, \ldots, n_s)$. Writing $\prod''$, we obtain the generating function for $U'(n_1, n_2, \ldots, n_s)$.

The generating functions for P, q', Q' are defined in a similar way.

2.2. By definition it is clear that

$$q_r(n_1, n_2, \ldots, n_s) = \sum_{j=1}^{r} p_j(n_1, n_2, \ldots, n_s), \tag{2.2.1}$$

$$q'_r(n_1, n_2, \ldots, n_s) = \sum_{j=1}^{r} p'_j(n_1, n_2, \ldots, n_s), \tag{2.2.2}$$

$$Q_r(n_1, n_2, \ldots, n_s) = \sum_{j=1}^{r} P_j(n_1, n_2, \ldots, n_s), \tag{2.2.3}$$

$$Q'_r(n_1, n_2, \ldots, n_s) = \sum_{j=1}^{r} P'_j(n_1, n_2, \ldots, n_s). \tag{2.2.4}$$

LEMMA I. *For all r,*

$$P_r(n_1, n_2, \ldots, n_s) = q_r(n_1-r, n_2-r, \ldots, n_s-r), \tag{2.2.5}$$

$$P'_r(n_1, n_2, \ldots, n_s) = q'_r(n_1-r, n_2-r, \ldots, n_s-r). \tag{2.2.6}$$

PROOF. Corresponding to a partition of $(n_1, n_2, \ldots, n_s)$ say $(a_{11}, a_{21}, \ldots, a_{s1}) + (a_{12}, a_{22}, \ldots, a_{s2}) + \cdots + (a_{1r}, a_{2r}, \ldots, a_{sr})$ we obtain a partition $(a_{11}-1, a_{21}-1, \ldots, a_{s1}-1) + (a_{12}-1, a_{22}-1, \ldots, a_{s2}-1) + \cdots + (a_{1r}-1, a_{2r}-1, \ldots, a_{sr}-1)$ of $(n_1-r, n_2-r, \ldots, n_s-r)$ by subtracting the vector $(1, 1, 1, \ldots, 1)$ from each vector. If the vectors of the first set have all positive components, then those of the second set have nonnegative components (some of them may be zero-vectors). This sets up a 1-1 correspondence between the partitions of the first type and those of the second type, completing the proof. The proof of 2.2.6 is similar.

Now we show that the generating functions for $q_r(n_1, n_2, \ldots, n_s)$ and $p'_r(n_1, n_2, \ldots, n_s)$ are closely related. Then in the future we need consider only $q_r(n_1, n_2, \ldots, n_s)$, results about the other functions follow from these.

Let $\alpha_1, \alpha_2, \alpha_3, \ldots$ be an infinite sequence of numbers with $|\alpha_k| < 1$ for all k and $\sum_{k=1}^{\infty} |\alpha_k| < \infty$.

Let

$$\prod (1-\alpha_k y)^{-1} = 1 + \sum_{r=1}^{\infty} B(r) y^r. \tag{2.2.7}$$

Put

$$S(m) = \sum \alpha_k^m. \tag{2.2.8}$$

By logarithmically differentiating 2.2.7 and equating coefficients of y^r we have

$$rB(r) = \sum_{m=1}^{r} S(m) B(r-m). \tag{2.2.9}$$

In particular if the α_i's are replaced by $x_1^{k_1}x_2^{k_2}\cdots x_s^{k_s}$, then

$$S(m) = \sum x_1^{mk_1}x_2^{mk_2}\cdots x_s^{mk_s} = \prod_{i=1}^{s}(1-x_i^m)^{-1}.$$

Thus 2.2.9 becomes

$$r\phi_r(x) = \sum_{m=1}^{r}\phi_{r-m}(x)\Big/\prod_{i=1}^{s}(1-x_i^m). \tag{2.2.10}$$

Similarly

$$r\psi_r(x) = \sum_{m=1}^{r}(-1)^{m-1}\psi_{r-m}(x)\Big/\prod_{i=1}^{s}(1-x_i^m), \tag{2.2.11}$$

where $\phi_0(x_1, x_2, \ldots, x_s)=\psi_0(x_1, x_2, \ldots, x_s)=1$. Using these recurrences we can express ϕ_r and ψ_r as rational functions of x_i whose denominators are

$$\prod_{m=1}^{r}\prod_{i=1}^{s}(1-x_i^m).$$

Hence we can write

$$\phi_r(x) = \lambda_r(x)\Big/\prod_{m=1}^{r}\prod_{i=1}^{s}(1-x_i^m), \tag{2.2.12}$$

where

$$\lambda_r(x) = \sum \lambda(m_1, m_2, \ldots, m_s)x_1^{m_1}x_2^{m_2}\cdots x_s^{m_s} \tag{2.2.13}$$

is a polynomial of degree $\binom{r}{2}$ in each of x_i.

If we take the logarithm and exponential of the expression

$$\prod(1-\alpha_k y)^{-1} = 1+\sum B(r)y^r$$

we obtain

$$1+\sum B(r)y^r = \exp\left(\sum_{m=1}^{\infty} S(m)/my^m\right),$$

or

$$B(r) = \sum_{(r)}\prod\{S(m)/m\}^{h_m}/h_m!, \tag{2.2.14}$$

where the summation is over all partitions of $r=\sum mh_m$ and the product is over all $m\in M$.

Similarly from

$$\prod(1+\alpha_k y) = 1+\sum_{r=1}^{\infty} A(r)y^r,$$

we obtain

$$A(r) = (-1)^r\sum_{(r)}(-1)^{h_m}\prod\{S(m)/m\}^{h_m}/h_m!. \tag{2.2.15}$$

In our case $S(m)=\prod_{i=1}^{s}(1-x_i^m)^{-1}$ and

$$\phi_r(x_1, x_2, \ldots, x_s) = \sum_{(r)} \prod \left\{m \prod_{i=1}^{s} (1-x_i^m)\right\}^{-h_m} \Big/ h_m!, \tag{2.2.16}$$

$$\psi_r(x_1, x_2, \ldots, x_s) = (-1)^r \sum_{(r)} \prod (-1)^{h_m} \left\{m \prod_{i=1}^{s} (1-x_i^m)\right\}^{-h_m} \Big/ h_m!, \tag{2.2.17}$$

where the ranges of summation and product are as above. It follows that

$$\phi_r(x_1, x_2, \ldots, x_s^{-1}) = (-1)^r x_s^r \psi_r(x_1, x_2, \ldots, x_s), \tag{2.2.18}$$

$$\psi_r(x_1, x_2, \ldots, x_s^{-1}) = (-1)^r x_s^r \phi_r(x_1, x_2, \ldots, x_s), \tag{2.2.19}$$

$$\phi_r(x_1, x_2, \ldots, x_{s-1}^{-1}, x_s^{-1}) = (x_{s-1}x_s)^r \phi_r(x_1, x_2, \ldots, x_s), \tag{2.2.20}$$

$$\psi_r(x_1, x_2, \ldots, x_s) = x_s^{\binom{r}{2}} \lambda_r(x_1, x_2, \ldots, x_s^{-1}) \Big/ \prod_{m=1}^{r} \prod_{i=1}^{s} (1-x_i^m), \tag{2.2.21}$$

and

$$p_r'(n_1, n_2, \ldots, n_s) = \sum \lambda_r\left(m_1, m_2, \ldots, \binom{r}{2} - m_s\right) \prod_{i=1}^{s} q_r(n_i - m_i), \tag{2.2.22}$$

$$q_r(n_1, n_2, \ldots, n_s) = \sum \lambda_r(m_1, m_2, \ldots, m_s) \prod_{i=1}^{s} q_r(n_i - m_i), \tag{2.2.23}$$

where the sums are over all $m_i \leq \binom{r}{2}$.

(Here $q_r(n)$ is the number of partitions of n into almost r parts.) Thus it is possible to find $q_r(n_1, n_2, \ldots, n_s)$ and the other functions if $\lambda_r(m_1, m_2, \ldots, m_s)$ are known. Alternatively the $q_r(n_1, n_2, \ldots, n_s)$ are determined by using the recurrence (2.2.10). A computational procedure based on (2.2.10) will be discussed in §9.

By using generating functions it is easy to show the following.

THEOREM 2.I. *The absolute value of the excess of the number of partitions of the vector* (n_1, n_2) *into an even number of distinct parts over those into an odd number of distinct parts is unbounded for* $n_1 \to \infty$ *and* $n_2 \geq 2$ *fixed.*

In one dimension this excess is bounded, as is seen from the Euler identity

$$\prod_{i=1}^{\infty} (1-x^i) = \sum (-1) x^{\lambda(\lambda/2)(3\lambda \pm 1)},$$

where the right-hand side has coefficients 0, ± 1.

THEOREM 2.II. *The following identity holds*:

$$\begin{aligned} &\sum_{m=1}^{n_1+n_2} (-1)^m \sum_{s=1}^{m} (-1)^s \binom{m}{s} \binom{s+n_1-1}{n_1} \binom{s+n_2-1}{n_2} \\ &\quad = 2^{n_2-1} \sum_{r=0}^{\mathrm{Min}(n_1,n_2)} 2^{n_1-r} \binom{_1n}{r} \binom{n_2}{r} \end{aligned} \tag{2.2.24}$$

PROOF. The generating function for the number of compositions $C_{m,n}$ of n into exactly m parts in the one-dimensional case is $C_m(x) = \sum C_{m,n}x^n = (x+x^2+\cdots)^m = x^m(1-x)^{-m}$ and therefore

$$C_{m,n} = \binom{n-1}{m-1}.$$

In the two-dimensional case the generating function is

$$\{1/(1-x_1)(1-x_2)-1\}^m = C_m(x_1, x_2) = \sum C_m(n_1, n_2)x_1^{n_1}x_2^{n_2}.$$

The coefficient of $x_1^{n_1}x_2^{n_2}$ is given by

$$(-1)^m \sum_{s=1}^{m} (-1)^s \binom{m}{s}\binom{s+n_1-1}{n_1}\binom{s+n_2-1}{n_2}.$$

This is the number of compositions of (n_1, n_2) into exactly m parts (vectors with nonnegative components). By summing from $m=1$ to n_1+n_2 we obtain the number of compositions of (n_1, n_2) which is given by

$$2^{n_2-1} \sum_{r=0}^{\mathrm{Min}(n_1,n_2)} 2^{n_1-r}\binom{n_1}{r}\binom{n_2}{r}$$

as shown by MacMahon [**21**, p. 171]. This proves the identity.

3.1. **Recurrence relations.** Call Δ the G.C.D. of the set of k's. All sums in t below are over $\{t : t|\Delta\}$.

THEOREM 3.I. *u, u', U, U' satisfy the recurrence relations*

$$n_1u(n_1, \ldots, n_s) = \sum_{k_i \ge 0} \left(u(n_1-k_1, \ldots, n_s-k_s) \sum_t k_1/t\right); \tag{3.1.1}$$

$$n_1u'(n_1, \ldots, n_s) = \sum_{k_i \ge 0} \left(u'(n_1-k_1, \ldots, n_s-k_s) \sum_t (-1)^{t-1}k_1/t\right); \tag{3.1.2}$$

$$n_1U(n_1, \ldots, n_s) = \sum_{k_i > 0} \left(U(n_1-k_1, \ldots, n_s-k_s) \sum_t k_1/t\right); \tag{3.1.3}$$

$$n_1U'(n_1, \ldots, n_s) = \sum_{k_i > 0} \left(U'(n_1-k_1, \ldots, n_s-k_s) \sum_t (-1)^{t-1}k_1/t\right). \tag{3.1.4}$$

PROOF OF (3.1.1). By (2.1.6):

$$\prod{}' (1-x_1^{k_1}x_2^{k_2}\cdots x_s^{k_s})^{-1} = \sum u(n_1, \ldots, n_s)x_1^{n_1}\cdots x_s^{n_s}.$$

Hence

$$\log \sum u(n_1, \ldots, n_s)x_1^{n_1}\cdots x_s^{n_s} = -\sum \log(1-x_1^{k_1}\cdots x_s^{k_s}) = \sum x_1^{mk_1}x_2^{mk_2}\cdots x_s^{mk_s}/m.$$

Differentiating partially with respect to x_1, we get

$$\sum n_1u(n_1, \ldots, n_s)x_1^{n_1}\cdots x_s^{n_s} = \sum u(n_1, \ldots, n_s)x_1^{n_1}\cdots x_s^{n_s} \sum k_1x_1^{mk_1}\cdots x_s^{mk_s}.$$

Equating coefficients, we obtain

$$n_1 u(n_1, \dots, n_s) = \sum_{k_i \geq 0} \left(u(n_1 - k_1, \dots, n_s - k_s) \sum_t k_1/t \right).$$

The proofs of (3.1.2), (3.1.3), (3.1.4) are similar.

3.2. THEOREM 3.II. *For all r, q_r, p'_r satisfy the following recurrence relations:*

$$rq_r(n_1, n_2, \dots, n_s) = \sum_{j=1}^{r} \sum_m q_{r-j}(n_1 - jm_1, \dots, n_s - jm_s), \tag{3.2.1}$$

$$rp'_r(n_1, n_2, \dots, n_s) = \sum_{j=1}^{r} \sum_m (-1)^{j-1} p'_{r-j}(n_1 - jm_1, \dots, n_s - jm_s), \tag{3.2.2}$$

where the sums are over all m_i such that $0 \leq m_i \leq [n_i/j]$, $i = 1(1)s$.

REMARK. Similar recurrence relations hold for q'_r, Q_r, Q'_r, p_r, P_r, P'_r.

PROOF. From (2.2.10) it is clear that the generating functions $\phi_r(x_1, \dots, x_s)$, $\psi_r(x_1, x_2, \dots, x_s)$ of q_r and p'_r satisfy recurrence relations

$$r\phi_r(x_1, \dots, x_s) = \sum_{j=1}^{r} \phi_{r-j}(x_1, \dots, x_s) \Big/ \prod_{i=1}^{s} (1 - x_i^j),$$

$$r\psi_r(x_1, \dots, x_s) = \sum_{j=1}^{r} (-1)^{j-1} \psi_{r-j}(x_1, \dots, x_s) \Big/ \prod_{i=1}^{s} (1 - x_i^j).$$

We can expand each of the denominators as a power series in x_i and the theorem then follows by equating coefficients of $x_1^{n_1} x_2^{n_2} \cdots x_s^{n_s}$ on both sides.

4.1. **Multipartitions and multipermutations.** In §2, it was shown that $q_r(n_1, \dots, n_s)$, $p'_r(n_1, \dots, n_s)$ depend on $\lambda_r(m_1, \dots, m_s)$ and $q_r(n)$, the number of partitions of n into almost r parts. In this section we show that the quantities $\lambda_r(m_1, \dots, m_s)$ can be evaluated with the aid of certain r-tuples whose components are permutations.

Let

$$\pi = \begin{pmatrix} 1 & 2 & 3 & \cdots & r \\ t_1 & t_2 & t_3 & \cdots & t_r \end{pmatrix}$$

be a permutation on r marks. We define two characteristic numbers $m_1 = m_1(\pi)$ and $m_2 = m_2(\pi)$ by the equations

$$m_1 = \sum_{t_i = t_j + 1, i < j} (r - t_j), \qquad m_2 = \sum_{t_{j+1} < t_j} (r - j).$$

For example, if

$$\pi = \begin{pmatrix} 1 & 2 & 3 & 4 & 5 & 6 & 7 & 8 & 9 \\ 7 & 1 & 3 & 5 & 2 & 9 & 8 & 4 & 6 \end{pmatrix}$$

then

$$m_1 = (9-6) + (9-4) + (9-8) + (9-2) = 16,$$

$$m_2 = (9-1) + (9-4) + (9-6) + (9-7) = 18.$$

We note that if

$$\pi^{-1} = \begin{pmatrix} 1 & 2 & 3 & 4 & \cdots & r \\ \sigma_1 & \sigma_2 & \sigma_3 & \sigma_4 & \cdots & \sigma_r \end{pmatrix}$$

is the inverse of π, then

$$m_1 = \sum_{\sigma_{i+1} < \sigma_i} (r-i).$$

THEOREM 4.I. *For all $r>0$, $\lambda_r(m_1, m_2)$ is the number of permutations on r marks having characteristic numbers m_1, m_2.*

REMARK. This shows that $\lambda_r(m_1, m_2) \geq 0$.

PROOF. First let $r=2$, and let the two parts be (a_1, b_1), (a_2, b_2), where $a_1 \leq a_2$ and if $a_1 = a_2$ then $b_1 \leq b_2$. We have two cases:

I. $b_1 \leq b_2$;

II. $b_1 > b_2$.

In the first case $a_1 \leq a_2$, $b_1 \leq b_2$; so the number of partitions of this type is $q_2(n_1)q_2(n_2)$. In the second case $b_2 < b_1$ implies $b_2 \leq b_1 - 1$, and $a_1 < a_2$ implies $a_1 \leq a_2 - 1$. Moreover $a_1 + (a_2 - 1)$ is a partition of $n_1 - 1$, and $b_1 + (b_2 - 1)$ is a partition of $n_2 - 1$. Therefore the number of partitions of this type is $q_2(n_1-1)q_2(n_2-1)$. Hence $q_2(n_1, n_2) = q_2(n_1)q_2(n_2) + q_2(n_2-1)$. Comparing this with formula (2.2.23), we see that $\lambda_2(0, 0) = \lambda_2(1, 1) = 1$, and $\lambda_2(m_1, m_2) = 0$ for all other m_1, m_2.

Now suppose $r \geq 3$. Let $(n_1, n_2) = (a_1, b_1) + (a_2, b_2) + \cdots + (a_r, b_r)$ be a partition of (n_1, n_2), where the parts (a_j, b_j) are supposed to be lexicographically ordered, i.e. $a_i \leq a_{i+1}$, and if $a_i = a_{i+1}$, then $b_i \leq b_{i+1}$.

There are r mutually exclusive possibilities for the b_i, namely

$$\begin{array}{llllll} b_1 & \leq b_2 & \leq \cdots \leq & b_{r-1} & \leq b_r, \\ b_1 & \leq b_2 & \leq \cdots \leq & b_r & < b_{r-1}, \\ \vdots & \vdots & & \vdots & \vdots \\ b_r & < b_{r-1} & < \cdots < & b_2 & < b_1. \end{array}$$

With each of these cases we associate a permutation π by requiring that $b_{t_j} < b_{t_{j+1}}$ whenever $t_j > t_{j+1}$ and $b_{t_j} \leq b_{t_{j+1}}$ otherwise. For example if $r=4$ and $b_3 \leq b_4 < b_1 \leq b_2$, we put

$$\pi = \begin{pmatrix} 1 & 2 & 3 & 4 \\ 3 & 4 & 1 & 2 \end{pmatrix}.$$

Now the conditions

$$b_{t_1} \leq b_{t_2} \leq \cdots \leq b_{t_j} < b_{t_{j+1}} \leq \cdots \leq b_{t_r}$$

can be written in the form

$$b_{t_1} \leq b_{t_2} \leq \cdots \leq b_{t_j} \leq b_{t_{j+1}} - 1 \leq \cdots \leq b_{t_r} - 1,$$

thereby converting the strict inequality $b_{t_j} < b_{t_{j+1}}$ into the nonstrict inequality $b_{t_j} \leq b_{t_{j+1}} - 1$. Clearly we introduce $r-j$ minus ones in this process. If there are strict inequalities at different places we can repeat the procedure and finally obtain a partition of $n_2-\sum_{t_j>t_j+1}(r-j)$. Similarly strict inequalities in the a's can be replaced by nonstrict inequalities. Corresponding to a given arrangement of the b_i we infer strict inequalities at certain places in the arrangement of the a_i. To make this clear we consider an example.

Let $b_{t_1}\, b_{t_2}\, b_{t_3}\, b_{t_4}\, b_{t_5}$ be $b_3\, b_1\, b_2\, b_5\, b_4$. This implies by definition of π that $b_3<b_1$, $b_5<b_4$.

Now $b_3<b_2$ implies $a_2<a_3$, and $b_5<b_4$ implies $a_4<a_5$.

Thus

$$b_3 < b_1 \leq b_2 \leq b_5 < b_4, \quad m_2 = 5,$$
$$a_1 \leq a_2 < a_3 \leq a_4 < a_5, \quad m_1 = 4;$$

with every partition of this form we can associate as above a partition of n_1-4 and a partition of n_2-5.

In general we see that the strict inequalities in the b_i which follow from π give rise to strict inequalities in the a_i at positions corresponding to those of the inverse permutation of π. There will be a number of arrangements of the b_i and a_i to which there correspond preassigned values of m_1, m_2 and we denote this number by $\lambda_r(m_1, m_2)$. Clearly $\sum \lambda_r(m_1, m_2)=r!$, the total number of permutations on r marks.

We have

$$m_2 = \sum_{t_j+1<t_j} (r-j), \qquad m_1 = \sum_{\sigma_i+1<\sigma_i} (r-i),$$

which in terms of the permutation π means

$$m_1 = \sum_{t_i=t_j+1,\, i<j} (r-t_j).$$

Now $\lambda_r(m_1, m_2)$ is the number of permutations on r marks for which these characteristic numbers are m_1, m_2 and as shown above, such a permutation gives rise to $q_r(n_1-m_1)q_r(n_2-m_2)$ partitions of (n_1, n_2). Hence

$$q_r(n_1, n_2) = \sum \lambda_r(m_1, m_2) q_r(n_1-m_1) q_r(n_2-m_2).$$

The maximum value of m_1 or m_2 is $\binom{r}{2}$ corresponding to strict inequalities throughout for the arrangement of the a_i or of the b_i, i.e. $\lambda_r(m_1, m_2)=0$ if m_1 or $m_2>\binom{r}{2}$. Thus the summation extends over $0\leq m_i\leq\binom{r}{2}$, $i=1, 2$.

Hence the above expression for $q_r(n_1, n_2)$ can be written as

$$q_r(n_1, n_2) = \sum \lambda_r(m_1, m_2) q_r(n_1-m_1) q_r(n_2-m_2).$$

The same equation follows from

$$\sum q_r(n_1, n_2) x_1^{n_1} x_2^{n_2} = \sum \lambda_r(m_1, m_2) x_1^{m_1} x_2^{m_2} \Big/ \prod_{i=1}^{r} (1-x_1^i)(1-x_2^i).$$

Thus $\lambda_r(m_1, m_2)$ can be obtained by analytic or by combinatorial methods for $s=2$.

For $s \geq 3$, the analytic methods are essentially unchanged, but the combinatorial character of $\lambda_r(m_1, m_2, \ldots, m_s)$ becomes more complex.

We will show that for $r > 0$, and $s \geq 3$, $\lambda_r(m_1, m_2, \ldots, m_s)$ depend on ordered sets of permutations (multipermutations).

First let $s=3$ and $(n_1, n_2, n_3)=(a_1, b_1, c_1)+(a_2, b_2, c_2)+\cdots+(a_r, b_r, c_r)$ where the components of the vectors on the right are arranged lexicographically, i.e. according to the size of the a_i, for equal a_i according to the size of the b_i, and for equal a_i, b_i according to the size of the c_i.

Thus the a_i are arranged in ascending order, and to arrange the b_i and c_i in ascending order we use permutations π, π'. These are defined by the requirement that $b_{t_j} < b_{t_{j+1}}$ whenever $t_j > t_{j+1}$, $b_{t_j} \leq b_{t_{j+1}}$ otherwise, and $c_{t'_j} < c_{t'_j}$ whenever $t'_j > t'_{j+1}$, $c_{t'_j} \leq c_{t'_{j+1}}$ otherwise. This already determines strict inequalities at certain places among the b_i and c_i. Now as before the arrangement of the b's implies strict inequalities at certain places among the a's. Indeed whenever $b_{t_{j+1}} < b_{t_j}$, we must have $a_{t_j} < a_{t_{j+1}}$. Similarly whenever $c_{t'_{j+1}} < c_{t'_j}$, we must have either $b_{t'_j} < b_{t'_{j+1}}$ or $a_{t'_j} < a_{t'_{j+1}}$. The system of inequalities so arising are not mutually exclusive, and so we must apply Sylvester's principle of exclusion and inclusion.

If there are l such systems, each with its own m_1, m_2, m_3 and thus contributing $+1$ to $\lambda_r(m_1, m_2, m_3)$, then any two of the systems have in common a new system, the partitions corresponding to which have been counted twice, and thus these $\binom{l}{2}$ systems each contribute -1 to some $\lambda_r(m_1, m_2, m_3)$; similarly there are $\binom{l}{3}$ systems requiring a correcting contribution of $+1$, and so on. Ultimately each pair of permutations contributes $+1$ to $\sum \lambda_r(m_1, m_2, m_3)$ because

$$l-\binom{l}{2}+\binom{l}{3}-\cdots+(-1)^{l-1}\binom{l}{l} = 1.$$

Since there are $(r!)^2$ possible pairs of permutations, we have $\sum \lambda_r(m_1, m_2, m_3)=(r!)^2$.

The following example will serve to explain the situation.

Let $b_{t_1}\, b_{t_2}\, b_{t_3}$ be $b_1\, b_2\, b_3$ and $c_{t'_1}\, c_{t'_2}\, c_{t'_3}$ be $c_3\, c_2\, c_1$; by definition of π', $c_3 < c_2 < c_1$.

Thus there are four possibilities for the arrangement of a_i, b_i namely, A, B, C, D, where

A: $a_1 < a_2 < a_3,\ b_1 \leq b_2 \leq b_3;\ (m_1, m_2, m_3)=(3, 0, 3)$.
B: $a_1 \leq a_2 \leq a_3,\ b_1 < b_2 < b_3;\ (m_1, m_2, m_3)=(0, 3, 3)$.
C: $a_1 < a_2 \leq a_3,\ b_1 \leq b_2 < b_3;\ (m_1, m_2, m_3)=(2, 1, 3)$.
D: $a_1 \leq a_2 < a_3,\ b_1 < b_2 \leq b_3;\ (m_1, m_2, m_3)=(1, 2, 3)$.

The $\binom{4}{2}$ cases common to A and B, A and C, etc. are

AB: $a_1 < a_2 < a_3,\ b_1 < b_2 < b_3;\ (m_1, m_2, m_3)=(3, 3, 3)$.
AC: $a_1 < a_2 < a_3,\ b_1 \leq b_2 < b_3;\ (m_1, m_2, m_3)=(3, 1, 3)$.
AD: $a_1 < a_2 < a_3,\ b_1 < b_2 \leq b_3;\ (m_1, m_2, m_3)=(3, 2, 3)$.
BC: $a_1 < a_2 \leq a_3,\ b_1 < b_2 < b_3;\ (m_1, m_2, m_3)=(2, 3, 3)$.
BD: $a_1 \leq a_2 < a_3,\ b_1 < b_2 < b_3;\ (m_1, m_2, m_3)=(1, 3, 3)$.
CD: $a_1 < a_2 < a_3,\ b_1 < b_2 < b_3;\ (m_1, m_2, m_3)=(3, 3, 3)$.

The $\binom{4}{3}$ cases common to ABC, etc. are

ABC: $a_1 < a_2 < a_3$, $b_1 < b_2 < b_3$; $(m_1, m_2, m_3) = (3, 3, 3)$.
ABD: $a_1 < a_2 < a_3$, $b_1 < b_2 < b_3$; $(m_1, m_2, m_3) = (3, 3, 3)$.
ACD: $a_1 < a_2 < a_3$, $b_1 < b_2 < b_3$; $(m_1, m_2, m_3) = (3, 3, 3)$.
BCD: $a_1 < a_2 < a_3$, $b_1 < b_2 < b_3$; $(m_1, m_2, m_3) = (3, 3, 3)$.

Lastly the case common to all is

ABCD: $a_1 < a_2 < a_3$, $b_1 < b_2 < b_3$; $(m_1, m_2, m_3) = (3, 3, 3)$,

and thus this pair of permutations contributes $+1$ to each of $\lambda_3(3, 0, 3)$, $\lambda_3(0, 3, 3)$, $\lambda_3(2, 1, 3)$, $\lambda_3(1, 2, 3)$; $\lambda_r(3, 3, 3)$, and -1 to each of $\lambda_3(3, 1, 3)$, $\lambda_3(3, 2, 3)$, $\lambda_3(2, 3, 3)$, $\lambda_3(1, 3, 3)$.

By considering all $(r!)^2$ possible pairs of permutations and their contribution to $\lambda_r(m_1, m_2, m_3)$ it is thus possible to evaluate all of the $\lambda_r(m_1, m_2, m_3)$.

The argument is quite general and applicable for $s > 3$. For $s = 4$ we have four sequences a_i, b_i, c_i, d_i, and we again define a permutation π'' such that $d_{t_j''} < d_{t_{j+1}''}$ whenever $t_j'' > t_{j+1}''$, $d_{t_j''} \leq d_{t_{j+1}''}$ otherwise.

After c_i we proceed to the arrangement of d_i; thus there are strict inequalities at certain places among the d_i which follow from the definition of π''.

Now note that whenever $d_{t_{j+1}''} < d_{t_j''}$, either $c_{t_j''} < c_{t_{j+1}''}$ or $b_{t_j''} < b_{t_{j+1}''}$ or $a_{t_j''} < a_{t_{j+1}''}$. Either one of these inequalities already holds, or there will be a set of l' new systems of inequalities. Again by the principle of exclusion and inclusion, any two of the new systems have in common a system, the partitions corresponding to which have been counted twice and so on. These new systems must be considered with each of the systems of inequalities which resulted from the arrangement of c_i. Again the contribution of each triple of permutations to $\sum \lambda_r(m_1, m_2, m_3, m_4)$ is $+1$ for the same reason and

$$\sum \lambda_r(m_1, m_2, m_3, m_4) = (r!)^3.$$

It was observed by taking random triples of permutations that in general most triples give rise to only one possibility of arrangement of the a_i, b_i, c_i, d_i.

A natural conjecture was to prove that the $\lambda_r(m_1, \ldots, m_s)$ are nonnegative. Carlitz **[2]** noted that one or more of these coefficients are negative when $s = 2$, $r = 4$, but Fine **[12]** points out that he was mistaken. Wright **[35]** confirmed the conjecture (i) for $r = 2, 3$ and all s and (ii) for $r = 4, 5$ and $s = 2$. Our combinatorial approach implies the truth of this statement for $s = 2$ and all r.

For s-tuples the first components are arranged in ascending order and to arrange the remaining $s-1$ components we use $s-1$ permutations $\pi_1, \pi_2, \ldots, \pi_{s-1}$. Let π_s be the identity permutation. Now write the set of permutations as $\pi_0, \pi_1, \ldots, \pi_{s-1}$, π_0 and form $\zeta_i = \pi_{i-1}^{-1}\pi_i$, $i = 1, \ldots, s$. Gordon **[14]** has shown, by extending our argument, that $m_1, \ldots, m_s$ are certain characteristic numbers of $\zeta_1, \ldots, \zeta_s$ respectively, and used this fact to prove the above conjecture.

5.1. **Asymptotic results.**

THEOREM 5.I. *For* $n_i \to \infty$ *and* r *fixed* (*more generally for* $r = o(\min(n_i^{1/3}))$),

$$(5.1.1) \quad q_r(n_1, n_2, \ldots, n_s) \sim r!^{-1} \prod_{i=1}^{s} \binom{n_i + r - 1}{r-1} \sim \left(\prod_{i=1}^{s} n_i\right)^{r-1} r!^{-1}(r-1!)^{-s}.$$

PROOF. As shown in §2

$$(5.1.2) \qquad q_r(n_1, n_2, \ldots, n_s) = \sum \lambda_r(m_1, \ldots, m_s) q_r(n_1 - m_1) \cdots q_r(n_s - m_s)$$

where $q_r(n)$ denotes the number of partitions of n into at most r parts. Erdös and Lehner [10] have shown that

$$(5.1.3) \qquad q_r(n) \sim r!^{-1}\binom{n+r-1}{r-1} \quad \text{as } n \to \infty \text{ and } r = o(n^{1/3}).$$

This can also be expressed as

$$q_r(n) = n^{r-1}/r!\,r-1!\{1 + o(1)\}.$$

Therefore

$$(5.1.4) \quad q_r(n_i - m_i) = n_i^{r-1}/r!(r-1)!\{1 + o(1)\} \quad \text{for } n_i \to \infty,\ r = (n_i^{1/3}),\ m_i \leq r^2,$$

because $(n_i - m_i)^{r-1} = n_i^{r-1} - (r-1)m_i n_i^{r-2} + (r-1)(r-2)/2m_i^2 n_i^{r-3} - \cdots$, if $r = \varepsilon_i n_i^{1/3}$, where $\varepsilon_i \to 0$ as $n_i \to \infty$. The kth term after the first in this expansion is $\leq r^{3k} n_i^{r-k-1} = \varepsilon^{3k} n_i^{r-1}$, where $\varepsilon = \max \varepsilon_i$, and the sum of the absolute value of these terms is

$$\leq \sum_{k=1}^{\infty} n_i^{r-1} \varepsilon^{3k} = n_i^{r-1}\varepsilon^3/(1-\varepsilon^3),$$

this proves (5.1.4).

We can substitute this in (5.1.2) and we get

$$q_r(n_1, n_2, \ldots, n_s) \sim \left(\prod_{i=1}^{s} n_i\right)^{r-1} A,$$

where

$$A = \frac{1}{(r!\,r-1!)^s} \sum \lambda_r(m_1, \ldots, m_s) = r!^{s-1}\{r!(r-1)!\}^{-s} = r!^{-1}(r-1!)^{-s},$$

i.e.

$$q_r(n_1, n_2, \ldots, n_s) \sim r!^{-1}(r-1!)^{-s} \prod_{i=1}^{s} n_i^{r-1}.$$

It is easy to see that

$$\prod_{i=1}^{s} \binom{n_i + r - 1}{r-1}$$

is asymptotic to $(r-1!)^{-s} \prod_{i=1}^{s} n_i^{r-1}$. This completes the proof of (5.1.1).

The number of compositions of $(n_1, n_2, \ldots, n_s)$ into at most r nonnegative parts, i.e., partitions taking order into account, is

$$k_r(n_1, n_2, \ldots, n_s) = \prod_{i=1}^{s} \binom{n_i+r-1}{r-1},$$

which is clearly $\leq r!q_r(n_1, n_2, \ldots, n_s)$, because some parts (vectors) may be equal. Theorem 5.I shows that

$$k_r(n_1, \ldots, n_s) \sim r!q_r(n_1, \ldots, n_s) \quad \text{as the } n_i \to \infty \text{ and } r = o(\min n_i^{1/3}).$$

REMARK 1. The asymptotic behavior of p_r, p'_r, P_r, P'_r, q'_r, Q_r, Q'_r follows from that of q_r, e.g.,

$$(5.1.5) \quad P_r(n_1, n_2, \ldots, n_s) \sim r!^{-1} \prod_{i=1}^{s} \binom{n_i-1}{r-1} \sim r!^{-1}(r-1!)^{-s} \prod_{i=1}^{s} n_i^{r-1}$$

and similar results for the others.

REMARK 2. The following results follow as a particular case of (5.1.5).

Assume without loss of generality that $n_1 \geq n_2 \geq \cdots \geq n_s$. Let $n_1, \ldots, n_{s-1} \to \infty$ and suppose $n_s = o(n_i^{1/3})$ for $1 \leq i \leq s-1$. Then

$$(5.1.6) \qquad U(n_1, n_2, \ldots, n_s) \sim \prod_{i=1}^{s-1} n_i^{n_s-1} n!_s^{-1}(n_s-1!)^{1-s}.$$

Under the same conditions

$$(5.1.7) \qquad U'(n_1, n_2, \ldots, n_s) \sim \prod_{i=1}^{s-1} n_i^{n_s-1} n!_s^{-1}(n_s-1!)^{1-s}.$$

The proof of (5.1.6) follows from (5.1.4), using the fact that $r = n_s$ in this case. The proof of (5.1.7) is similar.

5.2. THEOREM 5.II. *For fixed* $r \geq 3$ *and* $n_i \to \infty$, $1 \leq i \leq s$, $q_r(n_1, n_2, \ldots, n_s)$, $p_r(n_1, n_2, \ldots, n_s)$, $q'_r(n_1, n_2, \ldots, n_s)$, $p'_r(n_1, n_2, \ldots, n_s)$ *are equal to*

$$\left(\prod_{i=1}^{s} n_i\right)^{r-1} r!^{-1}(r-1!)^{-s}A, \quad \textit{where } A = 1 + \binom{r}{2} \sum_{i=1}^{s} \frac{1}{n_i} + o\left(\frac{1}{\min n_i}\right).$$

(5.2.2) $Q_r(n_1, n_2, \ldots, n_s)$, $Q'_r(n_1, n_2, \ldots, n_s)$, $P_r(n_1, n_2, \ldots, n_s)$, $P'_r(n_1, n_2, \ldots, n_s)$

are equal to

$$\left(\prod_{i=1}^{s} n_i\right)^{r-1} r!^{-1}(r-1!)^{-s}B,$$

where

$$B = 1 - \binom{r}{2} \sum_{i=1}^{s} \frac{1}{n_i} + o\left(\frac{1}{\min n_i}\right).$$

To prove Theorem 5.II we require the following lemma.

LEMMA. *For every* m_i, $\lambda_r(m_1, m_2, \ldots, m_s)$ *satisfy*

$$(5.2.3) \quad \sum m_i\lambda_r(m_1, \ldots, m_s) = \sum \left(\binom{r}{2} - m_i\right)\lambda_r(m_1, \ldots, m_s) = \tfrac{1}{2}(r!)^{s-1}\binom{r}{2},$$

where the sum is over all m_i *such that* $0 \le m_i \le \binom{r}{2}$.

PROOF OF LEMMA. Let

$$\sum_{m_1, m_2, \ldots, m_{s-1} = 0}^{\binom{r}{2}} \lambda_r(m_1, m_2, \ldots, m_s) = S_{m_s},$$

By symmetry of the λ's

$$S_{m_s} = S_{\binom{r}{2} - m_s}$$

Hence

$$\sum m_s\lambda_r(m_1, m_2, \ldots, m_s) = \sum \left(\binom{r}{2} - m_s\right)\lambda_r(m_1, m_2, \ldots, m_s)$$

$$= S_1 + 2S_2 + 3S_3 + \cdots + \binom{r}{2} S_{\binom{r}{2}}$$

but

$$S_1 = S_{\binom{r}{2}} = 1, \qquad S_j = S_{\binom{r}{2}-1}.$$

Thus if $\binom{r}{2}$ is odd,

$$\sum m_s\lambda_r(m_1, \ldots, m_s) = \binom{r}{2}\left\{1 + S_1 + S_2 + \cdots + S_{\frac{1}{2}(\binom{r}{2}-1)}\right\}$$

$$= \binom{r}{2} \cdot \tfrac{1}{2}(r!)^{s-1},$$

because the inner sum is one half of the sum of all the λ's.

Again for $\binom{r}{2}$ even

$$\sum m_s\lambda_r(m_1, \ldots, m_s) = \binom{r}{2}\left\{1 + S_1 + S_2 + \cdots + S_{\frac{1}{2}\binom{r}{2}-1}\right\} + \frac{1}{2}\binom{r}{2} S_{\frac{1}{2}\binom{r}{2}}$$

$$= \frac{1}{2}\binom{r}{2}\left\{1 + S_1 + S_2 + \cdots + S_{\binom{r}{2}}\right\} = \frac{1}{2}\binom{r}{2}(r!)^{s-1}.$$

This completes the proof of the Lemma.

PROOF OF THEOREM 5.II. We apply the following result of Glaisher and Sylvester ([**13**], [**27**]), also proved by Rieger [**24**] and Wright [**36**].

$$(5.2.4) \qquad P_r(n) = n^{r-1}/r!\,r-1!\{1 + r(r-1)(r-3)/4n + O(n^{-2})\}$$

for fixed $r \ge 3$ and $n \to \infty$.

Using this we have

$$q_r(n) = n^{r-1}/r!\,r-1!\{1 + r(r-1)(r+1)/4n + O(n^{-2})\},$$

and

$$q_r(n-m_1) = n^{r-1}/r!r-1!+n^{r-2}/r!r-2!(r^2+r-4m_1)/4+O(n^{r-3}).$$

Now

$$q_r(n_1, n_2, \ldots, n_s) = \sum \lambda_r(m_1, m_2, \ldots, m_s)q_r(n_1-m_1)\cdots q_r(n_s-m_s)$$

$$= \sum \lambda_r(m_1, m_2, \ldots, m_s) \prod_{i=1}^{s} (n_1^{r-1}/r!r-1!+n_i^{r-2}/r!r-2!(r^2+r-4m_1)/4+O(n_i^{r-3}))$$

coefficient of $(n_1n_2\cdots n_s)^{r-1}$ is

$$\sum \lambda_r(m_1, \ldots, m_s)/r!r-1!^s = r!^{-1}r-1!^{-s}.$$

The coefficient of $n_1^{r-2}(n_2n_3\cdots n_s)^{r-1}$ is

$$r!^{-s}r-2!^{-1}r-1!^{1-s} \sum (r^2+r-4m_1)/4\lambda_r(m_1, \ldots, m_s)$$
$$= r!^{-s}(r-2!)^{-1}(r-1!)^{1-s}\{r!^{s-1}(r^2+r)/4-r!^{s-1}(r^2-r)/4\}$$
$$= r-2!^{-1}r-1!^{-s}/2.$$

By symmetry the coefficient of $(n_1n_2\cdots n_{i-1}n_{i+1}\cdots n_s)^{r-1}n_i^{r-2}$ is the same. This completes the proof of (5.2.4) for $q_r(n_1, \ldots, n_s)$, $p_r'(n_1, n_2, \ldots, n_s)$.

But $p_r(n_1, n_2, \ldots, n_s)=q_r(n_1, n_2, \ldots, n_s)-q_{r-1}(n_1, n_2, \ldots, n_s)$, and so the first $s+1$ terms in (5.2.4) remain the same; similarly (5.2.1) holds for $q_r'(n_1, n_2, \ldots, n_s)$ which is equal to $\sum_{j=1}^{r} p_j'(n_1, n_2, \ldots, n_s)$. Now

$$P_r(n_1, n_2, \ldots, n_s) = q_r(n_1-r, n_2-r, \ldots, n_s-r)$$
$$= r!^{-1}r-1!^{-s} \prod_{i=1}^{s} (n_i-r)^{r-1}+\tfrac{1}{2}r-2!^{-1}(r-1!)^{-s}$$
$$\cdot \sum_{i=1}^{s} \{(n_1-r)\cdots(n_{i-1}-r)(n_{i+1}-r)\cdots(n_s-r)\}^{r-1}(n_i-r)^{r-2}$$
$$+\text{terms of lower order}$$
$$= r!^{-1}(r-1!)^{-s} \prod_{i=1}^{s} n_i^{r-1}+(r-1!)^{-s}r!^{-1}$$
$$\cdot \sum (n_1n_2\cdots n_{i-1}n_{i+1}\cdots n_s)^{r-1}n_i^{r-2}(-r(r-1)+r(r-1)/2)$$
$$+\text{terms of lower order}$$
$$= r!^{-1}(r-1!)^{-s} \prod_{i=1}^{s} n_i^{r-1}\cdot B.$$

This proves (5.2.2) for $P_r(n_1, n_2, \ldots, n_s)$, and for $P_r'(n_1, \ldots, n_s)$, Q_r, Q_r' the proof is similar.

REMARK. From (5.2.2) we obtain the following result for $U(n_1, n_2, \ldots, n_s)$ and $U'(n_1, n_2, \ldots, n_s)$.

For a fixed n_s and $n_i \to \infty$, $1 \leq i \leq s-1$,

(5.2.5) $U(n_1, n_2, \ldots, n_s)$ is equal to

$$n_s!-1(n_s-1!)^{1-s}\prod_{i=1}^{s-1} n_i^{n_s-1}\cdot C, \quad \text{where } C = 1-\binom{n_s}{2}\sum_{i=1}^{s-1}\frac{1}{n_i}+\cdots,$$

(5.2.6) $U'(n_1, n_2, \ldots, n_s)$ is equal to

$$n_s!-1(n_s-1!)^{1-s}\prod_{i=1}^{s-1} n_i^{n_s-1} D, \quad \text{where } D = 1-\binom{n_s}{2}\sum_{i=1}^{s-1}\frac{1}{n_i}+\cdots.$$

The result is a particular case of (5.2.2) for $r=n_s$.

Sylvester and Glaisher obtained the following formula. For $r\geq 7$

$$P_r(n) = \frac{1}{r!}\left\{\frac{n^{r-1}}{r-1!}+\frac{r(r-3)}{4(r-2)!}n^{r-2}+\frac{r(9r^3-58r^2+75r-2)}{288\,(r-3)!}n^{r-3}\right.$$
$$\left.+\frac{r^2(r-1)(r-3)(3r^2-19r+2)}{1152(r-4)!}n^{r-4}+\cdots\right\}. \tag{5.2.7}$$

We obtain, for r, $s\geq 3$,

$$q_r(n_1, \ldots, n_s) = \frac{\prod n_i^{r-1}}{r!(r-1!)^s}\left\{1+\frac{r(r-1)}{2}\left(\sum_{i=1}^{s}\frac{1}{n_i}\right)+\frac{r^2(r-1)^2}{4}\left(\sum_{i\neq j}\frac{1}{n_in_j}\right)\right.$$
$$+\frac{r(r-1)(3r^2-7r+2)}{24}\left(\sum_{i=1}^{s}\frac{1}{n_i^2}\right)+\frac{r^2(r-1)^2(r-2)(r-3)}{48}\left(\sum_{i=1}^{s}\frac{1}{n_i^3}\right)$$
$$+\frac{r^2(r-1)(^23r^2-7r+2)}{48}\left(\sum_{i\neq j}\frac{1}{n_in_j^2}\right) \tag{5.2.8}$$
$$\left.+\left(\frac{r^3(r-1)^3}{8}+\mu r\left(\frac{r-1}{2}\right)^{s+1}\right)\left(\sum_{i\neq j\neq k}\frac{1}{n_in_jn_k}\right)+\cdots\right\},$$

where

$$\mu = 1 \quad \text{for } s = 3,$$
$$= 0 \quad \text{for } s > 3.$$

PROOF. This follows from

$$q_r(n_1, \ldots, n_s) = \frac{1}{r!}\prod_{i=1}^{s}\binom{n_i+r-1}{r-1}+\frac{1}{2^{s+1}(r-2!)^{s+1}}\prod_{i=1}^{s}n_i^{r-2} \tag{5.2.9}$$
$$+\text{terms of lower order.}$$

The generating function $\phi_r(x_1, \ldots, x_s)$ satisfies

$$\phi_r(x_1, \ldots, x_s) = \sum_{(r)}\prod (h_m!)^{-1}(m\beta_s(m))^{-h_m} \tag{5.2.10}$$

where the sum is over all partitions of $r=\sum mh_m$ and

$$\beta_s(m) = \prod_{i=1}^{s}(1-x_i^m). \tag{5.2.11}$$

The terms of the highest orders come from the coefficient of $x_1^{n_1}\cdots x_s^{n_s}$ in

$$\frac{1}{r!}\prod (1-x_i)^{-r}+\frac{1}{2(r-2)!}\prod_{i=1}^{s}(1-x_i^2)^{-1}(1-x_i)^{-(r-2)}. \tag{5.2.12}$$

Now notice that $q_r(n_1,\ldots,n_s)=\sum_{t_i=1}^{r}(\beta(t_1,\ldots,t_s,n_1,\ldots,n_s)n_1^{t_1-1}\cdots n_s^{t_s-1})$ are semipolynomials in the sense of Wright [35] and the coefficients $\beta(t_1,\ldots,t_s,n_1,\ldots,n_s)$ are independent of $n_1,\ldots,n_s$ because $\max t_i+\min t_i>r$ in our case. This proves the result.

THEOREM 5.III. *For any $\varepsilon>0$, if the n_i are sufficiently large,*

$$\begin{aligned} r!^{-1}\prod_{i=1}^{s}\binom{n_i+r-1}{r-1} &\le q_r(n_1,n_2,\ldots,n_s)\\ &\le r!^{-1}(r-1)!^{-s}\prod_{i=1}^{s}\left(n_i+\frac{r}{2}+\varepsilon\right)^{r-1}. \end{aligned} \tag{5.2.13}$$

$$\begin{aligned} r!^{-1}\prod_{i=1}^{s}\binom{n_i-1}{r-1} &\le P_r(n_1,n_2,\ldots,n_s)\\ &\le r!r-1!^{-s}\prod_{i=1}^{s}\left(n_i-\frac{r}{2}+\varepsilon\right)^{r-1}. \end{aligned} \tag{5.2.14}$$

$$\begin{aligned} r!^{-1}\prod_{i=1}^{s}\binom{n_i-1}{r-1} &\le p_r(n_1,n_2,\ldots,n_s)\\ &\le r!^{-1}(r-1!)^{-s}\prod_{i=1}^{s}\left(n_i+\frac{r}{2}+\varepsilon\right)^{r-1}. \end{aligned} \tag{5.2.15}$$

REMARK 1. For $r\le 5$ we will give exact expressions for q_r, P_r, P_r' in §7.

REMARK 2. Bounds for Q_r, Q_r', q_r', p_r', P_r' follow from the bounds for q_r, P_r, p_r.

PROOF OF (5.2.14). $P_r(n_1,\ldots,n_s)$ is the number of partitions of $(n_1,\ldots,n_s)$ into exactly r parts, and there can be at most $r!$ possible arrangements of the parts corresponding to a given partition. Hence $r!P_r(n_1,n_2,\ldots,n_s)$ is greater than or equal to the number of compositions of $(n_1,n_2,\ldots,n_s)$ into exactly r parts, which is

$$\prod_{i=1}^{s}\binom{n_i-1}{r-1}.$$

This implies

$$P_r(n_1,n_2,\ldots,n_s)\ge r!^{-1}\prod_{i=1}^{s}\binom{n_i-1}{r-1}.$$

From (5.2.2) it follows by comparing the coefficients of the remaining s terms that

$$P_r(n_1,\ldots,n_s)\le r!^{-1}(r-1!)^{-s}\prod_{i=1}^{s}\left(n_i-\frac{r}{2}+\varepsilon\right)^{r-1}. \tag{5.2.16}$$

Now (5.2.13) follows from (5.2.14) replacing n_i by n_i+r, $i=1,2,\ldots,s$.

The proof of (5.2.15) is similar.

5.3. Here we define partition functions under restrictions of a different type.

DEFINITION. Let $u^{(r)}(n_1, n_2, \ldots, n_s)$ denote the number of all partitions of $(n_1, n_2, \ldots, n_s)$ with the restriction that the parts (vectors) have at most r nonzero components and $u'^{(r)}(n_1, n_2, \ldots, n_s)$ the number of such partitions into distinct parts.

The generating functions of these can be obtained as in §2, but because of their complexity, it is not possible to obtain an asymptotic result even for $u^{(2)}(n_1, n_2, n_3)$ for $n_1, n_2, n_3 \to \infty$. We prove

THEOREM 5.IV. *If $n_1 \to \infty$ and $n_i = o(n_1^{1/4})$, $2 \leq i \leq s$, then*

$$u^{(r)}(n_1, n_2, \ldots, n_s) \sim (6n_1/\pi^2)^{\sum_{i=2}^{s} n_i/2} \exp\{\pi\sqrt{2n_1/3}\}/n_1 \cdot 4\sqrt{3} \prod_{i=2}^{s} n_i!, \tag{5.3.1}$$

$$u'^{(r)}(n_1, n_2, \ldots, n_s) \sim (12n_1/\pi^2)^{\sum_{i=2}^{s} n_i/2} \exp\{\pi\sqrt{n_1/3}\}/4 \cdot 3^{1/4} n_1^{3/4} \prod_{i=2}^{s} n_i! \tag{5.3.2}$$

where $r \geq 2$.

REMARK 1. It follows from the definition that

$$u^{(1)}(n_1, n_2, \ldots, n_s) = p(n_1)p(n_2), \ldots, p(n_s).$$

PROOF OF THEOREM 5.IV. First we prove the result for

$$u^{(2)}(n_1, n_2, n_3) \quad \text{for } n_1 \to \infty,\ n_2, n_3 \text{ (fixed)} \geq 3.$$

The generating function for $u^2(n_1, n_2, n_3)$ is defined by

$$\begin{aligned} &\prod_{i=1}^{\infty} \{(1-x_1^i)(1-x_2^i)(1-x_3^i)\}^{-1} \prod_{i,j,k \geq 1}^{\infty} \{(1-x_1^i x_2^j)(1-x_1^j x_3^k)(1-x_2^i x_3^k)\}^{-1} \\ &\qquad = \sum_{n_1 n_2 n_3 = 0}^{\infty} u^{(2)}(n_1, n_2, n_3) x_1^{n_1} x_2^{n_2} x_3^{n_3}. \end{aligned} \tag{5.3.3}$$

We make use of the identity

$$\begin{aligned} &\{(1-a)(1-ax)(1-ax^2)(1-ax^3)\cdots\}^{-1} \\ &\qquad = 1 + \frac{a}{1-x} + \frac{a^2}{(1-x)(1-x^2)} + \frac{a^3}{(1-x)(1-x^2)(1-x^3)} + \cdots, \\ &\qquad\qquad |a| < 1,\ |x| < 1, \end{aligned} \tag{5.3.4}$$

and notice that the asymptotic behavior of the function in (5.3.3) as $n_1 \to \infty$ and n_2, n_3 are fixed is controlled by the leading term which comes from the product

$$\begin{aligned} &\prod_{i=1}^{\infty} (1-x_1^i)^{-1} \cdot \frac{x_2^{n_2}}{(1-x_1)(1-x_1^2)\cdots(1-x_1^{n_2})} \\ &\qquad \cdot \frac{x_3^{n_3}}{(1-x_1)(1-x_1^2)\cdots(1-x_1^{n_3})}. \end{aligned} \tag{5.3.5}$$

For fixed n_2, n_3, $u^{(2)}(n_1, n_2, n_3)$ being an increasing function of n_1, we can apply Ingham's Tauberian theorem [19] to (5.3.4), to get

$$u^{(2)}(n_1, n_2, n_3) \sim (6n_1/\pi^2)^{(n_2+n_3)/2} \exp\{\pi\sqrt{2n_1/3}\}/4\sqrt{3}\, n_1 n_2! n_3!.$$

Using Nanda's argument [23], it is easily shown that this result is also true for $n_2, n_3 = o(n_1^{1/4})$, and by using an inductive proof along the lines of Robertson [25], it follows that (5.3.1) holds for all finite s and $n_1 \to \infty$, $n_i = o(n_1^{1/4})$, $2 \le i \le s$. Thus (5.3.1) gives the asymptotic behavior of $u^{(2)}(n_1, n_2, \ldots, n_s)$, and it also holds for $r = s$ as proved by Robertson.

Again for $u^{(r)}(n_1, n_2, \ldots, n_s)$, $3 \le r \le s-1$, the asymptotic behavior of its generating function as $n_1 \to \infty$ and $n_i = o(n_1^{1/4})$, $2 \le i \le s$, is controlled by the leading term which comes from the product

$$\prod_{i=1}^{\infty} (1-x_1^i)^{-1} \prod_{j=2}^{s} \frac{x_j^{n_j}}{(1-x_1)(1-x_1^2)\cdots(1-x_1^{n_j})},$$

where the second product has the maximum number of factors $(1-x_1^i)^{-1}$ and the coefficient of $x_1^{n_1}, x_2^{n_2}, \ldots, x_s^{n_s}$ in this product is

$$\sim (6n_1/\pi^2)^{(1/2)\sum_{i=1}^{s} n_i} \exp\{\pi\sqrt{2n_1/3}\}/4n_1\sqrt{3} \prod_{i=1}^{s} n_i!,$$

thus (5.3.1) is valid for $2 \le r \le s$. The proof of (5.3.2) is similar.

DEFINITION. Let $c^{(r)}(n_1, n_2, \ldots, n_s)$ denote the number of all partitions of $(n_1, n_2, \ldots, n_s)$ with the restriction that parts (vectors) have exactly r nonzero components, $c'^{(r)}(n_1, n_2, \ldots, n_s)$ the number of such partitions into distinct parts. It is clear from (5.3.1), (5.3.2) and generating functions of $c^{(r)}(n_1, \ldots, n_s)$, $c'^{(r)}(n_1, \ldots, n_s)$ that as $n_1 \to \infty$ and $n_i = o(n_1^{1/4})$, $2 \le i \le s$ and $r \ge 2$:

$$c^{(r)}(n_1, \ldots, n_s) = o(u^{(r)}(n_1, \ldots, n_s)), \qquad c'^{(r)}(n_1, \ldots, n_s) = o(u'^{(r)}(n_1, \ldots, n_s)).$$

We will obtain asymptotic expressions for $c^{(2)}(n_1, n_2, \ldots, n_s)$ and $c'^{(2)}(n_1, n_2, \ldots, n_s)$, as $n_1 \to \infty$ and $n_i = o(n_1^{1/4})$, $2 \le i \le s$.

THEOREM 5.V. *For $n_1 \to \infty$ and $n_i = o(n_1^{1/4})$*

$$(5.3.6)\qquad \begin{aligned} c^{(2)}(n_1, n_2, \ldots, n_s) &\sim \prod_{i=2}^{s} n_i!^{-1} \binom{n_1 - 1}{\sum_{i=2}^{s} n_i - 1} \\ &\sim n_1^{(\sum_{i=2}^{s} n_i - 1)} \left(\sum_{i=2}^{s} n_i - 1\right)!^{-1} \prod_{i=2}^{s} n_i!^{-1}. \end{aligned}$$

$$(5.3.7)\qquad c'^{(2)}(n_1, \ldots, n_s) \sim \prod_{i=2}^{s} n_i!^{-1} \binom{n_1 - \sum_{i=2}^{s} \binom{n_i}{2}}{\sum_{i=2}^{s} n_i - 1}.$$

Proofs of (5.3.6) and (5.3.7) are similar to those of (5.3.1) and (5.3.2).

6.1. **Properties of** $q_r(n_1, n_2, \ldots, n_s)$, $\lambda_r(m_1, m_2, \ldots, m_s)$.

THEOREM 6.I. *For $m_i < M$, $i = 1, 2, \ldots, s$,*

$$\lambda_r(m_1, m_2, \ldots, m_s) \le \{(2M)!\}^{s-1}.$$

PROOF. First we prove the theorem for $s=2$. Recall that

$$m_2 = \sum_{t_{j+1}<t_j} (r-j).$$

For any permutation π, let $j_1<j_2<\cdots<j_\alpha$ be integers such that $t_{j_i+1}<t_{j_i}$ $(i=1, 2, \ldots, \alpha)$.

Let $r-j_\alpha=k_1$, $r-j_{\alpha-1}=k_2, \ldots, r-j_1=k_\alpha$ then $m_2=k_1+k_2+\cdots+k_\alpha$, where $k_1<k_2<k_3<\cdots<k_\alpha$.

If m_2 is fixed and $r\to\infty$, the number of such partitions of m_2 is bounded, but if m_2 and r both tend to ∞ the number of such partitions is unbounded. We show that the number of permutations with fixed $m_1, m_2\leq M$ is bounded by $(2M)!$.

Assume that $r\geq 2M$, and let

$$\pi = \begin{pmatrix} 1 & 2 & \cdots & r \\ t_1 & t_2 & \cdots & t_r \end{pmatrix}$$

be any permutation. Suppose $t_1=1$, $t_2=2, \ldots, t_k=k$, $t_{k+1}=k'>k+1$ and let $t_{k''}=k+1$, $k''>k+1$. Then $t_{k''}<t_{k+1}$, which implies $k''>r-M$.

Let

$$\sigma = \begin{pmatrix} 1 & 2 & 3 & \cdots & r \\ \sigma_1 & \sigma_2 & \sigma_3 & \cdots & \sigma_r \end{pmatrix}$$

be the inverse of

$$\pi = \begin{pmatrix} 1 & 2 & 3 & \cdots & r \\ t_1 & t_2 & t_3 & \cdots & t_r \end{pmatrix};$$

then $\sigma_{k'}<\sigma_{k+1}$, which implies $k'>r-M$.

Thus the $(r-M-k)$ numbers $t_{k+1}, t_{k+2}, \ldots, t_{r-M}$ are all different and greater than $r-M$, but less than or equal to r; hence their number is at most M, and therefore $r-M-k\leq M$, or $k\geq r-2M$ thus $t_i=i$ for $i\leq r-2M$ and $\lambda_r(m_1, m_2)$ $\leq(2M)!$, the number of permutations on the remaining $2M$ marks.

Now suppose $s=3$. As in §4, let

$$(n_1, n_2, n_3) = (a_1, b_1, c_1)+(a_2, b_2, c_2)+\cdots+(a_r, b_r, c_r)$$

and define π and π' so that $b_{t_j}<b_{t_{j+1}}$ whenever $t_j>t_{j+1}$, $b_{t_j}\leq b_{t_{j+1}}$ otherwise, and $c'_{t_j}<c'_{t_{j+1}}$ whenever $t'_j>t'_{j+1}$, $c_{t'_j}\leq c_{t'_{j+1}}$ otherwise. Consider $\lambda_r(m_1, m_2, m_3)$ for $m_i<M$.

As in the case $s=2$ the permutation π must satisfy $t_i=i$ for $i\leq r-2M$.

Notice that $m_3<M$ implies $t'_1<t'_2<\cdots<t'_{r-M}$. We show that $t'_1=1$. If not suppose, $t'_1=k\geq 2$, but $k\leq r-2M$, because the permutation is strictly increasing up to t'_{r-M}. This implies $c_k<c_{k-1}$, thus either $b_{k-1}<b_k$ or $a_{k-1}<a_k$; which is not the case; hence $t'_1=1$.

Similarly we show that

$$t'_i = i \qquad \text{for } i \leq r-2M.$$

If not assume that

$$t_1' = 1, \ldots, t_l' = l, \qquad t_{l+1}' = l' > l+1,$$

where $l+1 < r-2M$. This implies $c_{l+1} < c_l$ and thus either $a_l < a_{l+1}$ or $b_l < b_{l+1}$ which is not the case. Hence $t_i' = i$ for $i \le r-2M$, and $\lambda_r(m_1, m_2, m_3) \le ((2M)!)^2$, the number of pairs of permutations on the remaining $2M$ marks.

For higher dimensions all the $(s-1)$-tuples of permutations satisfy

$$t_i = t_i' = t_i'' = \cdots = t_i^{(s-2)} = i \quad \text{for } i \le r-2M,$$

and therefore

$$\lambda_r(m_1, m_2, \ldots, m_s) \le ((2M)!)^{s-1}.$$

6.2. From the proof we see that $\lambda_r(m_1, m_2)$ is constant for $m_1 \le M$, $m_2 \le M$ and $r \ge 2M$; therefore it suffices to consider permutations on $2M$ marks only in order to find

$$\lambda(m_1, m_2) = \operatorname*{limit}_{r \to \infty} \lambda_r(m_1, m_2).$$

These can be obtained by finding the partitions of m_1, m_2 into distinct parts. For example if $m_1 = m_2 = 4$, the partitions of m_1, m_2 in question are

$$\begin{array}{ll} 4, & 4 \\ 4, & 1+3 \\ 1+3, & 4 \\ 1+3, & 1+3. \end{array}$$

To these correspond 9 permutations on 8 marks

$$\left.\begin{array}{cccccccc} 1 & 2 & 3 & 5 & 4 & 6 & 7 & 8 \\ 1 & 2 & 5 & 6 & 3 & 4 & 7 & 8 \\ 1 & 5 & 6 & 7 & 2 & 3 & 4 & 8 \\ 5 & 6 & 7 & 8 & 1 & 2 & 3 & 4 \end{array}\right\} \text{ corresponding to 4, 4;}$$

$$\begin{array}{ccccccccl} 1 & 2 & 6 & 8 & 3 & 4 & 5 & 7 & \quad\text{corresponding to 4, } 1+3; \\ 1 & 2 & 5 & 6 & 7 & 3 & 8 & 4 & \quad\text{corresponding to } 1+3\text{, 4;} \end{array}$$

$$\left.\begin{array}{cccccccc} 1 & 2 & 3 & 4 & 6 & 5 & 8 & 7 \\ 1 & 2 & 3 & 4 & 8 & 6 & 7 & 5 \\ 1 & 2 & 3 & 6 & 8 & 4 & 7 & 5 \end{array}\right\} \text{ corresponding to } 1+3,\ 1+3.$$

Thus $\lambda_r(4, 4) = 9$ for all $r \ge 8$.

6.3. THEOREM 6.II. *The $\lambda_r(m_1, m_2)$ enjoy the following properties*:

$$\lambda_r(1, m_2) = \begin{cases} 1, & \textit{for } 1 \le m_2 \le r-1, \\ 0, & \textit{for } m_2 \ge r. \end{cases} \tag{6.3.1}$$

$$\lambda_r(2, m_2) = \begin{cases} \lambda_{r-1}(2, m_2), & \textit{for } 1 \le m_2 \le r-3, \\ 0, & \textit{for } m_2 > 2r-4, \\ \lambda_{r-1}(2, m_2)+1, & \textit{for } r-3 < m_2 \le 2r-4. \end{cases} \tag{6.3.2}$$

(6.3.3) $$\lambda_r(3, m_2) = \lambda_{r-1}(3, m_2), \quad \textit{for } 1 \leq m_2 \leq [\tfrac{r}{2}].$$

(6.3.4) $$\lambda_r(3, m_2+1)+\lambda_{r-1}(3, m_2)-\lambda_r(3, m_2)-\lambda_{r-1}(3, m_2+1) = 0.$$

If

$$\sum_{m_1=0}^{\binom{r}{2}} \lambda_r(m_1, m_2) = S_r(m_2),$$

then

$$S_r(m_2) = S_r(\tbinom{r}{2}-m_2).$$

(6.3.5) $$S_{r+1}(m_2) = \sum_{j=0}^{r} S_r(m_2-j).$$

(6.3.6) $$S_{r+1}(m_2) = \sum_{j=0}^{m_2} S_r(j), \quad \textit{for } m_2 \leq r.$$

(6.3.7) $$S_r(m_2)+S_r(m_2-r) = \sum_{j=0}^{m_2} S_{r-1}(j).$$

PROOF. The proofs of (6.3.1) to (6.3.4) follow from formulas for $\lambda_r(m_1, m_2)$, $m_2=0, 1, 2, 3$, $r \geq 2$ given in §8.

To prove (6.3.5), we notice that $S_r(m_2)$ is the number of permutations on r marks having characteristic number m_2. Consider permutations

$$\pi = \begin{pmatrix} 1 & 2 & 3 & \cdots & r \\ t_1 & t_2 & t_3 & \cdots & t_r \end{pmatrix}$$

on r marks having characteristic numbers $m_2, m_2-1, \ldots, m_2-r$. From these we can obtain permutations on $(r+1)$ marks having characteristic number m_2 in a unique way by the transformation $t_i \to t_i+1$ and by placing 1 at a suitable position. This is illustrated by the following example, where we have listed nine permutations on 4 marks with characteristic numbers $m_2=0, 1, 2$ and the corresponding permutations on 5 marks with characteristic number 2.

$$1\ 2\ 3\ 4 \longrightarrow 2\ 3\ 4\ 1\ 5;$$

$$\left.\begin{matrix} 1 & 2 & 4 & 3 \\ 1 & 3 & 4 & 2 \\ 2 & 3 & 4 & 1 \end{matrix}\right\} \longrightarrow \left.\begin{matrix} 2 & 3 & 5 & 1 & 4 \\ 2 & 4 & 5 & 1 & 3 \\ 3 & 4 & 5 & 1 & 2 \end{matrix}\right\};$$

$$\left.\begin{matrix} 2 & 4 & 1 & 3 \\ 3 & 4 & 1 & 2 \\ 1 & 4 & 2 & 3 \\ 2 & 3 & 1 & 4 \\ 1 & 3 & 2 & 4 \end{matrix}\right\} \longrightarrow \left.\begin{matrix} 1 & 3 & 5 & 2 & 4 \\ 1 & 4 & 5 & 2 & 3 \\ 1 & 2 & 5 & 3 & 4 \\ 1 & 3 & 4 & 2 & 5 \\ 1 & 2 & 4 & 3 & 5 \end{matrix}\right\}.$$

Formulae (6.3.6) and (6.3.7) follow from (6.3.5).

6.4. THEOREM 6.III. *The $q_r(n_1, n_2)$ enjoy the following properties*:

(6.4.1) $q_3(n, 2) = 2q_3(n, 1)$.

(6.4.2) $q_3(n, 3) = 3q_3(n, 1)+q_3(n-3, 0)$.

(6.4.3) $q_3(n, 4) = q_3(n, 1)+q_3(n, 3)+q_3(n-1, 0)+q_3(n-2, 0)$.

(6.4.4) $q_3(n, 5) = q_3(n, 1)+q_3(n, 4)+q_3(n-1, 0)+q_3(n-2, 0)+q_3(n-3, 0)$.

(6.4.5) $q_r(n, 1)-q_r(n-1, 1) = q_r(n, 0)-q_r(n-r, 0)$.

PROOF. The proofs of (6.4.1) to (6.4.4) follow immediately by substituting the values of $q_3(n, m)$ obtained in §7, while (6.4.5) follows from the identity

$$q_r(n, 1) = q_r(n)+q_r(n-1)+\cdots+q_r(n-r+1),$$

which is a consequence of (6.3.1).

7.1. **Formulas for $q_r(n_1, n_2, \ldots, n_s)$, $r\leq 5$.** Recall that $P_r(n_1, n_2, \ldots, n_s)$ is the number of partitions of $(n_1, n_2, \ldots, n_s)$ into exactly r parts with positive components. If we consider all possible arrangements, i.e., partitions in which the order of the vectors also matters, then we obtain compositions.

It is clear that the number of such compositions of $(n_1, n_2, \ldots, n_s)$ is

$$K_r'(n_1, n_2, \ldots, n_s) = \prod_{i=1}^{s} \binom{n_i-1}{r-1}.$$

Clearly any partition

$$(n_1, n_2, \ldots, n_s) = \sum_{i=1}^{r} (a_{i1}, a_{i2}, \ldots, a_{ir})$$

into positive parts can be written in the form

$$\sum_{j=1}^{k} r_j(b_{j1}, b_{j2}, \ldots, b_{js}),$$

where the vectors $(b_{j1}, b_{j2}, \ldots, b_{js})$ are distinct, and the coefficients r_j are the multiplicities with which they appear among the $(a_{i1}, a_{i2}, \ldots, a_{is})$. Of course we must have

$$r = \sum_{j=1}^{k} r_j.$$

We denote by $P_r(n_1, n_2, \ldots, n_s; r_1, r_2, \ldots, r_k)$ the number of partitions of $(n_1, n_2, \ldots, n_s)$ into positive parts with fixed multiplicities $r_1, r_2, \ldots, r_k$. Then

$$\begin{aligned} K_r'(n_1, n_2, \ldots, n_s) &= \prod_{i=1}^{s} \binom{n_i-1}{r-1} \\ &= \sum \frac{r!}{r_1!r_2!\cdots r_k!} P_r(n_1, n_2, \ldots, n_s; r_1, r_2, \ldots, r_k) \end{aligned} \tag{7.1.1}$$

where the summation is extended over all partitions of r.

From this we can easily derive

$$P_r(n_1, n_2, \ldots, n_s) = \sum_{r_1, r_2, \ldots, r_k} P_r(n_1, n_2, \ldots, n_s; r_1, r_2, \ldots, r_k).$$

We can also obtain explicit formulas for

$$P'_r(n_1, n_2, \ldots, n_s) = P_r(n_1, n_2, \ldots, n_s; 1, 1, \ldots, 1).$$

All products are over $i=1(1)s$.

THEOREM 7.I.

(7.1.2) $P_1(n_1, n_2, \ldots, n_s) = 1.$

(7.1.3) $P_2(n_1, n_2, \ldots, n_s) = \left[\frac{1}{2}\left(1+\prod (n_i-1)\right)\right].$

(7.1.4) $P_3(n_1, n_2, \ldots, n_s) = \left[\frac{1}{6}\left\{\prod \binom{n_i-1}{2}+3\prod [(n_i-1)/2]+2\right\}\right].$

(7.1.5) $q_1(n_1, n_2, \ldots, n_s) = 1.$

(7.1.6) $q_2(n_1, n_2, \ldots, n_s) = \left[\frac{1}{2}\left(1+\prod (n_i + 1)\right)\right].$

(7.1.7) $q_3(n_1, n_2, \ldots, n_s) = \left[\frac{1}{6}\left\{\prod \binom{n_i+2}{2}+3\prod [n_i/2+1]+2\right\}\right].$

(7.1.8) $P'_1(n_1, n_2, \ldots, n_s) = 1.$

(7.1.9) $P'_2(n_1, n_2, \ldots, n_s) = \left[\frac{1}{2}\left\{\prod (n_i-1)-1\right\}\right].$

(7.1.10) $P'_3(n_1, n_2, \ldots, n_s) = \left[\frac{1}{6}\left\{\prod \binom{n_i-1}{2}-3\prod [(n_i-1)/2]+2\right\}\right].$

Formulas for q'_r, Q_r, Q'_r, p_r, p'_r follow from these.

PROOF. If $r = 2$, (6.1.1) becomes

$$2P'_2(n_1, n_2, \ldots, n_s)+P_2(n_1, n_2, \ldots, n_s; 2) = \prod (n_i-1),$$

where

$$P_2(n_1, n_2, \ldots, n_s; 2) = 1, \quad \text{if } 2|\Delta,$$
$$= 0, \quad \text{if } 2\nmid\Delta,$$

where $\Delta=$G.C.D. of $n_1, \ldots, n_s$. Hence

$$P'_2(n_1, n_2, \ldots, n_s) = \left[\tfrac{1}{2}\left\{\prod (n_i-1)-1\right\}\right].$$

$$\begin{aligned} P(n_1, n_2, \ldots, n_s) &= P'_2(n_1, n_2, \ldots, n_s)+P_2(n_1, n_2, \ldots, n_s; 2) \\ &= \left[\tfrac{1}{2}\left\{\prod (n_i-1) + 1\right\}\right]. \end{aligned}$$

which implies

$$q_2(n_1, n_2, \ldots, n_s) = \left[\tfrac{1}{2}\left\{\prod (n_i+1)+1\right\}\right].$$

If $r=3$, (7.1.1) becomes

$$6P_3'(n_1, n_2, \ldots, n_s)+3P_3(n_1, \ldots, n_s; 2, 1)+P_3(n_1, n_2, \ldots, n_s; 3) = \prod \binom{n_i-1}{2}, \tag{7.1.11}$$

where

$$P_3(n_1, \ldots, n_s; 3) = \begin{matrix} 1, & \text{if } 3|\Delta, \\ 0, & \text{otherwise.} \end{matrix} = \text{call this number } \lambda, \text{ now}$$

$$P_3(n_1, n_2, \ldots, n_s; 2, 1) = \prod [\tfrac{1}{2}(n_i-1)]-\lambda \tag{7.1.12}$$

because the number of partitions of $(n_1, \ldots, n_s)$ into exactly three parts with at least two equal parts is equal to the number of partitions of $(n_1, \ldots, n_s)$ into two parts, one even. The number of even summands for $(n_1, \ldots, n_s)$ is given by

$$\prod \tfrac{1}{2}[(n_i-1)].$$

We have to subtract from these the cases where the second part is half of the first part, which correspond to three equal parts. This proves (7.1.12). Substituting this in (7.1.11), we get (7.1.4), (7.1.7) and (7.1.10).

For $r=4$, denote $(n_1, \ldots, n_s)$ by N_s. Then (7.1.1) becomes

$$24P_4(N_s; 1, 1, 1, 1)+12P_4(N_s; 1, 1, 2)+6P_4(N_s; 2, 2) +4P_4(N_s; 3, 1)+P_4(N_s; 4) = \prod \binom{n_i-1}{3}. \tag{7.1.13}$$

Now notice that

$$2P_4(N_s; 1, 1, 2)+2P_4(N_s; 2, 2)+2P_4(N_s; 1, 3)+P_4(N_s; 4) = \prod [\tfrac{1}{4}(n_i-2)^2]. \tag{7.1.14}$$

$$2P_4(N_s; 2, 2)+P_4(N_s; 4) = \begin{cases} \prod \left(\frac{n_i}{2}-1\right), & \text{if } 2|n_i,\ i = 1, \ldots, s, \\ 0, & \text{otherwise.} \end{cases} \tag{7.1.15}$$

$$P_4(N_s; 3, 1)+P_4(N_s; 4) = \prod \{[(n_i-4)/3]+1\}. \tag{7.1.16}$$

$$P_4(N_s, 4) = \begin{cases} 1, & \text{if } 4|n_i,\ i = 1, \ldots, s, \\ 0, & \text{otherwise.} \end{cases}$$

Substituting the values of $P_4(N_s; 4)$, $P_4(N_s; 3, 1)$, $P_4(N_s; 2, 2)$, $P_4(N_s; 2, 1, 1)$ in (7.1.13), we obtain

$$\begin{aligned} P_4'(n_1, n_2, \ldots, n_s) &= P_4(n_1, n_2, \ldots, n_s; 1, 1, \ldots, 1) \\ &\cdot \frac{1}{24}\left\{\prod \binom{n_i-1}{3} +8 \prod ([(n_i-4)/3]+1) \right. \\ &\left. +3\left(\prod (n_i/2-1), 0\right) -6 \prod [(n_i/2-1)^2]-6\eta_4(n_i)\right\} \end{aligned} \tag{7.1.17}$$

where

$$\eta_h(n_i) = 1, \quad \text{if } h|n_i,\ i = 1, \ldots, s,$$
$$= 0, \quad \text{otherwise,}$$

$$\prod\left(\frac{n_i}{2}-1\right), 0 = \begin{cases} \prod\left(\frac{n_i}{2}-1\right), & \text{if } 2|n_i,\ i = 1, \ldots, s, \\ 0, & \text{otherwise.} \end{cases}$$

$$P_4(n_1, n_2, \ldots, n_s)$$

(7.1.18)
$$= \frac{1}{24}\left\{\prod \binom{n_i-1}{3} + 8 \prod ([(n_i-4)/3]+1) + 3\left(\prod (n_i/2-1), 0\right) + 6 \prod [(n_i/2-1)^2] + 6\eta_4(n_i)\right\}$$

$$q_4(n_1, n_2, \ldots, n_s)$$

(7.1.19)
$$= \frac{1}{24}\left\{\prod \binom{n_i+3}{3} + 8 \prod ([n_i/3]+1) + 3\left(\prod (n_i/2+1), 0\right) + 6 \prod [(n_i/2+1)^2] + 6\eta_4(n_i)\right\}.$$

Wright [**35**] gives also

(7.1.20)
$$q_5(n_1, \ldots, n_s) = \frac{1}{120} \prod \binom{n_i+4}{4} + \frac{1}{12} \prod \frac{1}{2}\left\{\binom{n_i+3}{3} + [(n_i/4+1/2)^2]\right\} + \frac{1}{6} \prod \left[\frac{1}{6} n_i(n_i+5)+1\right] + \frac{1}{8} \prod \frac{1}{2} ([n_i/2]+1)([n_i/2]+2) + \frac{1}{4} \prod \left[\frac{1}{4}n_i+1\right] + \frac{1}{6} \prod \left\{\left[\frac{1}{6}n_i+\frac{2}{3}\right] + \eta_6(n_i)\right\} + \frac{1}{5}\eta_5(n_i).$$

8.1. **Formulas for $\lambda_r(m_1, m_2)$ for $m_2=0, 1, 2, 3, r \geq 2$.** Recall that m_1 and m_2 are given by

(8.1.1)
$$m_2 = \sum_{t_{j+1}<t_j} (r-j),$$

(8.1.2)
$$m_1 = \sum_{t_i=t_{j+1}, i<j} (r-t_j).$$

If $m_2=0$, then π is the identity permutation, which has $m_1=0$. Hence

(8.1.3)
$$\lambda_r(0, 0) = 1, \qquad \lambda_r(m_1, 0) = 0 \quad \text{if } m_1 \geq 1.$$

If $m_2=1$, then $t_1<t_2<\cdots<t_{r-1}>t_r$. For this permutation $m_1=r-t_r$, hence

(8.1.4)
$$\lambda_r(m_1, 1) = 1, \quad \text{if } 1 \leq m_1 \leq r-1,$$
$$\lambda_r(m_1, 1) = 0, \quad \text{if } m_1 \geq r.$$

If $m_2=2$, then $t_1<t_2<\cdots<t_{r-2}>t_{r-1}>t_r$.
There are three possibilities:

(1) If $t_r=r$, then $m_1=r-t_{r-1}$, giving a contribution of 1 to $\lambda_r(m_1, 2)$ for $2 \leq m_1 \leq r-1$.

(2) If $t_r=t_{r-1}+1$, then $m_1=r-t_r$, giving a contribution of 1 to $\lambda_r(m_1, 2)$ for $1 \leq m_1 \leq r-2$.

(3) If $t_{r-1}+1<t_r<r$, then $m_1=2r-t_{r-1}-t_r$, giving a contribution to $\lambda_r(m_1, 2)$ equal to the number of solutions of

$$t_r = 2r-t_{r-1}-m_1, \qquad 1 < t_{r-1}+1 < 2r-t_{r-1}-m_1 < r,$$

i.e., of

$$t_{r-1} \geq 1, \qquad t_{r-1} \leq r-1-m_1/2, \qquad t_{r-1} \geq r+1-m_1.$$

The number of solutions is clearly

$$= \max \{0, \min ([r-1-m_1/2], [m_1/2-1])\}.$$

Therefore

$$\lambda_r(m_1, 2) = \max \{0, \min ([r-1-m_1/2], [m_1/2+1])\}. \tag{8.1.5}$$

A more detailed analysis yields a formula for $\lambda_r(m_1, 3)$.
In particular we have the following results.

$$\text{For } r \geq 2 \max (m_1, 1), \qquad \lambda_r(m_1, 1) = 1. \tag{8.1.6}$$

$$\text{For } r \geq 2 \max (m_1, 2), \qquad \lambda_r(m_1, 2) = [m_1/2+1]. \tag{8.1.7}$$

$$\text{For } r \geq 2 \max (m_1, 3), \qquad \lambda_r(m_1, 3) = [m_1^2/12+m_1+1/4]. \tag{8.1.8}$$

9.1. **Evaluation of $q_r(n_1, n_2, \ldots, n_s)$ and $\lambda_r(m_1, m_2, \ldots, m_s)$.**

METHOD I. First we consider the case $s=2$. The method is quite general and suitable for high-speed computing for $s \leq 3$. It is easily extended to the case $s>3$ if the computing machine and coding used allow for arithmetic operations on variables having four or more subscripts. We use $\{(1-x_1^m)(1-x_2^m)\}^{-1}$ as an operator.

Consider a formal power series in two variables x_1, x_2 denoted by $\sum_{i,j=0}^{\infty} a_{ij}x_1^i x_2^j$ whose coefficients may be written as an infinite matrix (a_{ij}).

Let

$$\{(1-x_1^m)(1-x_2^m)\}^{-1}\left(\sum_{i,j=0}^{\infty} a_{ij}x_1^i x_2^j\right) = \sum_{i,j=0}^{\infty} b_{ij}x_1^i x_2^j.$$

It is clear that the matrix of coefficients (b_{ij}) is obtained from (a_{ij}) by applying the following row and column operations.

By the column operation

$$\begin{bmatrix} b'_{n0} \\ b'_{n1} \\ b'_{n2} \\ \vdots \end{bmatrix} = \begin{bmatrix} a_{n0} \\ a_{n1} \\ a_{n2} \\ \vdots \end{bmatrix} + \begin{bmatrix} a_{n-m,0} \\ a_{n-m,1} \\ a_{n-m,2} \\ \vdots \end{bmatrix} + \cdots + \begin{bmatrix} a_{n-jm,0} \\ a_{n-jm,1} \\ a_{n-jm,2} \\ \vdots \end{bmatrix}, \qquad j = \left[\frac{n}{m}\right], \tag{9.1.1}$$

we obtain the matrix (b'_{ij}).

Now apply to (b'_{ij}) the row operation:

$$[b_{0,n}, b_{1,n}, \ldots] \tag{9.1.2}$$
$$= [b'_{0,n}, b'_{1,n}, \ldots]+[b'_{0,n-m}, b'_{1,n-m}, \ldots]+\ldots+[b'_{0,n-jm}, b'_{1,n-jm}, \ldots].$$

We write $(b_{ij}) = (a_{ij})_{m,m}$.

Thus from (2.2.10) we get

$$r(q_r(i,j)) = (q_{r-1}(i,j))_{1,1}+(q_{r-2}(i,j))_{2,2}+\cdots+(q_0(i,j))_{r,r}, \tag{9.1.3}$$

where

$$(q_0(i,j)) = \begin{pmatrix} 1 & 0 & 0 & \cdots \\ 0 & 0 & 0 & \cdots \\ 0 & 0 & 0 & \cdots \end{pmatrix},$$

$$(q_1(i,j)) = \begin{pmatrix} 1 & 1 & 1 & \cdots \\ 1 & 1 & 1 & \cdots \\ 1 & 1 & 1 & \cdots \end{pmatrix}.$$

By this recursive technique $q_r(n_1, n_2)$ is easily computable, and the method is suitable for computing $q_r(n_1, n_2), q_r(n_1, n_2, n_3), \ldots$.

The operations involved, for $s=2$, are row and column additions, and divisions by r. The initial input is the matrix

$$\begin{pmatrix} 1 & 0 & 0 & \cdots \\ 0 & 0 & 0 & \cdots \\ 0 & 0 & 0 & \cdots \\ . & . & . & \end{pmatrix}.$$

For $s=3$ we must perform additions on three-dimensional arrays of numbers.

From $q_r(n_1, n_2)$ we can obtain $\lambda_r(m_1, m_2)$ by multiplying the formal power series,

$$\sum_{q_r} (n_1, n_2)x_1^{n_1}x_2^{n_2},$$

by

$$(1-x_1)(1-x_1^2)\cdots(1-x_1^r)(1-x_2)(1-x_2^2)\cdots(1-x_2^r).$$

This multiplication involves subtraction. First we multiply the formal power series by $(1-x_1)(1-x_2)$, then by $(1-x_1^2)(1-x_2^2), \ldots$, and finally by $(1-x_1^r)(1-x_2^r)$. Multiplication of $\sum_{a_{ij}} x_1^i x_2^j$ by $(1-x_1^m)$ gives a new matrix b_{ij} of coefficients such that

$$\begin{bmatrix} b_{n,0} \\ b_{n,1} \\ \vdots \end{bmatrix} = \begin{bmatrix} a_{n,0} \\ a_{n,1} \\ \vdots \end{bmatrix} - \begin{bmatrix} a_{n-m,0} \\ a_{n-m,1} \\ \vdots \end{bmatrix}. \tag{9.1.4}$$

Multiplication by $(1-x_2^m)$ will have a similar effect on rows.

For $s=3$ and higher dimensions, $q_r(n_1, n_2, \ldots, n_s)$ and $\lambda_r(m_1, m_2, \ldots, m_s)$ can be computed by multiplying formal power series in several variables by $\prod_{i=1}^{s} (1-x_i^m)^{\pm 1}$.

METHOD II. Here we first determine $\lambda_r(m_1, \ldots, m_s)$; then $q_r(n_1, \ldots, n_s)$ is easily computable as a coefficient in the product of two polynomials in s variables. First we consider $s=2$.

We need a scheme that generates all permutations on r marks. Lehmer **[20]** mentions that there are at least ten different methods of producing permutations automatically; of all these methods two are very fast, namely the Tompkins-Paige **[30]** cyclic method and a method for generating permutations by transposition by M. B. Wells **[31]**. We describe a method we used for our computations, but it is not as fast as the two mentioned above.

The permutations are generated by the following technique. The permutation

$$\pi = \begin{pmatrix} 1 & 2 & \ldots & r \\ t_{r-1} & t_{r-2} & \ldots & t_0 \end{pmatrix}$$

will be denoted by $(t_{r-1}, t_{r-2}, \ldots, t_0)$.

Let $(t_{r-1}, t_{r-2}, \ldots, t_{k'}, t_{k'-1}, \ldots, t_k, \ldots, t_1, t_0)$ be a permutation with $t_0 < t_1 < \cdots < t_{k'-1} > t_{k'}$ and let

$$\pi'' = (t_{r-1}, t_{r-2}, \ldots, t_{k'}, t_0, t_1, \ldots, t_{k'-1}).$$

Now find $t_0 - t_{k'}, t_1 - t_{k'}, \ldots$ and let k be the smallest integer such that $t_k - t_{k'} > 0$. Interchanging $t_k, t_{k'}$, we get the next permutation

$$\pi' = (t_{r-1}, \ldots, t_k, t_0, t_1, \ldots, t_{k-1}, t_{k'}, t_{k+1}, \ldots, t_{k'-1}).$$

When starting with the identity permutation it can be easily shown that this technique generates all the permutations on r marks.

While generating the new permutation π' from π we also find the corresponding change in m_1, m_2. We have

$$m_2(\pi') - m_2(\pi) = -\tfrac{1}{2}k'(k'-3) - \begin{cases} (k'+1), & \text{if } t_k > t_{k'+2} > t_{k+1}, \\ 0, & \text{otherwise.} \end{cases}$$

To find $m_1(\pi') - m_1(\pi)$ we notice that if

$$\pi = (t_{r-1}, \ldots, t_{k'}, t_{k'-1}, \ldots, t_r, \ldots, t_0),$$
$$\pi' = (t_{r-1}, \ldots, t_k, t_0, t_1, \ldots, t_{k-1}, t_{k'}, t_{k+1}, \ldots, t_{k'-1}),$$

we subtract $r - t_j$ from m_1 if $t_{j+1} - t_j = 1$, subtract $r - t_{k-1}$ from m_1 if $t_{k'} - t_{k-1} = 1$ and add $r - t_{k'}$ to m_1 if $t_k - t_{k'} = 1$.

In this way we generate all permutations and find the change in m_1, m_2 from one permutation to the next. This change is added to m_1, m_2 and a unit stored in the (m_1, m_2) location. When all permutations have been generated, the number of units stored in the (m_1, m_2) location will be the required number of permutations having characteristic numbers (m_1, m_2) and the table of values of $\lambda_r(m_1, m_2)$ can be printed out.

The values of $\lambda_r(m_1, m_2)$ for $r \leq 9$ were obtained by generating all permutations on r marks on the SWAC Computer and then the $q_r(n_1, n_2)$, $r \leq 9$, $n_1, n_2 \leq 29$ were evaluated on the IBM 709 as coefficients in the product of two polynomials in two variables. For $s \geq 2$ and large values of r Method I is more suitable.

The tables of $q_r(n_1, n_2)$ have been extended to $n_1, n_2 = 1(1)49$, $r = 2(1)98$, and certain values of $q_r(n_1, n_2, n_3)$, $\lambda_r(n_1, n_2, n_3)$ have also been computed. For more details of these methods and for numerical results the reader is referred to Cheema and Motzkin [**9**].

BIBLIOGRAPHY

1. F. C. Auluck, *On partitions of bipartite numbers*, Proc. Cambridge Philos. Soc. **49** (1953), 72–83. MR **14**, 726.

2. L. Carlitz, *The expansion of certain products*, Proc. Amer. Math. Soc. **7** (1956), 558–564. MR **19**, 29.

3. ———, *A problem in partition*, Duke Math. J. **30** (1963), 203–213. MR **26** #6143.

4. ———, *A note on the Jacobi theta formula*, Bull. Amer. Math. Soc. **68** (1962), 591–592. MR **27** #3561.

5. T. Chaundy, *Unrestricted plane partitions*, Quart. J. Math. Oxford **3** (1932), 76–80.

6. M. S. Cheema, *Tables of partitions of Gaussian integers, giving the number of partitions of $n+im$*, Math. Tables, vol. 1, Nat. Inst. Sci. India, New Delhi, 1956. MR **19**, 17.

7. ———, *Vector partitions and combinatorial identities*, Math. Comp. **18** (1964), 414–420. MR **29** #4697.

8. M. S. Cheema and T. S. Motzkin, *Vector partitions and permutation vectors*, Notices Amer. Math. Soc. **8** (1961), 167. Abstract #580-585.

9. ———, *Evaluation of multipartition numbers*, Univ. of California Press, Los Angeles, Calif. (to appear).

10. P. Erdös and J. Lehner, *The distribution of the number of summands in the partitions of a positive integer*, Duke Math. J. **8** (1941), 335–345. MR **3**, 69.

11. L. Euler, *Observationes analyticae variae de combinationibus*, Comment. Acad. Sci. Imp. Petrop. **13** (1741–48), 64–93 (1751).

12. N. J. Fine, Math. Reviews **19** (1958), 29.

13. J. W. L. Glaisher, *Formulae for partitions into given elements derived from Sylvester's theorem*, Quart. J. Math. Oxford **40** (1909), 275–348.

14. B. Gordon, *Two theorems on multipartite partitions*, J. London Math. Soc. **38** (1963), 459–464. MR **28** #1187.

15. ———, *Two new representations of the partition function*, Proc. Amer. Math. Soc. **13** (1962), 869–873. MR **25** #5005.

16. H. Gupta, *Partitions of j-partite numbers into k summands*, J. London Math. Soc. **33** (1958), 403–405. MR **20** #7007.

17. H. Gupta, A. E. Gwyther and J. C. P. Miller, *Tables of partitions*, Royal Soc. Math. Tables Vol. 4, (1958).

18. G. H. Hardy and S. Ramanujan, *Asymptotic formulae in combinatory analysis*, Proc. London Math. Soc. (2) **17** (1918), 75–115.

19. A. E. Ingham, *A Tauberian theorem for partitions*, Ann. of Math. (2) **42** (1941), 1075–1090. MR **3**, 166.

20. D. H. Lehmer, *Teaching combinatorial tricks to a computer*, Proc. Sympos. Appl. Math. vol. 10, Amer. Math. Soc., Providence, R.I., 1960, pp. 179–193. MR **22** #4127.

21. P. A. MacMahon, *Combinatory analysis*, Vols. I, II, Chelsea, New York, 1960. MR **25** #5003.

22. Gunter Meinardus, *Zur additiven Zahlentheorie in mehreren Dimensionen*. I, Math. Ann. **132** (1956), 333–346. MR **18**, 642.

23. V. S. Nanda, *Bipartite partitions*, Proc. Cambridge Philos. Soc. **53** (1957), 273–277. MR **19**, 16.

24. G. J. Rieger, *Über Partitionen*, Math. Ann. **138** (1959), 356–362. MR **21** #7188.

25. M. M. Robertson, *Asymptotic formulae for the number of partitions of a multi-partite number*, Proc. Edinburgh Math. Soc. (2) **12** (1960/61), 31–40. MR **23** #A2406.

26. ———, *Partitions of large multipartites*, Amer. J. Math. **84** (1962), 16–34. MR **25** #3919.

27. J. J. Sylvester, *On subvariants, i.e., semi-invariants to binary quantics of an unlimited order: Excursus on rational functions and partitions*, Amer. J. Math. **5** (1882), 119–136.

28. G. Szekeres, *An asymptotic formula in the theory of partitions*, Quart. J. Math. Oxford Ser. (2) **2** (1951), 85–108. MR **13**, 210.

29. J. A. Todd, *A table of partitions*, Proc. London Math. Soc. (2) **48** (1943), 229–242. MR **5**, 172; 328.

30. C. B. Tompkins, *Machine attacks on problems whose variables are permutations*, Proc. Sympos. Appl. Math., vol. 6, Amer. Math. Soc., Providence, R.I., 1956, pp. 195–211. MR **18**, 238.

31. M. B. Wells, *Generation of permutations by transposition*, Math. Comp. **15** (1961), 192–195. MR **23** #B553.

32. E. M. Wright, *Partitions of multi-partite numbers*, Proc. Amer. Math. Soc. **7** (1956), 880–890. MR **18**, 793.

33. ———, *The number of partitions of a large bi-partite number*, Proc. London Math. Soc. (3) **7** (1957), 150–160. MR **19**, 16.

34. ———, *Partitions of large bipartites*, Amer. J. Math. **80** (1958), 643–658. MR **20** #3111.

35. ———, *Partition of multipartite numbers into a fixed number of parts*, Proc. London Math. Soc. (3) **11** (1961), 499–510. MR **24** #A2573.

36. ———, *Partitions into k parts*, Math. Ann. **142** (1960/61), 311–316. MR **22** #12088.

University of Arizona

University of California, Los Angeles

SIMPLICIAL GEOMETRIES

HENRY H. CRAPO AND GIAN-CARLO ROTA

Introduction. We wish to draw attention to certain geometries which arise naturally in the study of simplicial complexes, and which carry the homology structure of the complex.

Connections between geometry and topology were once more noticeable than they are today. P. Alexandroff [**1**, p. 13] in formulating the fundamental problem of set-theoretic topology, stated: the set theoretic entities which permit a connection with the objects of polyhedral topology "deserve to be regarded as geometrical configurations, even if of the most general kind."

In support of Alexandroff, we shall show how the homology structure of any finite simplicial complex may be embodied in a sequence of geometric structures. We feel that, in this geometric interpretation, more powerful combinatorial methods for topology will be developed, in the spirit of Poincaré and the early Lefshetz. One immediate objective is to free combinatorial topology from its coefficients, in order better to reveal the intrinsic structure of simplicial complexes.

1. **Simplicial complexes.** A *simplicial complex* on a set T is a nonempty family S of subsets (*simplices*) of T, such that every subset (*face*) of a simplex is also a simplex.

For each simplicial complex S on an n-element set T and for each nonnegative integer $i \leq n$, let α_i be the number of faces of cardinality i, and let β_i be the Betti number [**1**], calculated for faces of cardinality i. (Our indexing differs by 1 from that conventionally used.) If $S=\{\varnothing\}$, $\beta_0=1$. Otherwise $\beta_0=0$.

We make essential use of the Euler-Poincaré formula. Note that the formula calculates an invariant of the simplicial complex, the Möbius function [**3**]:

$$\mu = \alpha_0-\alpha_1+\cdots+(-1)^n\alpha_n = \beta_0-\beta_1+\cdots+(-1)^n\beta_n.$$

Recall that for any face A, we have a Boolean algebra interval $[\varnothing, A]$, and $\mu[\varnothing, A] =(-1)^{|A|}$. In all nontrivial cases, $1-\mu=\alpha_1-\alpha_2+\cdots$ is the Euler characteristic.

For each fixed n-element set T, and any nonnegative integer $k\leq n$, let T_k be the set of k-element subsets of T (the *k-skeleton* of T). With each subset $A\subseteq T_k$ associate the simplicial complex

$$S(A) = T_0 \cup T_1 \cup \cdots \cup T_{k-1} \cup A$$

and write the corresponding face numbers and Betti numbers $\alpha_i(A)$, $\beta_i(A)$, $i=0,\ldots,k$.

Since all i-simplices are present, for $i \leq k-1$, all i-cycles bound, for $i \leq k-2$, hence only β_{k-1} and β_k may be nonzero. Indeed,

$$(-1)^{k-1}\mu = \beta_{k-1}-\beta_k = \binom{n-1}{k-1}-|A|.$$

We define the *rank* of a subset $A \subseteq T_k$ by the equivalent formulae

$$r(A) = \binom{n-1}{k-1}-\beta_{k-1}(A) = |A|-\beta_k(A).$$

2. **Combinatorial geometries.** The notion of a combinatorial geometry has already been recognized as a unifying concept in combinatorial theory [2]. Theories of mappings, extensions, orthogonality, connectivity, and coordinatization are now fairly well developed for these geometries. The best known examples of combinatorial geometries include those arising from linear and algebraic dependence, the classical geometries, partitions of type n, and matchings within a binary relation. Important work has been done on examples relating to linear inequalities and algebraic varieties.

A *combinatorial geometry* $G(P)$ is a set P (of *points*) together with a closure operator $A \to \bar{A}$ defined on all subsets of P, satisfying the Steinitz-MacLane *exchange property*:

$$\text{if } a \notin \bar{A} \text{ but } a \in \overline{A \cup b}, \text{ then } b \in \overline{A \cup a},$$

the *finite basis* property:

$$\text{each subset } A \subseteq P \text{ has a finite subset } A_f \subseteq A \text{ such that } \bar{A}_f = \bar{A}$$

and such that the empty set $\varnothing$, and all single-point sets are closed sets.

A closed set $A \subseteq P$ of points is called a *flat*, and has a well-defined *rank* λ in the lattice $L(P)$ of flats. The resulting *geometric lattices* are easily characterized as semimodular point lattices with no infinite chains.

Consider the definition of a combinatorial geometry. The exchange property is merely a convenient axiomatization of the fact:

> The points not on a flat A of rank k are partitioned into equivalence classes by the flats of rank $k+1$ which contain A.

In the present discussion, we shall be dealing with geometries on a finite set, where the finite basis property is trivially satisfied. The final condition is not an essential restriction of the theory. If it fails, we have a *pregeometry*.

For our present purposes, the characterization of the *geometric rank* function

$$r(A) = \lambda(\bar{A}),$$

defined for every subset $A \subseteq P$, provides the most convenient starting point.

A *Whitney rank function* [2], [4] is a nonnegative integer-valued function defined on all subsets of a set P, which is

normalized: $r(\varnothing) = 0$,
unit-increase: $r(A \cup a) = r(A)$ or $r(A)+1$,
semimodular: $r(A \cap B)+r(A \cup B) \leq r(A)+r(B)$,
finite basis: every subset $A \subseteq P$ has a finite subset $A_f \subseteq A$ such that
$r(A_f) = r(A)$,

for all subsets $A, B \subseteq P$ and all elements $a \in P$. If $r(A)=2$ for all two-element subsets $A \subseteq P$, then

$$a \in \bar{A} \text{ if and only if } r(A \cup a) = r(A)$$

defines a closure operator, and a geometry, on P.

3. Simplicial geometries.

THEOREM. *The function*

$$r(A) = \binom{n-1}{k-1} - \beta_{k-1}(A) = |A| - \beta_k(A)$$

is the geometric rank function of a geometry (for $k=0, 1$, a pregeometry) $G(T_k)$ of rank $\binom{n-1}{k-1}$ on the set T_k of $\binom{n}{k}$ points.

PROOF. Consider what happens when a single k-simplex a is added to a subset $A \subset T_k$. The equality

$$r(A) = r(A \cup a)$$

holds whenever some multiple of the boundary of the simplex a already bounds in A (so $\beta_k(A) = \beta_k(A \cup a)+1$ and $\beta_{k-1}(A) = \beta_{k-1}(A \cup a)$). Otherwise, we have

$$r(A)+1 = r(A \cup a),$$

because the boundary of the simplex a bounds a in the larger complex (so $\beta_k(A) = \beta_k(A \cup a)$ and $\beta_{k-1}(A) = \beta_{k-1}(A \cup a) - 1$). Evidently the function r is a normalized unit-increase function. Semimodularity is easily established in the special case

$$r(A \cup a) - r(A) \leq r(B \cup a) - r(B)$$

for subsets $B \subseteq A \subseteq P$, and thence extended to the general case. This completes the proof.

EXAMPLE ($n=5$).

	$G(T_0)$	$G(T_1)$	$G(T_2)$	$G(T_3)$	$G(T_4)$	$G(T_5)$
number of points	1	5	10	10	5	1
rank	0	1	4	6	4	1
corank	1	4	6	4	1	0

$k=0$ The single simplex $\varnothing$ is in the closure of the empty set of 0-simplices.
$k=1$ The five points, say, a, b, c, d, e, coincide.
$k=2$ The Desargues configuration (see Figure 1).

$k=3$ A geometry with five nontrivial planes (one for each 4-element set), in rank-6 space. A generic projection of this geometry into rank-4 space is indicated in Figure 2.

$k=4$ Five points in general position in rank-4 space.

$k=5$ A single point.

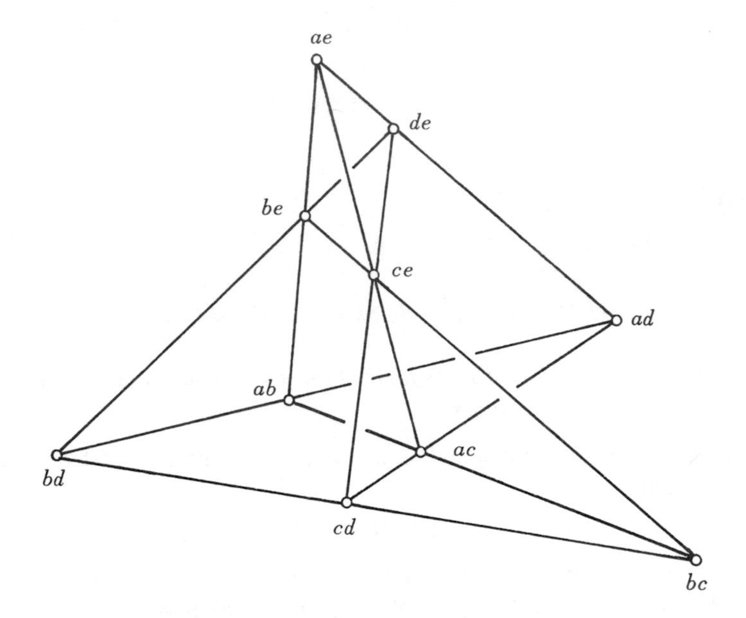

FIGURE 1

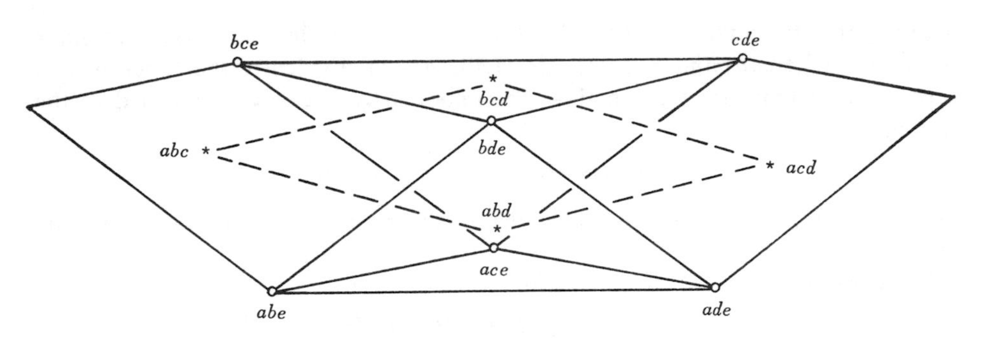

FIGURE 2

For $k=2$, the simplicial geometries are well known: the lattices $L(T_2)$ are isomorphic to the lattices of all partitions of the set T, and the geometries $G(T_2)$ are the geometries of complete linear graphs. For higher values of k, the simplicial geometries have not been studied, despite their relevance to problems of embedding and coloring.

THEOREM. *In a simplicial geometry $G(T_k)$ on an n-element set T, the set B of points (that is, of k-simplices) containing any fixed element $b \in T$ form a basis for the geometry.*

PROOF. The set B is the suspension, by b, of the complete skeleton $(T-b)_{k-1}$. Therefore $\beta_{k-1}(B)=0$, so $|B|=\binom{n-1}{k-1}-r(B)=r(T_k)$, and B is a basis.

4. **Orthogonality.** Given the geometric rank function r of a geometry $G(S)$ on a set S, the *nullity* of subsets $A \subseteq S$ is given by

$$n(A) = |A|-r(A),$$

and the *orthogonal rank* by

$$r^*(A) = n(S)-n(S-A).$$

The orthogonal rank is a Whitney rank function, and defines an orthogonal (pre-)*geometry* $G^*(S)$.

THEOREM. $G^*(T_k) \simeq G(T_{n-k})$.

PROOF. This is easily shown to be equivalent to Alexander duality:

$$\beta_k(T_k-A) = \beta_{n-k-1}(A'),$$

where A' is the set of complements in S of simplices in A. This completes the proof.

For simplicial subdivision of a sphere the notions of orthogonality of geometries and orthogonality of subdivisions may be shown to agree. Thus geometric orthogonality provides a basis for a unified theory of topological duality.

Combinatorial theory may derive numerous benefits from the intimate relationship between geometry and topology. Consider, for example, the fact that every proper 4-coloring of the vertices of a triangulation of a sphere is a simplicial map of the sphere onto a tetrahedron, and has a well-defined index.

REFERENCES

1. P. S. Aleksandrov, *Einfachste Grundbegriffe der Topologie*, Springer, Berlin, 1932; English transl., Dover, New York, 1961; Ungar, New York, 1965. MR **26** #6951; MR **33** #4916.

2. H. H. Crapo and G.-C. Rota, *On the foundations of combinatorial theory*. II: *Combinatorial geometries*, M.I.T. Press, Cambridge, Mass., 1970 (preliminary edition).

3. G.-C. Rota, *On the foundations of combinatorial theory*. I: *Theory of Möbius functions*, J. Wahrscheinlichkeitstheorie und Verw. Gebiete **2** (1966), 340–368. MR **30** #4688.

4. H. Whitney, *On the abstract properties of linear dependence*, Amer. J. Math. **57** (1935), 509–533.

UNIVERSITY OF WATERLOO, ONTARIO

MASSACHUSETTS INSTITUTE OF TECHNOLOGY

PROBLEMS AND RESULTS IN COMBINATORIAL ANALYSIS

P. ERDÖS

This review of some solved and unsolved problems in combinatorial analysis will be highly subjective. I will only discuss problems which I either worked on or at least thought about. The disadvantages of such an approach are obvious, but the disadvantages are perhaps counterbalanced by the fact that I certainly know more about these problems than about others (which perhaps are more important). I will mainly discuss finite combinatorial problems. I cannot claim completeness in any way but will try to refer to the literature in some cases; even so many things will be omitted. $|S|$ will denote the cardinal number of S; $c, c_1, c_2, \ldots$ will denote absolute constants not necessarily the same at each occurrence.

I. I will start with *some problems dealing with subsets of a set*. Let $|S|=n$. A well-known theorem of Sperner [**57**] states that if $A_i \subset S$, $1 \le i \le m$, is such that no A_i contains any other, then max $m = \binom{n}{[n/2]}$. The theorem of Sperner has many applications in number theory; as far as I know these were first noticed by Behrend [**2**] and myself [**8**].

I asked 30 years ago several further extremal problems about subsets which also have number theoretic consequences. Let $A_i \subset S$, $1 \le i \le m_1$, assume that there are no three distinct A's so that $A_i \cup A_j = A_r$. I conjectured that

$$\max m_1 = (1+o(1))\binom{n}{[n/2]},$$

but could not even prove max $m_1 = o(2^n)$. This latter result was proved by Sárközi and Szemerédi but was never published because it was superseded by the result of Kleitman [**44**], who first of all proved that max $m < 2^{3/2}\binom{n}{[n/2]}$ and recently (in this volume) that

$$\max m_1 < \binom{n}{[n/2]}\left(1+c\left(\frac{\log n}{n}\right)^{1/2}\right)$$

which is in fact stronger than my conjecture. It would be of interest to determine max m_1; maybe this question has no simple solution, but perhaps an asymptotic formula for

$$\max m_1 - \binom{n}{[n/2]}$$

is not quite hopeless.

The second problem I asked was: Let $A_i \subset S$, $1 \leq i \leq m_2$. Assume that there are not four distinct A's say A_i, A_r, A_s, A_t satisfying

$$A_i \cup A_r = A_s, \qquad A_i \cap A_r = A_t.$$

Kleitman proved $\max m_2 < c_1 2^n/n^{1/4}$ and I showed that $\max m_2 > c_2 2^n/n^{1/4}$. Presumably

$$\max m_2 = (c+o(1))2^n/n^{1/4}$$

but as far as I know this has not yet been proved. These results are not yet published (they will appear in Proc. Amer. Math. Soc.).

Here I would like to mention a question which goes back to Dedekind: How many families of subsets of S are there where no set of a family contains any other? Denote the number of such families by $f(n)$. There may not be a simple explicit formula for $f(n)$, but Kleitman, sharpening previous results of several authors, proved (not yet published)

$$\log f(n) = (1+o(1))\binom{n}{[n/2]} \log 2.$$

It would be interesting to give an asymptotic formula for $f(n)$ but this is probably rather difficult.

Kleitman proved several other conjectures of mine involving subsets, some of which have not yet been published.

Rota observed that Dilworth's theorem **[6]** implies Sperner's theorem and many other results in combinatorial analysis.

Ko, Rado and I **[26]** proved that if $n \geq 2k$, $A_i \subset S$, $|A_i| = k$, $A_i \cap A_j \neq \varnothing$, $1 \leq i < j \leq m$, then

$$\max m = \binom{n-1}{k-1}.$$

Assume now $|A_i \cap A_j| \geq r$. Put $\max m = f(n, k, r)$. We proved that for $n > n_0(k, r)$,

$$f(n, k, r) = \binom{n-r}{k-r}. \tag{1}$$

Min **[26]** observed (in the same paper) that (1) is not true in general; the determination of $f(n, k, r)$ in general seems to be a difficult problem. We conjectured

$$f(4k, 2k, 2) = \frac{1}{2}\left(\binom{4k}{2k} - \binom{2k}{k}^2\right), \tag{2}$$

but we could not decide whether (2) is true.

We also observed that if $A_i \subset S$, $A_i \cap A_j \neq \emptyset$, $1 \leq i < j \leq m$, then max $m = 2^{n-1}$. It does not seem to be easy to determine the number of families $A_i \subset S$, $A_i \cap A_j \neq \emptyset$, $1 \leq i < j \leq 2^{n-1}$. We could not even get an asymptotic formula for the number of these families.

Let $|S| = n$, $A_i \subset S$, $1 \leq i \leq k$. What is the smallest value of k so that there should always be three A's any two of which have the same union? This question like some other problems in this chapter has connections with number theory. (More generally we can ask what is the smallest value of $k = k_r$ so that there always are r A's any two of which have the same union.)

II. **Some geometric problems.** Let $z_i \geq 1$, $1 \leq i \leq n$. Consider all the sums $\sum_{i=1}^{n} \varepsilon_i z_i$, $\varepsilon_i = \pm 1$. I [9] proved as an easy application of Sperner's theorem that the number of sums which fall into the interior of an interval of length 2 is at most $\binom{n}{[n/2]}$, with equality if $z_i = 1$, $1 \leq i \leq n$. I conjectured that if the z_i are complex numbers satisfying $|z_i| \geq 1$, then every circle of radius 1 contains at most $\binom{n}{[n/2]}$ sums $\sum_{i=1}^{n} \varepsilon_i z_i$ (this would sharpen a result of Littlewood and Offord); more generally I conjectured that the above result may remain true if the z_i are vectors in Hilbert space or even in a Banach space.

Katona [40] and Kleitman [43] independently and almost simultaneously proved my conjecture in the plane by giving an interesting generalization of Sperner's theorem, and Kleitman [43] also proved that my conjecture holds in k-dimensional space if $n > n_0(k)$, but the general conjecture has not yet been settled. Sárközi, Szemerédi and I have the following conjecture: Let $|z_i| \leq 1$, $1 \leq i \leq n$, then there are at least $c2^n/n$ summands $\sum_{i=1}^{n} \varepsilon_i z_i$, $\varepsilon_i = \pm 1$, which are of absolute value $\leq \sqrt{2}$ (it is easy to see that $\sqrt{2}$ cannot be diminished; let an odd number of z's be 1 and an odd number i). The order of magnitude $c2^n/n$ is easily seen to be best possible if true. Analogous conjectures can easily be made for higher dimensions.

Sárközi and Szemerédi proved that if $-1 \leq z_i \leq 1$ then there are at least $\binom{n}{[n/2]}$ sums $\sum_{i=1}^{n} \varepsilon_i z_i$, $\varepsilon_i = \pm 1$, which are less than 1 in absolute value. It is easy to see that $\binom{n}{[n/2]}$ is best possible.

Sárközi and Szemerédi further observed that our conjecture in the plane is true if the following purely combinatorial result holds: Let $|S| = n$, $A_i \subset S$, $1 \leq i \leq k$, $B_j \subset S$, $1 \leq j \leq l$ (the A's and B's are all distinct). Assume $|A_{i_1} \cap A_{i_2}| \geq 2$, $1 \leq i_1 \leq i_2 \leq k$; $|B_{j_1} \cap B_{j_2}| \geq 2$, $1 \leq j_1 < j_2 \leq l$; $|A_i \cap B_j| \geq 1$, $1 \leq i \leq k$; $1 \leq j \leq l$. Then

$$k + l \leq 2^{n-1} - c2^n/n.$$

Miss E. Klein raised in 1932 the following problem: Let $f(n)$ be the smallest integer with the property that from $f(n)$ points in the plane one can always select vertices of a convex n-gon. Miss Klein proved that $f(4) = 5$, Makai and Turán proved $f(5) = 9$. Szekeres and I [31], [32] proved

$$2^{n-2} < f(n) \leq \binom{2n-4}{n-2}$$

(our proof of the lower bound contained an inaccuracy which was corrected by Kalbfleisch). It seems likely that $f(n) = 2^{n-2} + 1$ but this is not known for $n \geq 6$.

Let there be given 2^n points in the plane. Szekeres and I **[32]** proved that these points always determine an angle greater than $\pi(1-1/n)$, an earlier result of Szekeres **[58]** states that to every ε there are 2^n points so that every angle is less than $\pi(1-1/n)+\varepsilon$. Thus for 2^n points the problem of minimizing the maximum of the greatest angle is completely solved. It is not impossible that if $n>n_0$ then already $2^{n-1}+1$ points always determine an angle $>\pi(1-1/n)$, but we only proved that 2^n-1 points always determine an angle $\geq\pi(1-1/n)$. In higher dimensions sharp results are known only for special values of n, thus Danzer and Grünbaum **[5]** proved that 2^n+1 points in n-dimensional space always determine an angle $>\pi/2$.

Sylvester conjectured and Gallai first proved that if we have n points in the plane not all on a line then there is at least one line which goes through exactly two of the points. Denote by $f(n)$ the minimum number of such lines. N. G. de Bruijn and I conjectured that $f(n)\to\infty$ as $n\to\infty$. This was proved by Motzkin **[47]**. Kelly and Moser **[41]** proved that $f(n)\geq 3n/7$ and this is best possible for $n=7$. Motzkin conjectured that $f(n)\geq[n/2]$ and showed that for infinitely many n this is best possible.

Let there be given n points not all on a line, I observed that it easily follows from Gallai's result that these points determine at least n lines. G. Dirac conjectures that one of the n points is such that it is connected with the other points by more than cn distinct lines.

Assume now that the n points are such that not more than $n-k$ of them are on a line. I conjectured that these points determine at least ckn lines. If k is fixed and $n>n_0(k)$ then Kelly and Moser **[41]** determined the minimum number of lines which these points determine.

Let there be given n points in the plane, not all on a circle. I conjectured that these points determine always at least $\binom{n-1}{2}$ circles. B. Segre disproved this conjecture for $n=8$, but Elliott **[7]** proved it for $n>n_0$.

One can pose the following general problem: Let $a_1,\ldots,a_n$ be n elements, $A_1,\ldots,A_t$, $t>1$, be sets whose elements are the a's. Assume $|A_i|\geq r$, $1\leq i\leq t$, and that each r-tuple is contained in precisely one of the A's. Put $\min t=f(n;r)$. Hanani, Szekeres, de Bruijn and I **[3]** proved that $f(n;2)=n$ and Hanani proved (see Erdös, **[16]**)

$$c_1n^{3/2}<f(n;3)<c_2n^{3/2}.$$

Thus for $r=2$ the combinatorial and geometric problem has the same solution (in the geometric problem the A's are the lines joining the points) but for $r=3$ this is no longer the case (for $r=3$ the r's are circles). The cases $r>3$ have not been investigated.

Further geometric problems and results of a combinatorial nature can be found in **[10]**, **[12]**, **[42]**. Many very interesting problems on combinatorial geometry are found in the lithographed notes of Croft. Further I would like to refer to two books, Hadwiger and Debrunner **[35]** and **[42]**.

III. A well-known *theorem of Ramsey* **[50]** states that if $|S|\geq\aleph_0$ and we split the i-tuples of S into two classes then there is an infinite set all of whose i-tuples are in the same class. Many extensions and generalizations of Ramsey's theorem have been published in the last few years (see my remarks under Ramsey **[50]**). Here we will only be connected with the finite version of Ramsey's theorem. Denote by $f(i; k, l)$ the smallest integer so that if $|S|=f(i; k, l)$ and we split the i-tuples of S into two classes then there either is a subset of k elements all whose i-tuples are in the first class or a subset of l elements all whose i-tuples are in the second class. Ramsey was the first who obtained upper bounds for $f(i; k, l)$. Szekeres and I proved **[31]**

$$f(2; k, l) \leq \binom{k+l-1}{k-1} \tag{3}$$

and I **[11]** proved that

$$f(2; k, k) > 2^{k/2}. \tag{4}$$

It would be very nice to prove that

$$\lim_{k=\infty} f(2; k, k)^{1/k}$$

exists and to determine its value.

(4) was proved not by an explicit construction but by a simple probabilistic reasoning. It would be very desirable to obtain a good lower bound for (4) by an explicit construction.

I **[15]** proved by a more complicated probabilistic reasoning

$$f(2; 3, l) > cl^2/(\log l)^2 \tag{5}$$

and Graver and Yackel **[33]** recently showed that

$$f(2; k, l) < cl^{k-1} \log\log l/\log l. \tag{6}$$

My method which I used to prove (5) very likely will also give (k fixed, $l \to \infty$)

$$f(2; k, l) > c_1 l^{k-1}/(\log l)^{c_2} \tag{7}$$

but I have not worked out the formidable details.

Very little is known about the exact values of $f(2; k, l)$. Trivially $f(2; 2, l)=l$ and $f(2; k, 2)=k$. Further we have (see Graver and Yackel, **[33]**)

$$f(2; 3, 3) = 6, \quad f(2; 3, 4) = 9, \quad f(2; 3, 5) = 14, \quad f(2; 3, 6) = 18,$$
$$f(2; 3, 7) = 23, \quad f(2; 4, 4) = 18.$$

As far as I know nothing is known about the exact values of $f(i; k, l)$ for $i \geq 3$. Hajnal, Rado and I **[25]** proved that $f(i; k, l)$ is less than an $(i-1)$-times iterated exponential and greater than an $(i-2)$-times iterated exponential.

We can generalize the Ramsey numbers by division into more than two classes—even less is known about these than about division into two classes (Greenwood and Gleason **[34]**).

The following question which is related to Ramsey's theorem is perhaps of some interest: Let K_n be the complete graph of n vertices. Denote its vertices by $X_1, \ldots, X_n$. To the edge (X_i, X_j) we make correspond $\varepsilon_{i,j}$ where $\varepsilon_{i,j} = \pm 1$. Put

$$F(n) = \min \max \sum \varepsilon_{i,j}$$

where the maximum is taken over the edges of all the complete subgraphs of K_n and the minimum over all the $2^{C_{n,2}}$ (C is the binomial coefficient) choices of the $\varepsilon_{i,j}$. I can only prove [18]

$$n/4 < F(n) < cn^{3/2}.$$

It would be desirable to obtain better estimates for $F(n)$.

ADDED IN PROOF. J. Spencer and I proved $F(n) > c_1 n^{3/2}$.

IV. **Miscellaneous combinatorial problems.** Miller [46] in the course of some investigations in set theory introduced the following concept: A family of sets $\{A_\alpha\}$ is said to have property B if there is a set S which has a nonempty intersection with every A_α and does not contain any of the A_α's.

Hajnal and I [23] continued Miller's investigations and also asked the following question about finite sets: What is the smallest integer $m(n)$ for which there is a family of sets $\{A_k\}$, $|A_k| = n$, $1 \le k \le m(n)$, which does not have property B? Trivially $m(2) = 3$. It is not difficult to see that $m(3) = 7$. The value of $m(4)$ is unknown.

Schmidt [55] and I [19] proved

$$2^n(1+4/n)^{-1} < m(n) < n^2 2^{n+1}.$$

It would be of interest to give an asymptotic formula for $m(n)$ and to compute $m(4)$. Perhaps no simple formula for $m(n)$ exists.

Gallai asked: Does there exist a family of sets $\{A_k\}$, $1 \le k \le m_1(n)$, not having property B and satisfying $|A_k| \le n$ and $|A_{k_1} \cap A_{k_2}| \le 1$, $1 \le k_1 \le k_2 \le m_1(n)$? A priori it is not obvious that $m_1(n)$ is finite, but Hajnal and I proved that $m_1(n) < c^n$ for every n [24]. We cannot prove that $\lim_{n=\infty} m_1(n)^{1/n}$ exists.

Rado and I investigated the following question: A family of sets is called a Δ-system if every two members of the family have the same intersection. Denote by $F(k, l)$ the smallest integer for which if $\{A_i\}$, $1 \le i \le F(k, l)$, is a family of sets each having k elements, then it always contains a subfamily A_{i_r}, $1 \le r \le l$, which is a Δ-system. We proved

$$(l-1)^k < F(k, l) < k!(l-1)^k\left(1 - \frac{1}{2!(l-1)} - \cdots - \frac{k-1}{k!(l-1)^{k-1}}\right).$$

We believe that

$$F(k, l) < c^k l^k \tag{8}$$

holds. (8) would have many number-theoretic applications but is also of great intrinsic interest.

We investigated the problem of Δ-systems also if $k + l \ge \aleph_0$. But this set-theoretical problem is much simpler than the combinatorial one and we have determined $F(k, l)$ in this case [29].

I conjectured that if $a_1, \ldots, a_{2^k}$ is a sequence of length 2^k where a_i is one of the integers $0, 1, \ldots, k-1$ then there are always two consecutive blocks containing each of the integers the same number of times. This is obvious for $k=2$ and de Bruijn proved it for $k=3$. But for $k=4$ de Bruijn and I disproved it. Later Croft constructed a sequence of length 50 for $k=4$ without such consecutive blocks and he suggested that for $k=4$ there probably is an infinite sequence without two such consecutive blocks. With the help of the Atlas computer Churchhouse constructed such a sequence of length about 1700 which gives a strong support to the conjecture of Croft.

Steiner conjectured that if $n=6k+1$ or $6k+3$ then there exists a system of triplets of n elements so that every pair is contained in one and only one triplet of our system. It is obvious that if n is not of the above form then such a system does not exist.

Steiner's conjecture was first proved by Reiss **[51]**. More generally the following question can be asked: For which values of n is there a system of combinations taken s at a time formed from n elements so that every r-tuple is contained in one and only one of our s-tuples. Hanani **[37]**, **[38]** settled the cases $r=3$, $s=4$, $r=2$, $s=4$ and $r=2$, $s=5$. The general problem seems very difficult.

Finally I would like to call attention to an old conjecture of van der Waerden which seems surprisingly difficult: The permanent of an n by n doubly stochastic matrix is $\geq n!/n^n$. Equality only if all elements of the matrix are $1/n$.

V. **Some problems in combinatorial number theory.** Van der Waerden **[59]** proved the following theorem: If we split the integers into two classes at least one of them contains arbitrarily long arithmetic progressions. Here we are more concerned with the following finite form of van der Waerden's theorem: Let $f(n)$ be the smallest integers so that if we split the integers not exceeding $f(n)$ into two classes at least one of them contains an arithmetic progression of n terms. Van der Waerden's proof gives a very poor upper bound for $f(n)$. Sharpening a previous result of Rado and myself, Schmidt proved $f(n) > 2^{n-c(n\log n)^{1/2}}$ **[54]**. I understand that recently Belrekamp proved $f(n) > 2^n$ (Canad. Math. Bull. **11** (1968), 409–414). It would be very desirable to obtain better lower and especially upper bounds for $f(n)$. Undoubtedly $\lim_{n=\infty} f(n)^{1/n}$ exists—I expect the limit to be infinite. One could try to estimate $f(n, m)$ where $f(n, m)$ is the smallest integer so that if we split the integers not exceeding $f(n, m)$ into two classes either the first class contains an arithmetic progression of n terms or the second an arithmetic progression of m terms. Also one can consider splittings into more than two classes. R. Schneider and R. Ecks have certain results in these directions.

Define $f_\varepsilon(n)$ to be the smallest integer so that if $g(m) = \pm 1$, $1 \leq m \leq f_\varepsilon(n)$, is any number-theoretic function then there is an arithmetic progression of n terms

$$0 < a < a+d < \cdots < a+(n-1)d \leq f_\varepsilon(n)$$

for which

$$\left| \sum_{k=0}^{n-1} g(a+kd) \right| > \varepsilon n.$$

I proved $f_\varepsilon(n) > (1+\eta)^n$, $\eta = \eta(\varepsilon)$ [18]. The proof is probabilistic and similar to my proof with Rado. I would guess that $f_\varepsilon(n) < (1+\eta_1)^n$ and perhaps $\eta_1 \to 1$ as $\varepsilon \to 0$ but I cannot disprove

$$\lim_{n=\infty} f_\varepsilon(n)^{1/n} = \infty$$

for every $\varepsilon > 0$.

Roth [52], proved that if $g(m) = \pm 1$, $1 \le m \le n$, there always is an arithmetic progression $1 \le a < \cdots < a+kd \le n$ for which (for every $\varepsilon > 0$ if $n > n_0(\varepsilon)$)

$$\left| \sum_{l=0}^{k-1} g(a+ld) \right| > n^{1/4-\varepsilon} \tag{9}$$

and he conjectured that in (9) $n^{1/4-\varepsilon}$ can be replaced by $n^{1/2-\varepsilon}$. I proved [22] that there is a constant C and a $g(m) = \pm 1$ so that for every progression

$$\left| \sum_{l=0}^{k-1} g(a+ld) \right| < Cn^{1/2} \tag{10}$$

and I conjecture that (10) holds for every $C > 0$ if $n > n_0(C)$.

ADDED IN PROOF. This conjecture was just proved by J. Spencer.

An old conjecture of mine states that if $g(m) = \pm 1$, $1 \le m \le \infty$, then to every c there is a d and an m so that

$$\left| \sum_{k=1}^{m} g(kd) \right| > c. \tag{11}$$

The proof of (11) seems to present great difficulties.

Let $a_1 < \cdots < a_k$ be k distinct real numbers. Denote by $f(n; a_1, \ldots, a_k)$ the number of solutions of

$$n = \sum_{i=1}^{k} \varepsilon_i a_i, \qquad \varepsilon_i = 0 \text{ or } 1.$$

Moser and I [21] proved that

$$f(n; a_1, \ldots, a_k) < c2^k/k^{3/2}(\log k)^{3/2} \tag{12}$$

and we conjectured that in (12) $(\log k)^{3/2}$ can be omitted. This conjecture was proved by Sárközi and Szemerédi [53].

Moser and I further conjectured that if $k = 2l+1$ then

$$f(n; (a_1, \ldots, a_{2l+1})) \le f(n; -l, -l+1, \ldots, -1, 0, 1, \ldots, l-1, l). \tag{13}$$

As far as I know (13) has not yet been proved. Van Lint [45] found an asymptotic formula for $f(n; -l, \ldots, l)$.

We further conjectured that the number of solutions of

$$n = \sum_{i=1}^{k} \varepsilon_i a_i, \qquad \sum_{i=1}^{k} \varepsilon_i = t$$

is for every t less than $c2^k/k^2$.

Donald Newman conjectured that for every n and m there is a function $h_{n,m}$ having the following properties: $h_{n,m}$ is defined for $1 \le i \le n$ and

$$h_{n,m}(i) \neq h_{n,m}(j), \qquad 1 \le i < j \le n, \quad m < h_{n,m}(i) \le m+n,$$
$$(i, h_{n,m}(i)) = 1, \qquad 1 \le i \le n.$$

Baines and Daykin [**1**] proved that for $n=m$ such a function exists, but the general case is not yet settled. One would think that Hall's theorem can be applied here, but this seems to present great difficulties.

Schur [**56**] proved the following result: Denote by $H(n)$ the least integer so that if we split the integers from 1 to $H(n)$ into two classes, the equation $x+y=z$ is solvable in at least one class. Schur proved $H(n) \le [en!]$. It would be very interesting to decide whether $\lim_{n=\infty} H(n)^{1/n}$ is finite or not.

Saunders in his dissertation written under Ore proved the following result: To every k there is an $F(k)$ so that if we split the integers from 1 to $F(k)$ into two classes there are always k integers $a_1 < \cdots < a_k$ so that all the $2^k - 1$ sums

$$(14) \qquad \sum_{i=1}^{k} \varepsilon_i a_i, \qquad \varepsilon_i = 0 \text{ or } 1 \quad (\text{not all } \varepsilon_i = 0),$$

are in the same class. Graham and Rothschild then asked the following question: Split the set of all the integers into two classes. Does there then exist an infinite sequence $a_1 < \cdots$ so that all the sums

$$(15) \qquad \sum_{i=1}^{\infty} \varepsilon_i a_i, \qquad \varepsilon_i = 0 \text{ or } 1,$$

are in the same class, where in (15) not all the ε_i are 0 but only a finite number of them are different from 0?

I do not know the answer to this question. It easily follows from Ramsey's theorem that there is an infinite sequence where all the sums

$$\sum_{i=1}^{\infty} \varepsilon_i a_i, \quad \varepsilon_i = 0 \text{ or } 1, \qquad \sum_{i=1}^{\infty} \varepsilon_i = t,$$

are in the same class. But I cannot decide the following question: Does there exist an infinite sequence $a_1 < \cdots$ where all the a_i, $1 \le i < \infty$, and all the sums $a_i + a_j$, $1 \le i < j < \infty$, belong to the same class. This would of course follow from Graham's conjecture. The following weakening of Graham's conjecture also does not seem to be completely trivial: There is an infinite sequence $a_1 < \cdots$ so that all the sums

$$\sum_{i=1}^{\infty} \varepsilon_i a_i, \quad \varepsilon_i = 0 \text{ or } 1, \qquad \sum_{i=1}^{\infty} \varepsilon_i = t, \quad 1 \le t < \infty,$$

belong to the same class but the class may depend on t.

Finally I would like to ask the following question which I could not decide even if we assume the continuum hypothesis. Split the real numbers into two classes. Does there then exist a set of power $\aleph_1$, $\{a_\alpha\}$, $1 \leq \alpha < \omega_1$, so that all the sums

$$a_{\alpha_1} + a_{\alpha_2}, \qquad 1 \leq \alpha_1 < \alpha_2 < \omega_1,$$

belong to the same class?

For further problems of combinatorial number theory I refer to **[22]** and my first paper on extremal problems in number theory **[17]**. See also **[20]**.

VI. Finally I would like to call attention to some curious results and *problems* of Czipszer, Hajnal and myself **[4]** *which are partly graph-theoretic and partly analytic*—in fact they deal with Tauberian theorems.

Let G be an infinite graph whose vertices are the integers and $g(n)$ the number of edges of G both vertices of which do not exceed n. A monotone path of length k is a sequence of integers $i_1 < \cdots < i_{k+1}$ where i_j and i_{j+1}, $j = 1, \ldots, k$, are joined by an edge. We conjecture that if for every $\varepsilon > 0$ and $n > n_0(\varepsilon)$

$$g(n) > n^2\left(\frac{1}{4} - \frac{1}{4k} + \varepsilon\right)$$

then G contains infinitely many monotone paths of length k.

We proved this conjecture for $k=2$ and $k=3$, but could not settle the general case.

For $k=2$ we proved the following stronger theorem: Assume that for $n > n_0(\varepsilon)$

$$g(n) > \frac{n^2}{8} + \left(\frac{1}{32} + \varepsilon\right)\frac{n^2}{(\log n)^2},$$

then G contains infinitely many monotone paths of length k. This result is best possible since it fails if we only assume

$$g(n) = \frac{n^2}{8} + \frac{n^2}{32(\log n)^2} + o\left(\frac{n}{(\log n)^2}\right).$$

We further proved that there is a G for which

$$\liminf_{n=\infty} g(n)/n^2 > \tfrac{1}{4}$$

but G does not contain an infinite monotone path. On the other hand we showed that there is an $\alpha > 0$ so that if

$$\liminf_{n=\infty} g(n)/n^2 > \tfrac{1}{2} - \alpha$$

then G contains an infinite monotone path. We do not know the largest value of α which will insure this conclusion.

Several further problems of a combinatorial nature can be found in my two papers **[13]**, **[14]** on unsolved problems.

References

1. D. E. Daykin and M. J. Baines, *Coprime mappings between sets of consecutive integers*, Mathematika **10** (1963), 132–136. MR **29** #1176.

2. F. Behrend, *On sequences of numbers not divisible one by another*, J. London Math. Soc. **10** (1935), 42–44.

3. N. G. de Bruijn and P. Erdös, *On a combinatorial problem*, Nederl. Akad. Wetensch. Proc. **51** (1948), 1277–1279 = Indag. Math. **10** (1948), 421–423. MR **10**, 424.

4. J. Czipszer, P. Erdös and A. Hajnal, *Some external problems on infinite graphs*, Magyar Tud. Akad. Mat. Kutató Int. Közl. **7** (1962), 441–457. MR **27** #744.

5. L. Danzer and B. Grünbaum, *Über zwei Probleme bezüglich konvexer Körper von P. Erdös und von V. L. Klee*, Math. Z. **79** (1962), 95–99. MR **25** #1488.

6. R. P. Dilworth, *A decomposition theorem for partially ordered sets*, Ann. of Math. (2) **51** (1950), 161–166. MR **11**, 309.

7. P. D. T. A. Elliott, *On the number of circles determined by n points*, Acta Math. Acad. Sci. Hungar. **18** (1967), 181–188. MR **35** #4793.

8. P. Erdös, *On the density of some sequences of numbers*. II, J. London Math. Soc. **12** (1937), 7–11.

9. ———, *On a lemma of Littlewood and Offord*, Bull. Amer. Math. Soc. **51** (1945), 898–902. MR **7**, 309.

10. ———, *On sets of distances of n points*, Amer. Math. Monthly **53** (1946), 248–250. MR **7**, 471.

11. ———, *Some remarks on the theory of graphs*, Bull. Amer. Math. Soc. **53** (1947), 292–294. MR **8**, 479.

12. ———, *On some geometrical problems*, Mat. Lapok **8** (1957), 86–92. (Hungarian) MR **20** #6056.

13. ———, *Some unsolved problems*, Michigan Math. J. **4** (1957), 291–300. MR **20** #5157.

14. ———, *Some unsolved problems*, Magyar Tud. Akad. Mat. Kutató Int. Közl. **6** (1961), 221–254. MR **31** #2106.

15. ———, *Graph theory and probability*. II, Canad. J. Math. **13** (1961), 346–352. MR **22** #10925.

16. ———, *On some elementary geometrical problems*, Köz. Mat. Lapok **24** (1962), 193–201. (Hungarian)

17. ———, *Remarks on number theory*. IV. *Extremal problems in number theory*. I, Mat. Lapok **13** (1962), 228–255. (Hungarian) MR **33** #4020.

18. ———, *On combinatorial questions connected with a theorem of Ramsey and van der Waerden*, Mat. Lapok **14** (1963), 29–37. (Hungarian) MR **34** #7409.

19. ———, *On a combinatorial problem*. II, Acta Math. Acad. Sci. Hungar. **15** (1964), 445–447. MR **29** #4700.

20. ———, *Some recent advances and current problems in number theory*, Lectures on Modern Mathematics, vol. III. Wiley, New York, 1965, pp. 196–244. MR 31 #2191.

21. ———, *Extremal problems in number theory*, Proc. Sympos. Pure Math., vol. VIII, Amer. Math. Soc., Providence, R.I., 1965, pp. 181–189. MR **30** #4740.

22. ———, *Remarks on number theory*. V. *Extremal problems in number theory*. II, Mat. Lapok **17** (1966), 135–155. (Hungarian) MR **36** #133.

23. P. Erdös and A. Hajnal, *On a property of families of sets*, Acta Math. Acad. Sci. Hungar. **12** (1961), 87–123. MR **27** #50.

24. ———, *On chromatic number of graphs and set-systems*, Acta Math. Acad. Sci. Hungar. **17** (1966), 61–99, see 94–99. MR **33** #1247. We use probabilistic methods. Hales and Jewett [36] prove the finiteness of $m_1(n)$ in a completely different setting by a direct construction.

25. P. Erdös, A. Hajnal and R. Rado, *Partition relations for cardinal numbers*, Acta Math. Acad. Sci. Hungar. **16** (1965), 93–196. MR **34** #2475.

26. P. Erdös, Chao Ko and R. Rado, *Intersection theorems for systems of finite sets*, Quart. J. Math. Oxford Ser. (2) **12** (1961), 313–320. MR **25** #3829.

27. P. Erdös and R. Rado, *A combinatorial theorem*, J. London Math. Soc. **25** (1950), 249–255. MR **12**, 322.

28. ———, *Combinatorial theorems on classifications of subsets of a given set*, Proc. London Math. Soc. (3) **2** (1952), 417–439. MR **16**, 455.

29. P. Erdös and R. Rado, *Intersection theorems for systems of sets*, J. London Math. Soc. **35** (1960), 85–90. MR **22** #2554.

30. ———, *Intersection theorems for systems of sets*. II, J. London Math. Soc. **44** (1969), 467–479.

31. P. Erdös and G. Szekeres, *A combinatorial problem in geometry*, Compositio Math. **2** (1935), 463–470.

32. ———, *On some extremum problems in elementary geometry*, Ann. Univ. Sci. Budapest. Eötvös. Sect. Math. **3–4** (1960/61), 53–62. MR **24** #A3560.

33. J. E. Graver and James Yackel, *Some graph theoretic results associated with Ramsey's theorem*, J. Combinatorial Theory **4** (1968), 125–175; $f(2,3,6)$ was first determined by Gerson and Kalbfleisch. MR **37** #1278.

34. R. E. Greenwood and A. M. Gleason, *Combinatorial relations and chromatic graphs*, Canad. J. Math. **7** (1955), 1–7. MR **16**, 733.

35. H. Hadwiger and H. Debrunner, *Kombinatorische Geometrie in der Ebene*, Inst. Math., Univ. Genève, Geneva, 1960; English transl., Holt, Rinehart and Winston, New York, 1964. MR **22** #11210; MR **29** #1577.

36. A. Hales and R. I. Jewett, *Regularity and positional games*, Trans. Amer. Math. Soc. **106** (1963), 222–229. MR **26** #1265.

37. H. Hanani, *On quadruple systems*, Canad. J. Math. **12** (1960), 145–157. MR **22** #2558.

38. ———, *The existence and construction of balanced incomplete block designs*, Ann. Math. Statist. **32** (1961), 361–386. MR **29** #4161.

39. S. Hansen, *A generalization of a theorem of Sylvester on the lines determined by a finite point set*, Math. Scand. **16** (1965), 175–180. MR **34** #3411.

40. Gy. Katona, *On a conjecture of Erdös and a stronger form of Sperner's theorem*, Studia Sci. Math. Hungar. **1** (1966), 59–63. MR **34** #5690.

41. L. M. Kelly and W. O. J. Moser, *On the number of ordinary lines determined by n points*, Canad. J. Math. **1** (1958), 210–219. MR **20** #3494.

42. V. L. Klee (editor), *Convexity*, Proc. Sympos. Pure Math., vol. VII, Amer. Math. Soc., Providence, R.I., 1963.

43. D. Kleitman, *On a lemma of Littlewood and Offord on the distrbution of certain sums*, Math. Z. **90** (1965), 251–259. MR **32** #2336.

44. ———, *On a combinatorial problem of Erdös*, Proc. Amer. Math. Soc. **17** (1966), 139–141. MR **32** #2337.

45. J. H. van Lint, *Representation of* 0 *as* $\sum_{k=-N}^{N} \varepsilon_k k$, Proc. Amer. Math. Soc. **18** (1967), 182–184. MR **34** #5789.

46. E. W. Miller, *On a property of families of sets*, C.R. Soc. Sci. Varsovie **30** (1937), 31–38.

47. Th. Motzkin, *The lines and planes connecting the points of a finite set*, Trans. Amer. Math. Soc. **70** (1951), 451–464. This paper contains many further interesting problems some of which have been settled in the meantime. See, e.g., Hansen [**39**]. MR **12**, 849.

48. C. St. J. A. Nash-Williams, *On well-quasi-ordering transfinite sequences*, Proc. Cambridge Philos. Soc. **61** (1965), 33–39. MR **30** #3850.

49. R. Rado, *Studien zur Kombinatorik*, Math. Z. **36** (1933), 424–480.

50. F. P. Ramsey, *On a problem of formal logic*, Proc. London Math. Soc. (2) **30** (1929), 264–286, also Collected Papers, 82–111. See also Erdös and Rado [**28**]. For generalizations of Ramsey's Theorem see Erdös and Rado [**27**] and Nash-Williams [**48**]. I would further like to mention the following unpublished result of Galvin: Let F be a family of finite subsets of the integers so that every infinite set contains a set in F. Then every infinite subset S of the integers contains an infinite subset S_1, so that every $S_2 \subset S_1$, $|S_2| = \aleph_0$ has an initial segment in F.

51. M. Reiss, *Über eine Steinersche kombinatorische Aufgabe*, J. Reine Angew. Math. **56** (1859), 326–344.

52. K. F. Roth, *Remark concerning integer sequences*, Acta Arith. **9** (1964), 257–260. MR **29** #5806.

53. A. Sárközi and E. Szemerédi, *Über ein Problem von Erdös und Moser*, Acta Arith. **11** (1966), 205–208. MR **32** #102.

54. W. Schmidt, *Two combinatorial theorems on arithmetic progressions*, Duke Math J. **29** (1962), 129–140. MR **25** #1125.

55. ———, *Ein kombinatorisches Problem von P. Erdös*, Acta Math. Acad. Sci. Hungar. **15** (1964), 373–374. MR **29** #4701.

56. I. Schur, *Über die Kongruenz* $x^m+y^m \equiv z^m \pmod p$, Jber. Deutsch. Math.-Verein. **25** (1916), 114–117. See also Rado [**49**].

57. A. Sperner, *Ein Satz über Untermengen einer endlichen Menge*, Math. Z. **27** (1928), 544–548.

58. G. Szekeres, *On an extremum problem in the plane*, Amer. J. Math. **63** (1941), 208–210. MR **2**, 263.

59. B. L. van der Waerden, *Beweis einer Baudetschen Vermutung*, Nieuw Arch. Wisk. (2) **15** (1927), 212–216. See also R. Rado [**49**].

UNIVERSITY OF COLORADO

HUNGARIAN ACADEMY OF SCIENCES

[illegible]

MULTIROWED PARTITIONS WITH STRICT DECREASE ALONG COLUMNS[1] (NOTES ON PLANE PARTITIONS. IV)

BASIL GORDON

1. **Introduction.** For definitions and terminology, see [2]. We recall that if n is a nonnegative integer, $a_k(n)$ is the number of k-rowed partitions of n, while $b_k(n)$, $c_k(n)$, and $d_k(n)$ denote respectively the numbers of such partitions whose nonzero parts decrease strictly along rows, along columns, and along both. Put

$$A_k(x) = \sum_{n=0}^{\infty} a_k(n)x^n, \qquad B_k(x) = \sum_{n=0}^{\infty} b_k(n)x^n,$$

$$C_k(x) = \sum_{n=0}^{\infty} c_k(n)x^n, \quad \text{and} \quad D_k(x) = \sum_{n=0}^{\infty} d_k(n)x^n.$$

A classical theorem of MacMahon [4, p. 243] asserts that $A_k(x)=\prod_{\nu=1}^{\infty}(1-x^\nu)^{-\min(\nu,k)}$; on the other hand it was shown in [3] that

$$B_k(x) = \prod_{\nu=1}^{\infty} (1-x^\nu)^{-\min([(\nu+1)/2],[k/2]+2\{k\nu/2\})},$$

where $[\theta]$ and $\{\theta\}$ denote the integral and fractional parts of θ. In this note we shall consider $C_k(x)$ and $D_k(x)$. It turns out that there is a peculiar reciprocity between $a_k(n)$ and $d_k(n)$, and between $b_k(n)$ and $c_k(n)$, which enables one to express $C_k(x)$ and $D_k(x)$ as multiple series of k-rowed determinants. In the case of $C_k(x)$, the series in question can be summed explicitly in terms of the classical functions $P=\prod_{\nu=1}^{\infty}(1-x^\nu)^{-1}$, $Q=\prod_{\nu=1}^{\infty}(1-x^{2\nu-1})^{-1}$, and the "false ϑ-function"

$$F = \sum_{n=0}^{\infty} (-1)^n x^{\binom{n+1}{2}}.$$

However this has so far proved impossible for $D_k(x)$ when $k \geq 3$.

2. **A combinatorial lemma.** In this section we prove a lemma on standard tableaux which underlies the reciprocity between $a_k(n)$ and $d_k(n)$ mentioned in the

[1] The preparation of this paper was sponsored in part by NSF Grant No. GP-5497.

introduction. Let

$$m = m_1+m_2+\cdots+m_k \qquad (m_1 \geq m_2 \geq \cdots \geq m_k)$$

be a fixed partition, and consider its Ferrars graph $\mathcal{F}$. By a *standard tableau* we mean the configuration which results when the nodes of $\mathcal{F}$ are replaced by the integers $1, 2, \ldots, m$ in such a way that the entries in each row or column are monotonically increasing. For example, to the partition $3+2$ there correspond the standard tableaux

$$\begin{matrix}1\,2\,3\\4\,5\end{matrix}, \quad \begin{matrix}1\,2\,4\\3\,5\end{matrix}, \quad \begin{matrix}1\,2\,5\\3\,4\end{matrix}, \quad \begin{matrix}1\,3\,4\\2\,5\end{matrix}, \quad \text{and} \quad \begin{matrix}1\,3\,5\\2\,4\end{matrix}.$$

It is known (see for example [6, p. 385]) that the number of different standard tableaux corresponding to the partition $m=m_1+\cdots+m_k$ is

$$m!\prod_{i<j}(h_i-h_j)\Big/\prod_{i=1}h_i!,$$

where $h_i=m_i+k-i$. In what follows we fix a partition π, and consider the set $\mathcal{T}_\pi$ of all standard tableaux corresponding to π. For each such tableau T, let $P(T)$ be the set of integers in T which are "southwest" of their successors (i.e., the integers i which appear in a later row and earlier column than $i+1$). Let $Q(T)$ be the set of integers in T which are "northeast" of their successors. For example if the five tableaux listed above are denoted in order by T_1, T_2, T_3, T_4, T_5, we have $P(T_1)=\varnothing$, $P(T_2)=\{3\}$, $P(T_3)=\{4\}$, $P(T_4)=\{2\}$, $P(T_5)=\{2, 4\}$, and $Q(T_1)=\{3\}$, $Q(T_2)=\{2, 4\}$, $Q(T_3)=\{2\}$, $Q(T_4)=\{4\}$, $Q(T_5)=\varnothing$. This example illustrates the combinatorial lemma which we now state.

LEMMA 1. *Let the standard tableaux corresponding to the partition π be $T_1, T_2, \ldots, T_\nu$, where $\nu=|\mathcal{T}_\pi|$. Then the sets $P(T_1), \ldots, P(T_\nu)$ are a permutation of the sets $Q(T_1), \ldots, Q(T_\nu)$.*

PROOF. The proof is by induction on m. When $m=1$, there is only one partition, and only one standard tableau T; moreover $P(T)=Q(T)=\varnothing$. Now suppose $m>1$, and that the lemma has already been proved for all partitions of all integers less than m. For each partition π, let $\mathcal{P}_\pi$ be the class of all sets $P(T)$, where T runs through $\mathcal{T}_\pi$. Here we are using the word "class" in the sense of "class with multiplicities," i.e. each set M belongs to $\mathcal{P}_\pi$ with a multiplicity equal to the number of tableaux $T\in\mathcal{T}_\pi$ such that $M=P(T)$. Similarly, we let $\mathcal{Q}_\pi$ denote the class of all sets $Q(T)$, where T runs through $\mathcal{T}_\pi$. Our object is to prove that $\mathcal{P}_\pi=\mathcal{Q}_\pi$, and the idea of the proof is to obtain a recurrence satisfied by both $\mathcal{P}_\pi$ and $\mathcal{Q}_\pi$. Consider for definiteness a partition π: $m=m_1+m_2+m_3$ into three parts; it will be easily seen that the reasoning used in this case is perfectly general. If $m_1>m_2>m_3$, there are three possible positions for the integer m in any $T\in\mathcal{T}_\pi$, since it must be the final entry in one of the three rows. Let π_1, π_2, π_3 denote respectively the partitions $(m_1-1)+m_2+m_3$, $m_1+(m_2-1)+m_3$, and $m_1+m_2+(m_3-1)$; for any tableau $T\in\mathcal{T}_{\pi_i}$, let $T^{(i)}$ be the tableau obtained from T by placing the integer m at the end of its ith row. Thus, letting $\mathcal{T}_\pi^{(i)}=\{T^{(i)} \mid T\in\mathcal{T}_{\pi_i}\}$, we have $\mathcal{T}_\pi=\mathcal{T}_\pi^{(1)}\cup\mathcal{T}_\pi^{(2)}\cup\mathcal{T}_\pi^{(3)}$, where the union is disjoint. Hence $\mathcal{P}_\pi=\mathcal{P}_\pi^{(1)}\cup\mathcal{P}_\pi^{(2)}\cup\mathcal{P}_\pi^{(3)}$, where $\mathcal{P}_\pi^{(i)}=\{P(T)\mid T\in\mathcal{T}_\pi^{(i)}\}$. If $T\in\mathcal{T}_{\pi_3}$, we have $P(T^{(3)})=P(T)$, for $m-1$ cannot be southwest of m. Therefore $\mathcal{P}_\pi^{(3)}=\{P(T)\mid T\in\mathcal{T}_{\pi_3}\}=\mathcal{P}_{\pi_3}$. On the other hand, if $T\in\mathcal{T}_{\pi_2}$, we have $P(T^{(2)})=P(T)\cup\{m-1\}$ or $P(T^{(2)})=P(T)$ according as $m-1$ is in the third row of T or not. Let $\pi_{2,3}$ be the partition $m_1+(m_2-1)+(m_3-1)$, and for any tableau $V\in\mathcal{T}_{\pi_{2,3}}$, let $V^{(3)}$ be the tableau obtained from V by adjoining $m-1$ at the end of its third row. Then the tableaux $T\in\mathcal{T}_{\pi_2}$ for which $P(T^{(2)})=P(T)\cup\{m-1\}$ are precisely those of the form $T=V^{(3)}$, with $V\in\mathcal{T}_{\pi_{2,3}}$. From this it follows that $\mathcal{P}_\pi^{(2)}$ can be obtained by starting with the class $\{P(T)\mid T\in\mathcal{T}_{\pi_2}\}=\mathcal{P}_{\pi_2}$, and then adjoining $m-1$ to all sets of the subclass $\{P(V)\mid V\in\mathcal{T}_{\pi_{2,3}}\}=\mathcal{P}_{\pi_{2,3}}$. Finally, if $T\in\mathcal{T}_{\pi_1}$, then $P(T^{(1)})=P(T)$ or $P(T^{(1)})=P(T)\cup\{m-1\}$ according as $m-1$ is in the first row of T or not. The tableaux $T\in\mathcal{T}_{\pi_1}$ with $m-1$ in the third row are those of the form $T=V^{(3)}$ with $V\in\mathcal{T}_{\pi_{1,3}}$, using a notation which should be obvious by now. Moreover, for any such T we have $P(T)=P(V)$. Similarly, the tableaux $T\in\mathcal{T}_{\pi_1}$ with $m-1$ on the second row are those of the form $T=V^{(2)}$, where $V\in\mathcal{T}_{\pi_{1,2}}$. But now we have $P(T)=P(V)\cup\{m-2\}$ or $P(T)=P(V)$ according as $m-2$ is on the third row of V or not. The V's for which $m-2$ is on the third row are those of the form $V=W^{(3)}$, where $W\in\mathcal{T}_{\pi_{1,2,3}}$. Moreover for any such V we have $P(V)=P(W)$. The upshot of this discussion is that $\mathcal{P}_\pi^{(1)}$ can be obtained by starting from $\{P(T)\mid T\in\mathcal{T}_{\pi_1}\}=\mathcal{P}_{\pi_1}$, and adjoining $m-1$ to two (disjoint) subclasses, one of which is $\{P(V)\mid V\in\mathcal{T}_{\pi_{1,3}}\}=\mathcal{P}_{\pi_{1,3}}$, and the other of which is gotten by starting from $\{P(V)\mid V\in\mathcal{T}_{\pi_{1,2}}\}=\mathcal{P}_{\pi_{1,2}}$ and adjoining $m-2$ to the subclass $\{P(W)\mid W\in\mathcal{T}_{\pi_{1,2,3}}\}=\mathcal{P}_{\pi_{1,2,3}}$. Our results can be cast in the form of the following more symmetric construction of $\mathcal{P}_\pi$. First construct the classes $\mathcal{A}=\mathcal{P}_{\pi_{1,2,3}}$, $\mathcal{B}=\mathcal{P}_{\pi_{1,2}}\cup\mathcal{P}_{\pi_{1,3}}\cup\mathcal{P}_{\pi_{2,3}}$, and $\mathcal{C}=\mathcal{P}_{\pi_1}\cup\mathcal{P}_{\pi_2}\cup\mathcal{P}_{\pi_3}$ (where the unions are disjoint). $\mathcal{B}$ contains a subclass equal to $\mathcal{A}$, and to each set of this subclass we adjoin $m-2$, leaving the other sets of $\mathcal{B}$ unchanged. Call the resulting class $\mathcal{B}'$. Then $\mathcal{C}$ contains a subclass equal to $\mathcal{B}'$, and to each set of this subclass we adjoin $m-1$, leaving the other sets of $\mathcal{C}$ unchanged. The resulting class $\mathcal{C}'$ is equal to $\mathcal{P}_\pi$.

We can now go through the same analysis for the class $\mathcal{Q}_\pi$, and we obtain the same construction with $\mathcal{P}$ replaced by $\mathcal{Q}$ everywhere. Since $\mathcal{P}_{\pi_i}=\mathcal{Q}_{\pi_i}$, $\mathcal{P}_{\pi_{i,j}}=\mathcal{Q}_{\pi_{i,j}}$, and $\mathcal{P}_{\pi_{1,2,3}}=\mathcal{Q}_{\pi_{1,2,3}}$ by induction, it follows that $\mathcal{P}_\pi=\mathcal{Q}_\pi$.

If $m_1=m_2$ or $m_2=m_3$ the proof is even easier since then one or more of the subcases described above cannot occur.

3. **The reciprocity theorems.** Let π: $m=m_1+\cdots+m_k$ be a fixed partition, where $m_1\geq m_2\geq\cdots\geq m_k\geq 0$. (Note that we are allowing some of the m_i to vanish here.) Denote by $a(n; m_1,\ldots,m_k)$ or $a(n;\pi)$ the number of k-rowed partitions of n having exactly m_i nonzero parts on the ith row, and let $A(x; m_1,\ldots,m_k)=A(x;\pi)$ $=\sum_{n=0}^{\infty} a(n;\pi)x^n$ (where $a(0;\pi)=1$ or 0 according as $m=0$ or $m>0$). We define $B(x;\pi)$, $C(x;\pi)$, and $D(x;\pi)$ analogously, with strict decrease being required along rows for B, along columns for C, and along both for D. We first deal with the reciprocity between A and D. These functions can be calculated by the following

method, which we illustrate with the example $k=2$, $m_1=3$, $m_2=2$, for which the standard tableaux were tabulated earlier. Consider first the function $A(x; 3, 2)$. If the parts are denoted by $\begin{smallmatrix} a & b & c \\ d & e & \end{smallmatrix}$, there are five mutually exclusive and exhaustive possibilities for their ordering, viz. $a\geq b\geq c\geq d\geq e$, $a\geq b\geq d>c\geq e$, $a\geq b\geq d\geq e>c$, $a\geq d>b\geq c\geq e$, and $a\geq d>b\geq e>c$. It will be observed that these orderings are in 1:1 correspondence with the 5 standard tableaux, where the part marked 1 in the tableau is greatest, the part marked 2 is next greatest, etc. Whenever an integer in the tableau is southwest if its successor, a strict inequality occurs between the corresponding parts. Now the number of partitions $n=a+b+c+d+e$ with $a\geq b\geq c\geq d \geq e>0$ is generated by $x^5/(\mathbf{5})!$, where $(\boldsymbol{\nu})!=(1-x)(1-x^2)\cdots(1-x^\nu)$, while the number of partitions with $a\geq b\geq d>c\geq e>0$ is generated by $x^8/(\mathbf{5})!$ (since our inequalities are equivalent to $a-1\geq b-1\geq d-1\geq c\geq e>0$), etc. In general the number of partitions $n=n_1+n_2+\cdots+n_m$ with $n_1\geq n_2\geq\cdots\geq n_m>0$ and with certain of the inequalities being strict, is generated by $x^r/(\mathbf{m})!$, where $r=m+\sum_i i$, the summation being extended over all i such that the inequalities $n_i>n_{i+1}$ are strict. Hence if T is a given standard tableau corresponding to the partition $\pi: m=m_1+\cdots+m_k$, the number of partitions of n subject to the inequalities determined by T is generated by $x^{p(T)}/(\mathbf{m})!$, where $p(T)=m+\sum_{i\in P(T)} i$. It follows that

$$A(x; \pi) = \frac{1}{(\mathbf{m})!} \sum_{T\in\mathscr{T}_\pi} x^{p(T)}.$$

On the other hand, we can make a similar analysis of $D(x; \pi)$. In the example $m_1=3$, $m_2=2$ which was treated above, we now have the following five possibilities for the ordering of the parts: $a>b>c\geq d>e$, $a>b\geq d>c\geq e$, $a>b\geq d>e>c$, $a>d>b>c\geq e$, and $a>d>b\geq e>c$. Again there is a 1:1 correspondence between orderings and standard tableaux. But this time, whenever an integer in the tableau is not northeast of its successor, there is a strict inequality between the corresponding parts. It follows that the partitions satisfying the inequalities prescribed by a given tableau T are generated by $x^{q(T)}/(\mathbf{m})!$, where

$$q(T) = m+ \sum_{i=1;i\notin Q(T)}^{m-1} i = \binom{m+1}{2} - \sum_{i\in Q(T)} i.$$

Moreover

$$D(x; \pi) = \frac{1}{(\mathbf{m})!} \sum_{T\in\mathscr{T}_\pi} x^{q(T)}.$$

By Lemma 1, the classes $\{P(T) \mid T\in\mathscr{T}_\pi\}$ and $\{Q(T) \mid T\in\mathscr{T}_\pi\}$ coincide, so in particular the two sets of integers $\{\sum_{i\in P(T)} i \mid T\in\mathscr{T}_\pi\}$ and $\{\sum_{i\in Q(T)} i \mid T\in\mathscr{T}_\pi\}$ are the same. Therefore the sets $\{p(T)-m \mid T\in\mathscr{T}_\pi\}$ and

$$\left\{q(T)-\binom{m+1}{2} \,\middle|\, T\in\mathscr{T}_\pi\right\}$$

are negatives of each other. This implies that the functions

$$(\mathbf{m})!\, x^{-m}A(x; \pi) = \sum_{T\in\mathscr{T}_\pi} x^{P(T)-m}$$

and

$$(\mathbf{m})!\, x^{-\binom{m+1}{2}} D(x;\pi) = \sum_{T\in\mathscr{T}_\pi} x^{q(T)-\binom{m+1}{2}}$$

are interchanged by the map $x \to x^{-1}$. Thus

$$\begin{aligned}(\mathbf{m})!\, x^{-\binom{m+1}{2}} D(x;\pi) &= (1-x^{-1})(1-x^{-2})\cdots(1-x^{-m})x^m A(x^{-1};\pi)\\ &= (-1)^m x^{m-\binom{m+1}{2}}(\mathbf{m})!\, A(x^{-1};\pi),\end{aligned}$$

or finally

$$D(x;\pi) = (-x)^m A(x^{-1};\pi). \tag{1}$$

Now MacMahon [4, p. 240] has shown that if we put $\xi_\nu = x^\nu/(\boldsymbol{\nu})!$, and $h_i = m_i + k - i$ ($i = 1, \ldots, k$), then $A(x; m_1, \ldots, m_k)$ is equal to the determinant $\det(\alpha_{ij})$, where

$$\alpha_{ij} = x^{(i-j)(i-j+1)/2}\xi_{h_j-k+i}.$$

Substituting this into (1) and simplifying, we find that $D(x; m_1, \ldots, m_k) = \det(\delta_{ij})$, where

$$\delta_{ij} = x^{-(i-j)(i-j+1)/2}\eta_{h_j-k+i},$$

and where

$$\eta_\nu = x^{\binom{\nu+1}{2}}/(\boldsymbol{\nu})!.$$

(Both ξ_ν and η_ν vanish where $\nu < 0$.) Since the inequalities $m_1 \geq m_2 \geq \cdots \geq m_k$ are equivalent to $h_1 > h_2 > \cdots > h_k$, we have

$$D_k(x) = \sum_{h_1>\cdots>h_k} \det \frac{x^{-\frac{(i-j)(i-j+1)}{2}+\binom{h_j-k+i+1}{2}}}{(\mathbf{h}_j-\mathbf{k}+\mathbf{i})!}, \tag{2}$$

so that $D_k(x)$ can be expressed as a multiple series of determinants.

The reciprocity between B and C is similar to that between A and D, but the proof is simpler. For any partition π: $m = m_1 + \cdots + m_k$, let π': $m = m_1' + \cdots + m_l'$ be the conjugate partition. Then $c(n;\pi) = b(n;\pi')$, as is immediately seen by interchanging rows and columns of the partitions enumerated by $c(n;\pi)$. Now consider the example $m_1 = 3$, $m_2 = 2$ which was used earlier in connection with $A(x;\pi)$ and $D(x;\pi)$. Denoting the parts by $\begin{smallmatrix} a & b & c\\ d & e & \end{smallmatrix}$ as before, we find the following five possible orderings for the partitions enumerated by $b(n;3,2)$: $a>b>c\geq d>e$, $a>b\geq d>c\geq e$, $a>b\geq d>e>c$, $a\geq d>b>c\geq e$, and $a\geq d>b\geq e>c$. As before, these orderings are in 1:1 correspondence with the standard tableaux, where the part marked 1 in the tableau is greatest, the part marked 2 is next greatest, etc. But in the present case, whenever an integer i in the tableau is immediately to the left of $i+1$, or southwest of $i+1$, a strict inequality occurs between the corresponding parts. Hence, if $R(T)$ is the set of all such integers i, the number of partitions of n subject to the inequalities determined by T is generated by $x^{r(T)}/(\mathbf{m})!$, where $r(T) = m + \sum_{i\in R(T)} i$. From this we obtain $B(x;\pi) = (1/(\mathbf{m})!) \sum_{T\in\mathscr{T}_\pi} x^{r(T)}$. Now let $S(T)$ be the set of all integers i in the tableau T which occur immediately above $i+1$ or northeast of $i+1$. If T' is the tableau obtained by interchanging rows and columns of T (so that T' is associated with the partition π'), then clearly $R(T')=S(T)$. Hence

$$C(x;\pi) = B(x;\pi') = \frac{1}{(\mathbf{m})!}\sum_{T'\in\mathcal{T}_{\pi'}} x^{r(T')} = \frac{1}{(\mathbf{m})!}\sum_{T\in\mathcal{T}_\pi} x^{s(T)},$$

where $s(T)=m+\sum_{i\in S(T)} i$. Since $\{1, 2, \ldots, m-1\}$ is the disjoint union of $R(T)$ and $S(T)$, we have $r(T)-m+s(T)-m=\sum_{i=1}^{m-1} i=\binom{m}{2}$, or $\binom{m+1}{2}-r(T)=s(T)-m$. Therefore

$$B(x^{-1};\pi) = \frac{1}{(1-x^{-1})\cdots(1-x^{-m})}\sum_{T\in\mathcal{T}_\pi} x^{-r(T)}$$

$$= \frac{(-1)^m x^{\binom{m+1}{2}}}{(\mathbf{m})!}\sum_{T\in\mathcal{T}_\pi} x^{-r(T)}$$

$$= \frac{(-1)^m x^{-m}}{(\mathbf{m})!}\sum_{T\in\mathcal{T}_\pi} x^{s(T)} = (-x)^{-m}C(x;\pi).$$

Hence we have

$$C(x;\pi) = (-x)^m B(x^{-1};\pi), \tag{3}$$

which is entirely analogous to equation (1).

It was shown in [2] that if

$$\eta_\nu = x^{\binom{\nu+1}{2}}/(\boldsymbol{\nu})!$$

and $h_i=m_i+k-i$ $(i=1,\ldots,k)$, then $B(x;\pi)=\det(\beta_{i,j})$, where $\beta_{i,j}=\eta_{h_j-k+1}$. If we substitute this into (3), and use the fact that $\eta_\nu(x^{-1})=(-1)^\nu/(\boldsymbol{\nu})!$, we find easily that

$$C(x;\pi) = \det(\gamma_{i,j}), \tag{4}$$

where $\gamma_{i,j}=\xi_{h_j-k+i}$. (Recall that $\xi_\nu=x^\nu/(\boldsymbol{\nu})!$.) Hence

$$C_k(x) = \sum_{h_1>\cdots>h_k} \det \frac{x^{h_j-k+i}}{(\mathbf{h}_j-\mathbf{k}+\mathbf{i})!}. \tag{5}$$

4. **Summation of the series for $C_k(x)$.** The series (5) can be summed by applying Lemma 1 of [3], with $a_n=x^n/(\mathbf{n})!$, and $g_i=i-k$. We first have to calculate the sum $s=\sum a_n$ and the correlation coefficients $c_\nu=\sum a_n a_{n+\nu}$. Clearly $s=\sum_{n=0}^\infty x^n/(\mathbf{n})! =\sum_{m=0}^\infty p(m)x^m=P$. Since for any sequence $\{a_n\}$ we have the obvious relation $s^2=c_0+2\sum_{\nu=1}^\infty c_\nu$, it follows in the present case that

$$c_0+2\sum_{\nu=1}^\infty c_\nu = P^2. \tag{6}$$

Next observe that

$$xc_\nu-(1-x^{\nu+1})c_{\nu+1} = \sum_{n=0}^{\infty}\frac{x^{2n+\nu+1}}{(n)!\,(n+\nu)!}-\sum_{n=0}^{\infty}\frac{x^{2n+\nu+1}-x^{2n+2\nu+2}}{(n)!\,(n+\nu+1)!}$$

$$= \sum_{n=0}^{\infty}\frac{x^{2n+\nu+1}-x^{3n+2\nu+2}-x^{2n+\nu+1}+x^{2n+2\nu+2}}{(n)!\,(n+\nu+1)!}$$

$$= \sum_{n=0}^{\infty}\frac{x^{2n+2\nu+2}}{(n-1)!\,(n+\nu+1)!} = \sum_{n=0}^{\infty}\frac{x^{2n+2\nu+4}}{(n)!\,(n+\nu+2)!},$$

so that

$$xc_\nu = (1-x^{\nu+1})c_{\nu+1}+x^{\nu+2}c_{\nu+2}. \tag{7}$$

Summing these equations over all $\nu\geq 0$, we obtain $x\sum_{\nu=0}^{\infty}c_\nu=\sum_{\nu=0}^{\infty}c_{\nu+1}-xc_1$, or after using (6),

$$(1+x)c_0+2xc_1 = (1-x)P^2. \tag{8}$$

It is now convenient to introduce the quantity $F=(c_0+P^2)/2P^2$, so that $c_0=(2F-1)P^2$. Substituting this into (8), we find that $(1+x)(2F-1)P^2+2xc_1=(1-x)P^2$, from which it follows that $c_1=((1-(1+x)F)/x)P^2$. We now assert that

$$c_\nu = (-1)^\nu\frac{(1+x^\nu)(F-\sum_{\mu=0}^{\nu-1}(-1)^\mu x^{\binom{\mu+1}{2}})}{x^{\binom{\nu+1}{2}}}P^2-P^2. \tag{9}$$

When $\nu=0$ or 1, this reduces to the expressions we have just found for c_0 and c_1. It is therefore sufficient to verify that the expression on the right-hand side of (9) satisfies the recurrence (7). This is perfectly straightforward and will be omitted here. Since

$$c_\nu = \sum_{n=0}^{\infty}\frac{x^{2n+\nu}}{(n)!\,(n+\nu)!}$$

is a power series of the form $x^\nu+\cdots$, it follows from (9) that

$$F-\sum_{\mu=0}^{\nu-1}(-1)^\mu x^{\binom{\mu+1}{2}}$$

is a power series of the form

$$(-1)^\nu x^{\binom{\nu+1}{2}}+\cdots.$$

Letting $\nu\to\infty$, we see that

$$F = \sum_{\mu=0}^{\infty}(-1)^\mu x^{\binom{\mu+1}{2}}.$$

Thus F is a false θ-function in the sense of L. J. Rogers [**5**].

We note in passing that from equation (7) it follows that

$$\frac{xc_\nu}{c_{\nu+1}} = 1-x^{\nu+1}+\frac{x^{\nu+3}}{xc_{\nu+1}/c_{\nu+2}}.$$

Iteration of this leads to the continued fraction expansion

$$\frac{xc_\nu}{c_{\nu+1}} = 1-x^{\nu-1}+\frac{x^{\nu+3}}{1-x^{\nu+2}+}\frac{x^{\nu+4}}{1-x^{\nu+3}+}\cdots.$$

Thus the sum of this continued fraction can be expressed in terms of F by using equation (9). In particular we have

$$\frac{(2F-1)x^2}{1-(1+x)F} = 1-x+\frac{x^3}{1-x^2+}\frac{x^4}{1-x^3+}\cdots.$$

Since the left-hand side is obtained from F by a linear fractional transformation, we can derive from this a continued fraction expansion of F itself. The result is the remarkable identity

$$2F = 1+\frac{1-x}{1+x+}\frac{2x^2}{1-x+}\frac{x^3}{1-x^2+}\frac{x^4}{1-x^3+}\cdots. \tag{10}$$

Returning now to the main problem, we have to calculate the quantities $d_\nu = c_0 + 2(c_1+\cdots+c_{\nu-1})+c_\nu$. This is trivial from equation (9); we omit the details of the computation, and simply state the result, which is that

$$d_\nu = \frac{(-1)^\nu(1-x^\nu)\left(F-\sum_{\mu=0}^{\nu-1}(-1)^\mu x^{\binom{\mu+1}{2}}\right)}{x^{\binom{\nu+1}{2}}}P^2 \tag{11}$$

for $\nu>0$. We recall that by convention, $d_0=0$ and $d_{-\nu}=-d_\nu$. From Lemma 1 of **[3]** it follows that $C_k(x)$ is the Pfaffian of the skew-symmetric matrix $D_k=(d_{j-i})$ or

$$D'_k = \begin{bmatrix} 0 & P \\ -P & D_{k-1} \end{bmatrix}$$

according as k is even or odd. From this we see at once that $C_k(x)$ is of the form $x^{-\lambda_k}P^k\Phi_k(x, F)$, where λ_k is a nonnegative integer, and Φ_k is a polynomial with integer coefficients. Moreover the degree of Φ_k with respect to F is at most $[k/2]$, since each d_ν is linear in F. For example, direct calculation shows that $\lambda_1=0$, $\lambda_2=1$, $\lambda_3=3$, $\lambda_4=7$, while $\Phi_1=1$, $\Phi_2=(1-x)(1-F)$, $\Phi_3=(1-x)[(1+x^2)-(1+x+2x^2)F]$, and $\Phi_4=(1-x)(1-x^2)[(1-x)(1-x^3)F-(2-3x+2x^2+3x^4)F+1-2x+2x^2+x^4]$. A more thorough investigation of the polynomials $\Phi_k(x, F)$ will be undertaken elsewhere.

5. **Summation of the series for** $D_2(x)$. From equation (2) with $h_1=\alpha$, $h_2=\beta$, we have

$$D_2(x) = \sum_{\alpha>\beta\geq 0}\begin{vmatrix} \eta_{\alpha-1} & \eta_{\beta-1} \\ x^{-1}\eta_\alpha & \eta_\beta \end{vmatrix},$$

where

$$\eta_n = x^{\binom{n+1}{2}}/(\mathbf{n})!.$$

Thus

$$D_2(x) = \sum_{\alpha>\beta\geq 0} \left[\frac{x^{\binom{\alpha}{2}+\binom{\beta+1}{2}}}{(\alpha-1)!\,(\beta)!} - \frac{x^{\binom{\alpha+1}{2}+\binom{\beta}{2}-1}}{(\alpha)!\,(\beta-1)!}\right].$$

Denoting the correlation coefficient $\sum_{n=0}^{\infty} \eta_n\eta_{n+\nu}$ by γ_ν, we can write this equation in the form

$$D_2(x) = \sum_{\lambda\geq\mu\geq 0} \eta_\lambda\eta_\mu - x^{-1} \sum_{\lambda-2\geq\mu\geq 0} \eta_\lambda\eta_\mu$$

$$= \sum_{\nu=0}^{\infty} \gamma_\nu - x^{-1} \sum_{\nu=2}^{\infty} \gamma_\nu.$$

Since $Q^2 = (\sum_{n=0}^{\infty} \eta_n)^2 = \gamma_0 + 2\sum_{\nu=1}^{\infty} \gamma_\nu$, we have $D_2(x) = \frac{1}{2}(Q^2+\gamma_0) - \frac{1}{2}x^{-1}(Q^2-\gamma_0-2\gamma_1)$. It was shown in [3] that

$$\gamma_\nu + \gamma_{\nu+1} = x^{\binom{\nu+1}{2}} P; \tag{12}$$

in particular $\gamma_0+\gamma_1 = P$. Thus

$$D_2(x) = \tfrac{1}{2}(Q^2+\gamma_0) - \tfrac{1}{2}x^{-1}(Q^2+\gamma_0-2P). \tag{13}$$

In order to determine γ_0, we multiply equation (12) by $(-1)^\nu$ and sum over all $\nu\geq 0$. The series on the left telescopes, since $\gamma_\nu \to 0$ in the topology of formal power series. We therefore obtain

$$\gamma_0 = \sum_{\nu=0}^{\infty} (-1)^\nu x^{\binom{\nu+1}{2}} P = FP. \tag{14}$$

Substituting this into (15), and simplifying, we find that

$$2xD_2(x) = [2-(1-x)F]P - (1-x)Q^2. \tag{15}$$

We remark in passing that from (12) one can easily derive the continued fraction expansion

$$\frac{x^\nu \gamma_{\nu-1}}{\gamma_\nu} = 1 - x^\nu + \frac{x^{\nu+1}}{1-x^{\nu+1}+} \frac{x^{\nu+2}}{1-x^{\nu+2}+\cdots}.$$

In particular, putting $\nu = 1$ and using (14), we obtain

$$\frac{xF}{1-F} = 1 - x + \frac{x^2}{1-x^2+} \frac{x^3}{1-x^3+\cdots}.$$

This leads at once to the identity

$$F = \frac{1}{1+} \frac{x}{1-x+} \frac{x^2}{1-x^2+} \frac{x^3}{1-x^3+\cdots}, \tag{16}$$

which is closely related to (10).

From Gauss's identity

$$\sum_{n=0}^{\infty} x^{\binom{n+1}{2}} = \prod_{\nu=1}^{\infty} \frac{1-x^{2\nu}}{1-x^{2\nu-1}}$$

it follows that $Q^2 = GP$, where

$$G = \sum_{n=0}^{\infty} x^{\binom{n+1}{2}}.$$

Substituting this into (17) and simplifying, we find that

$$D_2(x) = P-(1-x)P \sum_{n=1}^{\infty} x^{\binom{2n+1}{2}-1}.$$

In other words

$$d_2(n) = p(n) - \sum_{\nu=1}^{\infty} p(n-(\nu+1)(2\nu-1)) + \sum_{\nu=1}^{\infty} p(n-\nu(2\nu+1)).$$

We note in conclusion that identities essentially equivalent to our formulae

$$\sum_{n=0}^{\infty} \frac{x^{2n}}{(\mathbf{n})!^2} = (2F-1)P^2$$

and

$$\sum_{n=0}^{\infty} \frac{x^{n(n+1)}}{(\mathbf{n})!^2} = FP$$

were developed by Auluck [**1**] in connection with a problem on the roughness of crystal surfaces.

References

1. F. C. Auluck, *On some new types of partitions associated with generalized Ferrars graphs*, Proc. Cambridge Philos. Soc. **47** (1951), 679–686. MR **13**, 536.

2. B. Gordon and L. Houten, *Notes on plane partitions.* I, J. Combinatorial Theory **4** (1968), 72–80. MR **36** #1339.

3. ———, *Notes on plane partitions.* II, J. Combinatorial Theory **4** (1968), 81–99. MR **36** #1339.

4. P. A. MacMahon, *Combinatory analysis*, vol. 2, Cambridge Univ. Press, New York, 1916.

5. L. J. Rogers, *On two theorems of combinatory analysis and some allied identities*, Proc. London Math. Soc. **16** (1917), 315–336.

6. H. Weyl, *Theory of groups and quantum mechanics*, Methuen, London, 1931.

University of California, Los Angeles

ROTA'S GEOMETRIC ANALOGUE TO RAMSEY'S THEOREM

R. L. GRAHAM AND B. ROTHSCHILD

1. Let $L=\{L_i \mid i=0, 1, 2, \ldots\}$ be a class of geometric lattices. For integers $k \geq 0$, $r \geq 0$, $t > 0$ consider the statement:

> $L(k, r, t)$. There is an integer $N=N_1(k, r, t)$, depending only on k, r, t, such that if $n \geq N$, and if the elements of L_n of rank r are colored with t colors, then there is an element x of rank k such that all the elements y of rank r with $y \leq x$ have the same color.

If we let $L_i = L(S_i)$, the subset lattice of a set S_i of i elements, then this statement, which we denote in this case by $S(k, r, t)$, becomes Ramsey's theorem for k, r, t.

Rota has conjectured that if one chooses the L_i to be $P_i(q)$, the lattice of subspaces of an i-dimensional vector space over $GF(q)$ (or equivalently, the lattice of projective subspaces of an $(i-1)$-dimensional projective space), then the corresponding statement, denoted in this case by $P_q(k, r, t)$, is also true. The conjecture is true for $r=1$ and any k, t and q. We will indicate part of the proof here. Details will appear elsewhere.

2. First we consider another statement, namely $A_q(k, r, t)$, by which we mean $L(k, r, t)$ with $L_i = A_i(q)$, the subspace lattice of an affine $(i-1)$-dimensional space over $GF(q)$. Using the well-known relationship between the affine and projective lattices (see the lemma below) we reduce P_q to A_q, of which the case $r=1$ is proved.

There are in fact three results we can obtain from the relationship, namely:

THEOREM 1. $P_q(k, r, t) \Rightarrow A_q(k, r, t)$.

THEOREM 2. $A_q(k+1, r+1, t) \Rightarrow P_q(k, r, t)$.

THEOREM 3. $\forall k\, A_q(k, r, t) \Rightarrow \forall k\, P_q(k, r, t)$.

Theorems 1 and 2 provide information about the relationship of the corresponding numbers N for the affine and projective cases. But it is clearly Theorem 3 that is necessary to reduce the projective to the affine problem for $r=1$, and thus we sketch a proof of Theorem 3 below.

3. LEMMA. *Let x be a dual atom of $P_n(q)$ (i.e. an $(n-2)$-dimensional hyperplane). Let $P_{n-1}=\{y \mid y\leq x\}$, $A_n=\{y \mid y\not\leq x$ or $y=0\}$. Then*:

(a) *P_{n-1} with the induced order is isomorphic to $P_{n-1}(q)$.*

(b) *A_n with the induced order is isomorphic to $A_n(q)$.*

(c) *For each $y\in P_{n-1}$, there is a $z\in A_n$ with $y\leq z$ and the rank of z one greater than the rank of y.*

(d) *For each $z\neq 0$, $z\in A_n$, the rank of $z\wedge x$ is one less than the rank of z.*

4. We now indicate a proof of Theorem 3. Assume $A_q(k, r, t)$ for all k. Let l_1 be a large integer. Then the lemma and $A_q(l_1, r, t)$ imply that if we color with t colors all the rank r elements of $P_n(q)$ for sufficiently large n, then (with A_n, x and P_{n-1} as in the lemma) there is an element of u_1 of rank l_1 in A_n such that all rank r elements y of A_n with $y\leq u_1$ have the same color. That is, $P_n(q)$ contains an element u_1 of rank l_1 such that when one divides $P_{l_1}=\{y \mid y\leq u_1\}$ into $P_{l_1-1}=\{y \mid y\leq u_1\wedge x\}$ and $A_{l_1}=\{y \mid y\leq u_1, y\not\leq u_1\wedge x$ or $y=0\}$ as in the lemma, then all the rank r elements of A_{l_1} have the same color. By the lemma P_{l_1-1} is isomorphic to $P_{l_1-1}(q)$ and A_{l_1} to $A_{l_1}(q)$. Hence we can apply these same arguments to P_{l_1-1} instead of $P_n(q)$.

So if we let l_2 be a large integer, and if l_1 is sufficiently large, then P_{l_1-1} contains an element u_2 of rank l_2 such that $P_{l_2}=\{y \mid y\leq u_2\}$ is isomorphic to $P_{l_2}(q)$, and it is divided into P_{l_2-1} and A_{l_2}, as in the lemma, with all rank r elements of A_{l_2} having the same color (but not necessarily the same as the color for A_{l_1}). (See Figure 1.)

We repeat this argument, say, $m=k_0(t-1)+1$ times, for an arbitrary k_0. Then this gives a sequence of pairs

$$(A_{l_1}, P_{l_1-1}), (A_{l_2}, P_{l_2-1}), \ldots, (A_{l_m}, P_{l_m-1}),$$

where $P_{l_i-1}\supseteq A_{l_{i+1}}\cup P_{l_{i+1}-1}$, and all the rank r elements of A_{l_i} have the same color (depending on i). But since there are only t colors, then one of them must occur k_0 times. So by renumbering, we obtain a sequence

$$(A_{m_1}, P_{m_1-1}), (A_{m_2}, P_{m_2-1}), \ldots, (A_{m_{k_0}}, P_{m_{k_0}-1})$$

with $P_{m_i-1}\supseteq A_{m_{i+1}}\cup P_{m_{i+1}-1}$, and with all the elements of any of the A_{m_i} of rank r having the same color.

Now we use part (c) of the lemma to find elements $a_1\in A_{m_{k_0}}$, $a_2\in A_{m_{k_0-1}}, \ldots$, $a_{k_0}\in A_{m_1}$ with $a_i>a_{i-1}$ for all i, and each a_i of rank i. Using part (d) of the lemma, we see that any y of rank r with $y\leq a_{k_0}$ is in A_{m_i} for some i. Hence all such y have the same color, and the element a_{k_0} establishes $P_q(k_0, r, t)$. Since k_0 was arbitrary, Theorem 3 is proved.

5. We note that if we consider $L(S_n)$ instead of $P_n(q)$ in the lemma, $L(S_{n-1})$ instead of P_{n-1}, and $L'(S_{n-1})$ instead of $A_n(q)$, where $L'(S_{n-1})$ is $L(S_{n-1})$ with an extra element appended below everything else, then the statements (a), (b), (c), (d) are still true. So the proof of Theorem 3 is still valid. But since coloring rank r

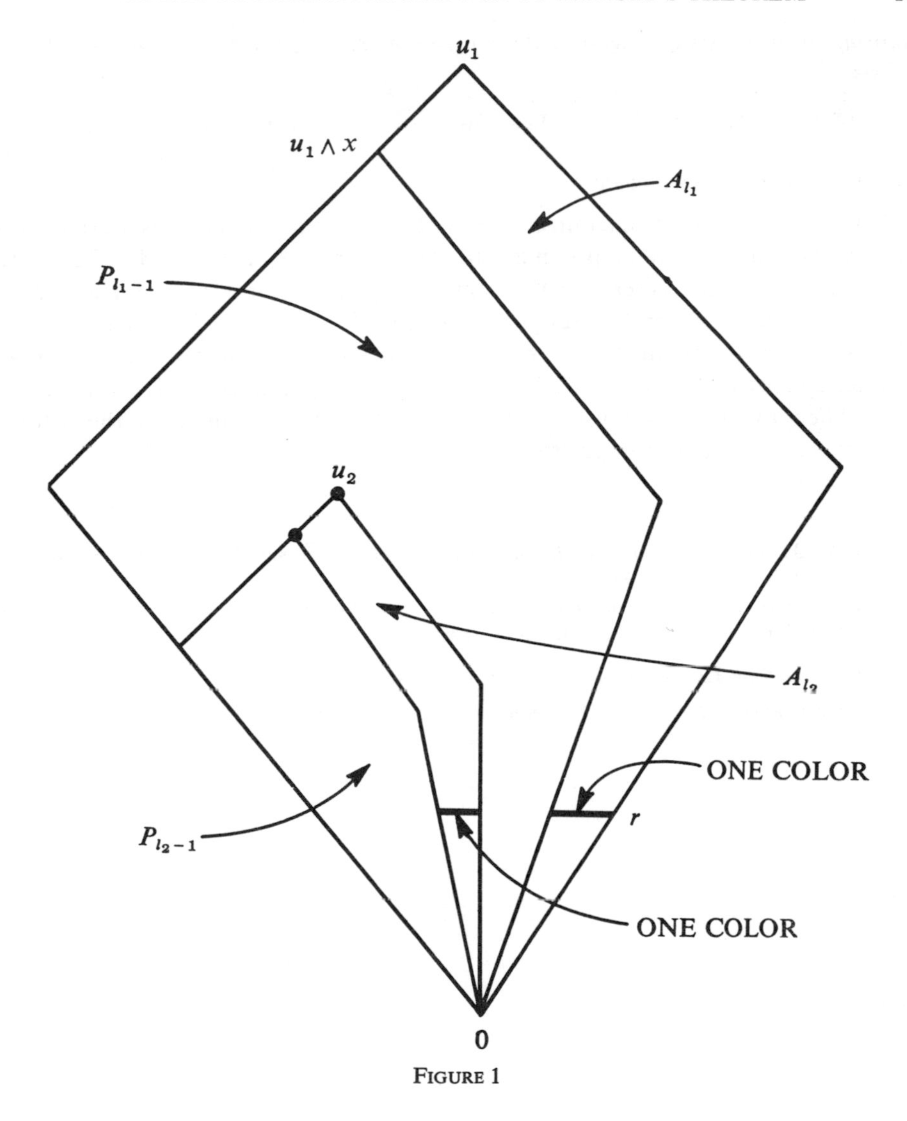

FIGURE 1

elements of $L'(S_{n-1})$ is equivalent to coloring rank $r-1$ elements of $L(S_{n-1})$, we obtain:

THEOREM 3′. $\forall k\, S(k, r-1, t) \Rightarrow \forall k\, S(k, r, t)$.

This is just the induction step in the proof of Ramsey's theorem.

6. Finally, we state the result from which one proves $A(k, 1, t)$ for all k.

THEOREM 4. *Let F be a finite set, and let $A=\{A_1, \ldots, A_r\}$ be a set of m-lists of elements of F. For each t there is a number $N=N(t, r, m)$ such that for $n \geq N$ and any* *coloring of the n-lists of F with t colors there are numbers* $m_1, m_2, \ldots, m_d$, $d \geq 2$, *and n-lists*

$$B_i = (x_{11}, \ldots, x_{1m_1}, A_i, x_{21}, \ldots, x_{2m_2}, A_i, \ldots, x_{d-1,m_{d-1}}, A_i, x_{d1}, \ldots, x_{dm_d}), \quad i = 1, 2, \ldots, r,$$

which all have the same color.

This result somewhat generalizes one of Hales and Jewett [**1**]. It uses arguments exactly like those used in proving van der Waerden's theorem. In fact both $A(k, 1, t)$ and van der Waerden's theorem are immediate corollaries of Theorem 4. To get $A(k, 1, t)$, we let $F = \mathrm{GF}(q)$ and $A = \{\text{all } (k-1)\text{-lists of } F\}$. Then the B_i of Theorem 4 are the points $(r=1)$ of an affine subspace of dimension $k-1$. To get van der Waerden's theorem, let $F = \{0, 1, 2, \ldots, l-1\}$, and let $m = 1$, $A = \{0, 1, \ldots l-1\}$. Then if we think of n-lists as representations of integers in base l, the B_i form a length l arithmetic subprogression.

References

1. A. W. Hales and R. I. Jewett, *Regularity and positional games*, Trans. Amer. Math. Soc. **106** (1963), 222–229. MR **26** #1265.

2. B. Rothschild, *A generalization of Ramsey's theorem and a conjecture of Rota*, Ph.D. Thesis, Yale University, New Haven, Conn., 1967.

Bell Telephone Laboratories, Murray Hill, New Jersey

Massachusetts Institute of Technology

COMBINATORIAL REPRESENTATIONS OF ABELIAN GROUPS

ALFRED W. HALES[1]

1. **Introduction.** In this paper we introduce an equivalence relation (*similarity*) on the class of rooted trees and discuss certain enumeration problems connected with this relation. The concept of similarity arose in the study of infinite abelian groups, so we begin by explaining the connection between trees and groups.

2. **Trees and groups.** Let T be a rooted tree, i.e. a connected graph with no cycles and with a distinguished vertex t (the root). (At this stage we put no restrictions on the cardinality of T.) If (i, j) is an edge of T, and if j lies "between" i and t, we assign the direction $i \to j$ to the edge (i, j). This makes T into a directed graph.

EXAMPLES.

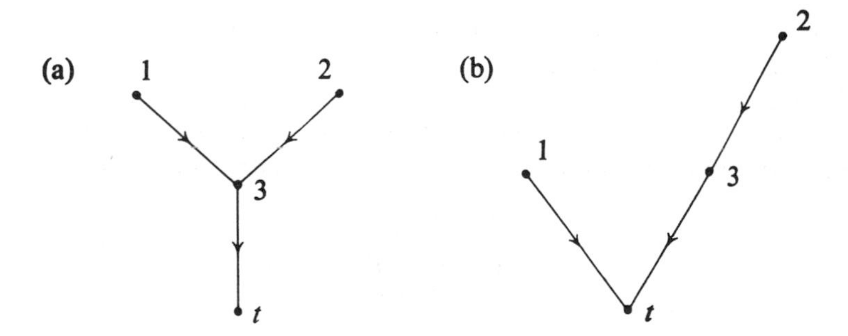

Now let p be a prime number. We define the *abelian group* T_p by giving generators and relations as follows: we take one generator x_i for each vertex i (other than t) of T; and for each directed edge $i \to j$ of T we take the relation $px_i = x_j$ (or, if $j = t$, the relation $px_i = 0$). It is clear that T_p is a p-primary abelian group, i.e. that every element of T_p has as order a power of p. More generally, if p is a prime element of a principal ideal domain R, we could define an R-module T_p in a manner analogous to the above.

[1] This work was supported in part by NSF Grant GP-5497.

Our interest in the above definition is chiefly due to the following theorems, obtained independently by Crawley and the author **[1]**:

(1) Any countable p-primary abelian group is isomorphic to T_p for some T.

(2) If $T_p \cong T'_p$ for some prime p, then $T_p \cong T'_p$ for *all* primes p (including prime elements in principal ideal domains).

Theorems (1) and (2) are of course generalizations of Ulm's Theorem, and from that viewpoint are chiefly of interest in the infinite case. In the finite case (1) is weaker than the fundamental theorem of abelian groups, which essentially says that T can be chosen to have all branches starting at the root (as in Example (b) above). Theorem (2), however, suggests a definition which leads to interesting enumeration problems in the finite case.

DEFINITION. Two rooted trees T and T' are *similar* (denoted by $T \sim T'$) if $T_p \cong T'_p$ for some (and hence all) p.

It is clear that similarity is an equivalence relation. There are various interesting questions arising about this relation in the infinite case, but we do not intend to discuss them here. From now on we restrict our attention to finite trees. The term "similar" can then be justified in the following way. We associate to a tree T (with $|T| = n+1$) an $n \times n$ "incidence matrix" using the vertices (other than t) of T as row and column labels and putting a 1 in the ij place if and only if $i \to j$ is an edge of T (0 entries otherwise). It is then easy to show that two trees are similar (in our sense) if and only if their associated incidence matrices are similar (in the classical sense). In fact, we obtain this result from theorem (2) above by taking R to be the ring of polynomials in one variable x, taking p to be the prime element x in R, and using the familiar correspondence between similarity of matrices and isomorphism of modules over R.

To illustrate this, look at Examples (a) and (b) above. In either case the group T_p is easily seen to be the direct sum of a cyclic group of order p and a cyclic group of order p^2, so the two trees are similar. The incidence matrix for the first is

$$\begin{pmatrix} 0 & 0 & 1 \\ 0 & 0 & 1 \\ 0 & 0 & 0 \end{pmatrix},$$

and for the second is

$$\begin{pmatrix} 0 & 0 & 0 \\ 0 & 0 & 1 \\ 0 & 0 & 0 \end{pmatrix},$$

and these two matrices are similar.

3. **Similarity classes.** Let T be a finite rooted tree with $|T| = n+1$. Then T_p has p^n elements, and by the fundamental theorem of abelian groups T_p is the direct sum of cyclic groups of orders $p^{a_1}, p^{a_2}, \ldots, p^{a_k}$. Here the integers $a_1, \ldots, a_k$ are uniquely determined (up to order) and sum to n. Thus the well-known 1-1 correspondence between isomorphism classes of abelian groups of order p^n and partitions of n

becomes, from our viewpoint, a 1-1 correspondence between similarity classes of trees with $n+1$ elements and partitions of n. In particular, the number of similarity classes of trees with $n+1$ elements is $p(n)$, the partition function of n.

During the rest of this paper we discuss the problem of determining the *size* of a similarity class of trees in terms of the associated partition. Thus the partition $3=2+1$ is associated with a similarity class containing two (isomorphism classes of) trees, namely those of Examples (a) and (b) above. In general this problem seems to be extremely difficult, and we are far from being able to give an explicit expression for the size of the similarity class in terms of the parameters $a_1, a_2, \ldots, a_k$. We can, however, say something about the appropriate generating function. Before doing this, we give a few simple isolated results without proof:

Partition	Size of Class	
$n=a+b$	$a-b+1$	$(a \geq b)$
$n=a+b+c$	$(a+b+1-2c)(a-b)+(a+1-c)$	$(a \geq b > c)$
	$\binom{a-b+2}{2}$	$(a \geq b = c)$
$n=a+1+1+\cdots+1$ (k times)	$\binom{a+k-1}{k}$	
$1=1$	1	
$3=1+2$	2	
$6=1+2+3$	7	
$10=1+2+3+4$	42	

In the general case we consider the partition $n=a_1+a_2+\cdots+a_k$. Let i_1 denote the number of ones among the a's, let i_2 denote the number of twos among the a's, etc. We denote by $f_{ni_1i_2\ldots}$ the size of the corresponding similarity class, and introduce the generating function

$$f(x, t_1, t_2, \ldots) = \sum_{n,i_1,i_2,\cdots=0}^{\infty} f_{ni_1i_2\ldots} x^n t_1^{i_1} t_2^{i_2} \cdots.$$

Although we cannot give an explicit formula for this function, we can obtain a "functional equation" for it by applying Pólya's theorem. In doing so we essentially mimic the technique of Harary and Prins [**2**]. For each $k \geq 1$, let $f^{(k)}(x, t_1, \ldots, t_k)$ denote $f(x, t_1, \ldots, t_k, 0, 0, \ldots)$, the generating function for the sizes of similarity classes corresponding to partitions involving only the numbers $1, 2, \ldots, k$. For example, we have $f^{(1)}(x, t_1) = 1 + xt_1 + x^2t_1^2 + x^3t_1^3 + \cdots$. In order to calculate $f^{(k+1)}$ in terms of $f^{(k)}$, we use the technique of Theorem 4 of [**2**] (although our partitions differ from theirs). A set of trees "counted" by $f^{(k)}$ gives rise to a tree "counted" by $f^{(k+1)}$ in the following way: take the disjoint union of the original set, add one new vertex to serve as the new root, and join it to each of the old roots. To obtain the partition associated with the new tree, take each of the old partitions and increase one of its largest parts by one. Then put these all together to obtain the new partition.

EXAMPLE. The partition associated with V is $1+1$. The partition associated with ⁞ is 2. These trees give rise to the tree:

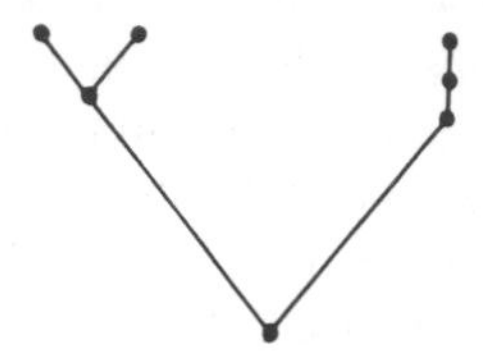

To get the partition for this tree, change $1+1$ to $1+2$, and 2 to 3. Putting these together gives $1+2+3$.

Now, using Pólya's theorem as in [2], we obtain the following result:

$$f^{(k+1)}(x, t_1, \ldots, t_{k+1}) = \exp \sum_{r=1}^{\infty} \frac{1}{r} f^{(k)*}(x^r, t_1^r, \ldots, t_k^r)$$

where * is the operation defined term by term on power series in the following manner: $(x^n t_1^{i_1} \cdots t_j^{i_j})^* = x^{n+1} t_1^{i_1} \cdots t_j^{i_j - 1} t_{j+1}$ if $i_j \geq 1$. Applying this recursion to the above formula for $f^{(1)}(x, t_1)$ and simplifying, we obtain

$$f^{(2)}(x, t_1, t_2) = \frac{1}{1-xt_1} \prod_{i=0}^{\infty} \frac{1}{1-x^{i+2}t_1^i t_2}.$$

On the other hand, letting k tend to infinity in the recursion gives the functional equation for f:

$$f(x, t_1, t_2, \ldots) = \exp \sum_{r=1}^{\infty} \frac{1}{r} f^*(x^r, t_1^r, t_2^r, \ldots).$$

As a check on this last result, we put $t_1 = t_2 = \cdots = 1$, which amounts to combining all similarity classes for a given n, i.e. counting the number of trees of a given size. Our functional equation then becomes

$$f(x, 1, 1, \ldots) = \exp \sum_{r=1}^{\infty} \frac{x^r}{r} f(x^r, 1, 1, \ldots)$$

which is one form of the well-known functional equation for the generating function which "counts" rooted trees.

REFERENCES

1. P. Crawley and A. W. Hales, *The structure of torsion abelian groups given by presentations*, Bull. Amer. Math. Soc. **74** (1968), 954–956. MR **38** #1163.

2. F. Harary and G. Prins, *The number of homeomorphically irreducible trees, and other species*, Acta Math. **101** (1959), 141–162. MR **21** #653.

UNIVERSITY OF CALIFORNIA, LOS ANGELES

DESIGNS WITH TRANSITIVE AUTOMORPHISM GROUPS[1]

MARSHALL HALL, JR.

1. **Introduction.** Some block designs have automorphism groups which are transitive on both the points and the blocks of the design. This paper deals with such designs. This interaction may lead from a known design to a new and interesting group or a known group may lead to a new design. The methods of this paper have led to the discovery of a symmetric block design with $v=56$, $k=11$, $\lambda=2$ which will be the subject of a separate paper, to be published jointly with Richard Lane and David Wales [**5**].

A design with a transitive automorphism group G can be considered as arising from the orbits of an appropriate intransitive subgroup S. Hence following D. G. Higman's [**3**] term we call it an orbital design. An orbital design always gives rise to a partially balanced block design in the sense of Bose and Shimamoto [**1**]. This relationship is the subject of §2.

The third section gives some examples illustrating this theory.

2. **Block designs with transitive automorphism groups.**

2.1. Let D be a block design with v points $a_1, \ldots, a_v$ and b blocks $B_1, \ldots, B_b$. Each block contains k distinct points, each point is in exactly r blocks, and every pair of distinct points a_i, a_j occurs together in exactly λ blocks. These parameters are known to satisfy the conditions:

$$bk = vr, \qquad r(k-1) = \lambda(v-1).$$

An automorphism α of D is a one-to-one mapping of the points onto themselves and of the blocks onto themselves preserving incidence. That is

$$a \to a^{\alpha} \in \{a_1, \ldots, a_v\}, \qquad B \to B^{\alpha} \in \{B_1, \ldots, B_b\}, \quad a \in B \text{ implies } a^{\alpha} \in B^{\alpha}.$$

Thus α is simultaneously a permutation of the points and of the blocks. The automorphisms of D form a group G, and we shall consider designs for which G is transitive on the points and also on the blocks.

If there are m blocks containing exactly the same points as a particular one, then by transitivity of G on the blocks it follows that every block is in a family of m blocks containing exactly the same points. If we identify blocks containing the

[1] This research was supported in part by ONR Contract N00014-67-A-0094-0010.

same set of points then in a natural way we may consider G to be an automorphism group of this new design, and G is transitive on its points and blocks. Without any real loss of generality we may assume that this has already been done, so that distinct blocks contain distinct sets of points. Hence the presentation of G as a permutation group on the set $\Omega=\{a_1, \ldots, a_v\}$ of points completely determines the action of G on the blocks.

The set of elements $S_1=\{x \mid x \in G, B_1x=B_1\}$ is a subgroup of G, called the stabilizer of B_1. Two elements in the same left coset S_1x of S_1 map B_1 onto the same block so that we may write

$$G = S_1+S_1x_2+\cdots+S_1x_b, \qquad [G:S_1] = b.$$

We may assume the coset representatives x_j numbered so that $B_1x_j=B_j$. Similarly let G_1 be the stabilizer of the point a_1 so that

$$G = G_1+G_1y_2+\cdots+G_1y_v$$

and the coset representatives numbered so that $a_1y_i=a_i$. The stabilizer of the block B_j is $S_j=x_j^{-1}S_1x_j$ and the stabilizer of the point a_i is $G_i=y_i^{-1}G_1y_i$.

2.2. The following theorem relates complexes in G to the design D.

THEOREM 1. *The elements of G mapping a point a_i onto a point of a block B_j are a complex F_{ij} consisting of double cosets G_iuS_j and $|F_{ij}|=k|G_i|$. The elements of G mapping a block B_j onto a block containing a_i are a complex E_{ji} consisting of double cosets S_jvG_i and $|E_{ji}|=r|S_j|$. These are inverse complexes, $E_{ji}=F_{ij}^{-1}$. If $i\neq j$ then $|E_{ii} \cap E_{ij}|=\lambda|S_i|$.*

PROOF. Let x be an element such that $a_ix \in B_j$. Then for $s \in G_i$, $t \in S_j$ we have $a_isx=a_ix \in B_j$ whence $(a_isx)t \in B_jt=B_j$ and so $a_i(sxt) \in B_j$. Hence the complex F_{ij} consists of double cosets G_iuS_j, and as B_j contains k points F_{ij} also consists of the k cosets of G_i mapping a_i onto the k points of B_j, whence $|F_{ij}|=k|G_i|$. In the same way the elements y for which $a_i \in B_jy$ form a complex E_{ji} consisting of double cosets S_jvG_i and $|E_{ji}|=r|S_j|$ as a_i is on r blocks. Since $a_ix \in B_j$ is equivalent to $a_i \in B_jx^{-1}$ it follows that $E_{ji}=F_{ij}^{-1}$.

Finally E_{ii} consists of the cosets of S_i corresponding to blocks containing a_i and E_{ij} of cosets of S_i corresponding to blocks containing a_j. As there are exactly λ blocks containing a_i and a_j it follows that $E_{ii} \cap E_{ij}$ contains exactly λ left cosets of S_i.

2.3. Let us now suppose that we are given G as a transitive permutation group on v letters. Let S_1^* be an intransitive subgroup of G. Choose $B_1\{a_1, \ldots, a_k\}$ as a set of letters consisting of full orbits of S_1^*. Then S_1^* stabilizes B_1. Now take S_1 as the full stabilizer of B_1. Here let $[G:S_1]=b$ and let

$$G = S_1+S_1x_2+\cdots+S_1x_b.$$

We now determine further blocks B_j by putting

$$B_j = B_1x_j, \qquad j = 2, \ldots, b. \tag{1}$$

Here if Δ is an orbit of S_1 contained in B_1, then Δx_j is an orbit of $S_j = x_j^{-1} S_1 x_j$ and $\Delta x_j \subseteq B_j$. Following D. G. Higman [3] we shall use the term "orbital" to describe a set of conjugate orbits and so the blocks $B_1, B_2, \ldots, B_b$ will form an *orbital design*. An orbital design will not in general be a block design. By construction every block B_j contains the same number of points (the letters determined by (1)). With G_1 the stabilizer of a_1 we have

$$G = G_1 + G_1 y_2 + \cdots + G_1 y_v,$$

and so if $B_{j_1}, B_{j_2}, \ldots, B_{j_r}$ are the full set of blocks containing the point a_1, then $B_{j_1} y_i, \ldots, B_{j_r} y_i$ will be the full set of blocks containing the point a_i. Hence in an orbital design every point occurs in r blocks. Counting incidences of points on blocks we have the familiar condition $bk = vr$. But if a_i and a_j are two distinct points the number of times that a_i and a_j occur together in blocks λ_{ij} will not be uniform for all pairs a_i, a_j. However we shall see that the number of distinct values of λ_{ij} is not large and is associated with the structure of G.

2.4. THEOREM 2. *If G is a transitive permutation group of rank $t+1$, then an orbital design derived from G is a partially balanced block design with at most t associate classes. Let G_1 have orbits $(a_1), \Delta_1, \Delta_2, \ldots, \Delta_t$. Let $\Delta_1, \Delta_2, \ldots, \Delta_u$ be the self-paired orbits and let the remaining orbits be $\Delta_{u+1}, \Delta'_{u+1}, \Delta_{u+2}, \Delta'_{u+2}, \ldots, \Delta_{u+v}, \Delta'_{u+v}$ where Δ_{u+j} and Δ'_{u+j} are paired orbits. Here $t = u + 2v$. We say that a_1 and a_i are sth associates if $s \leq u$ and $a_i \in \Delta_s$ or if $s = u+j$ and $a_i \in \Delta_{u+j}$ or Δ'_{u+j}. Also if for $x \in G$ we have $a_1 x = a_f$ and $a_i x = a_h$ we say that a_f and a_h are sth associates if a_1 and a_i are sth associates. Points a_f and a_h which are sth associates occur together in λ_s blocks where λ_s depends only on s.*

COROLLARY. *If G is doubly transitive then an orbital design is a block design.*

PROOF. We have already observed that in an orbital design each of the b blocks contains the same number k of points and each point is in the same number r of blocks.

We must still investigate the number of times λ_{ij} that distinct points a_i, a_j occur together in the blocks. If $B_{11}, B_{12}, \ldots, B_{1r}$ are the r blocks which contain the point a_1, then G_1 the stabilizer of a_1 will also permute these blocks among themselves. If $(a_1), \Delta_1, \Delta_2, \ldots, \Delta_t$ are the orbits of G_1 then if a_m and a_n are in the same orbit Δ_j there will be an $x \in G_1$ such that $a_1 x = a_1$ and $a_m x = a_n$. As $B_{11}x, B_{12}x, \ldots, B_{1r}x$ are merely $B_{11}, \ldots, B_{1r}$ in some order it follows that a_m and a_n occur the same number of times in these blocks and so $\lambda_{1m} = \lambda_{1n}$. Thus λ_{1j} depends only on the orbit of G_1 containing a_j.

We now turn to the theory of paired orbits. In Wielandt's book [4, p. 44] it is shown that if $\Delta_j(a_i)$ is an orbit of G_i, the stabilizer of a_i, then the set of letters $\{a_m\}$ with the property that for some $x \in G$ and $a_i x = a_m$ (putting $\Delta_j(a_m) = \Delta_j(a_i)x$) we have $a_i \in \Delta_j(a_m)$ forms an orbit of G_i, the *paired orbit* $\Delta'_j(a_i)$. We always have $(\Delta'_j)' = \Delta_j$. The orbits may be self-paired, and paired orbits have the same length, i.e. contain the same number of letters. Thus if $\Delta_j(a_1)$ is an orbit of G_1 and if $a_m \in \Delta'_j(a_1)$ then $a_1 \in \Delta_j(a_m)$. Hence $\lambda_{m1} = \lambda_j$ but $\lambda_{1m} = \lambda_{j'}$. As $\lambda_{1m} = \lambda_{m1}$ it follows that $\lambda_j = \lambda_{j'}$. Hence letters from paired orbits of G_1 occur equally often with a_1. If a_1 and a_i occur together in λ_s blocks then it follows that for $x \in G$, $a_1x=a_f$ and $a_ix=a_h$ also occur together in λ_s blocks. Thus the definition in the theorem of sth associates is such that letters a_i and a_j which are sth associates occur together in λ_s blocks, and $s=1, \ldots, u+v$ where $u+2v=t$. The theorem is now proved.

2.5. The intersection numbers $\mu_{ij}^{(\alpha)}$ are defined by D. G. Higman [3] by the rule

$$\mu_{ij}^{(\alpha)} = |\Delta_\alpha(b) \cap \Delta_i(a)|, \qquad b \in \Delta_j(a). \tag{2}$$

We must show that these numbers depend only on i, j, and α and not on the particular letters a, b. If $b' \in \Delta_j(a)$ then there is an $x \in G_a$ with $ax=a$ and $bx=b'$. Then $(\Delta_\alpha(b) \cap \Delta_i(a))x=(\Delta_\alpha(b') \cap \Delta_i(a))$ and so the number does not depend on the particular choice of b. Since the condition $b \in \Delta_j(a)$ is equivalent to $a \in \Delta_j'(b)$, by a similar argument the numbers do not depend on the particular choice of a.

THEOREM 3. *In an orbital design for any pair of points which are sth associates the number of points which are simultaneously jth associates of the first and uth associates of the second is* p_{ju}^s *and this number is independent of the pair of points with which we start. Furthermore* $p_{ju}^s=p_{uj}^s$.

PROOF. If a and b are sth associates then $b \in \Delta_s(a) \cup \Delta_s'(a)$. The points which are ith associates of a are the set $\Delta_i(a) \cup \Delta_i'(a)$ and those which are jth associates of b are the set $\Delta_j(b) \cup \Delta_j'(b)$. Thus

$$p_{ju}^s = |(\Delta_i(a) \cup \Delta_i'(a)) \cap (\Delta_j(b) \cup \Delta_j'(b))|, \qquad b \in \Delta_s(a) \cup \Delta_s'(a).$$

Since paired orbits are either identical or disjoint it follows that p_{ju}^s can be expressed in terms of the intersection numbers of (2) and so do not depend on the particular a and b. Higman's relation $\mu_{ij}^{(\alpha)}=\mu_{\alpha j'}^{(i)}$ is sufficient to prove $p_{ju}^s=p_{uj}^s$.

This now shows that orbital designs are partially balanced designs according to the definition of Bose and Shimamoto [1].

3. **Examples of designs with transitive groups.** A difference set in a group G regular on v points will give a symmetric block design D, but in this case both G_1 the stabilizer of a point and S_1, the stabilizer of a block are the identity, and so the the orbits consist of single points. Following Bruck [2] there may be multipliers of D, these being automorphisms of G which are also automorphisms of D. These automorphisms can be expressed as permutations in the holomorph of G, giving a larger group G^* more intimately related to the design than G itself.

For example with $b=v=16$, $r=k=6$, $\lambda=2$, let G be the elementary abelian group of order 16, $G=\langle a, b, c, d\rangle$, $a^2=b^2=c^2=d^2=1$. Then a difference set is $\{a, b, c, d, ab, cd\}$. But automorphisms α, β, γ, δ of G fix the difference set where:

$$\begin{array}{llll} a^\alpha = b & a^\beta = b & a^\gamma = c & a^\delta = a \\ b^\alpha = a & b^\beta = ab & b^\gamma = d & b^\delta = b \\ c^\alpha = c & c^\beta = c & c^\gamma = a & c^\delta = d \\ d^\alpha = d & d^\beta = d & d^\gamma = b & d^\delta = cd \end{array}$$

Thus in the larger group $G^* = \langle G, \alpha, \beta, \gamma, \delta\rangle$, the difference set is an orbit of the letter fixing the identity, and indeed G^* is a rank 3 group whose stabilizer has orbit lengths 1, 6, 9, the six letters of G_1 being the letters of the difference set.

The simple group G of order 6048 has a rank 4 representation on 36 letters in which the lengths of the orbits in the stabilizer are 1, 7, 7, 21. $G = \langle a, b\rangle$ and the stabilizer $G_1 = \langle a, d\rangle$ where

$$\begin{aligned} a = {} & (01)(02, 03, 04, 05, 06, 07, 08)(09, 10, 11, 12, 13, 14, 15) \\ & (16, 17, 18, 19, 20, 21, 22)(23, 24, 25, 26, 27, 28, 29) \\ & (30, 31, 32, 33, 34, 35, 36), \\ b = {} & (01, 02, 09, 16, 23, 20, 30, 17)(03, 35, 13, 29, 31, 24, 25, 11) \\ & (04, 26, 27, 07)(05, 21, 14, 10)(06, 19, 32, 36)(08, 18, 28, 22, 15, 12, 33, 34), \\ d = {} & (01)(02)(03, 05)(04, 25)(06)(07, 20)(08, 17)(16, 29)(18, 27) \\ & (19, 23)(21, 28)(22)(24)(26)(09, 14)(10)(11, 15)(12)(13)(30)(31) \\ & (32, 35)(33)(34, 36). \end{aligned}$$

Here in the stabilizer G_1, let us take as B_1 the fixed letter and the two 7 orbits:

$$B_1 = 01, 09, 10, 11, 12, 13, 14, 15, 30, 31, 32, 33, 34, 35, 36.$$

We find the orbital design is a symmetric block design D with $v = 36$, $k = 15$, $\lambda = 6$. The full automorphism group G^* of D is of order 12096 containing the further element

$$\begin{aligned} \alpha = {} & (01)(02, 26, 17, 04, 20, 07, 16, 27)(03, 18, 22, 08, 28, 29, 24, 25) \\ & (05, 06, 21, 19)(09, 30)(10, 32, 14, 36)(11, 34, 13, 33, 12, 35, 15, 31)(23). \end{aligned}$$

Here α induces an automorphism of G of order 2.

Bibliography

1. R. C. Bose and T. Shimamoto, *Classification and analysis of partially balanced incomplete block designs with two associate classes*, J. Amer. Statist. Assoc. **47** (1952), 151–184. MR **14**, 67.

2. R. H. Bruck, *Difference sets in a finite group*, Trans. Amer. Math. Soc. **78** (1955), 464–481. MR **16**, 1081.

3. D. G. Higman, *Intersection matrices for finite permutation groups*, J. Algebra **6** (1967), 22–42. MR **35** #244.

4. H. Wielandt, *Finite permutation groups*, Academic Press, New York, 1964. MR **32** #1252.

5. Marshall Hall, Jr., Richard Lane and David Wales, *Designs derived from permutation groups*, J. Combinatorial Theory **8** (1970), 12–22.

California Institute of Technology

TRUNCATED FINITE PLANES

HAIM HANANI

1. **Tactical configurations.**

DEFINITION 1. Given a set E of v elements, and given positive integers, k, l $(l \leq k \leq v)$ and λ, we denote by a *tactical configuration* $C[k, l, \lambda; v]$ a system of blocks (subsets of E), having k elements each and such that every subset of E having l elements is included in exactly λ blocks.

Tactical configurations $C[k, 2, \lambda; v]$ with $l=2$ are known as *balanced incomplete block designs* (BIBD).

THEOREM 1. *A necessary condition for the existence of a tactical configuration* $C[k, l, \lambda; v]$ *is that*

$$\lambda\binom{v-h}{l-h}\Big/\binom{k-h}{l-h} = \textit{integer}, \qquad h = 0, 1, \ldots, l-1. \tag{1}$$

PROOF. The left side of (1) is the number of blocks of $C[k, l, \lambda; v]$ that contain h fixed elements of E.

The condition of Theorem 1 is by far not sufficient [7]. In fact most of the work done on tactical configurations and especially BIBD's is to prove the existence or nonexistence of certain configurations satisfying the condition of Theorem 1 (see, e.g., [**6**, Chapters 10–15]).

A generalization of tactical configurations may be obtained by allowing the different blocks to have varying number of elements, as in the following definition.

DEFINITION 2. Let a set E having v elements be given, further let k, l $(l \leq k \leq v)$ and λ be positive integers and let $K_k = \{k_i\}_{i=1}^{n}$ be a finite nonempty set of integers $k_i \geq k$. We denote by a *C-design* $C[K_k, l, \lambda; v]$ a system of blocks (subsets of E) such that the number of elements in each block is some $k_i \in K_k$ and that every subset of E having l elements is included in exactly λ blocks.

If $K_k = \{k\}$ consists of one integer only we shall write $C[k, l, \lambda; v]$ instead of $C[\{k\}, l, \lambda; v]$. The C-designs $C[k, l, \lambda; v]$ are the tactical configurations introduced in Definition 1.

In the special case $l=2$, C-designs have been used extensively by Parker, Bose and Shrikhande [**9**], [**3**], [**2**] for construction of orthogonal Latin squares (see [**6**, p. 196]) and by the author [**7**] for construction of balanced incomplete block designs (see [**6**, p. 226]). For $l=3$, C-designs have been used by the author [**8**] for construction of special tactical configurations.

The following properties of the C-designs are easily established.

THEOREM 2. *For every* $k_i \in K_k$, $C[K_k, l, \lambda; k_i]$ *exists.*

THEOREM 3. *If* $K_k' \subset K_k$, *then from the existence of* $C[K_k', l, \lambda; v]$ *follows the existence of* $C[K_k, l, \lambda; v]$.

THEOREM 4. *If* λ *is a multiple of* λ', *then from the existence of* $C[K_k, l, \lambda'; v]$ *follows the existence of* $C[K_k, l, \lambda; v]$.

THEOREM 5. *If* $C[K_k', l, \lambda'; v]$ *exists and if for every* $k_i' \in K_k'$, $C[K_k, l, \lambda''; k_i']$ *exists, then also* $C[K_k, l, \lambda; v]$ *exists, where* $\lambda = \lambda'\lambda''$.

We shall make use mostly of the special case of Theorem 5, when $\lambda' = \lambda'' = 1$, namely

THEOREM 5*. *If* $C[K_k', l, 1; v]$ *exists and if for every* $k_i' \in K_k$, $C[K_k, l, 1; k_i']$ *exists, then also* $C[K_k, l, 1; v]$ *exists.*

DEFINITION 3. A set K_k such that a C-design $C[K_k, l, 1; v]$ exists for every $v \geq k$ will be denoted as a $K(k, l)$*-set.*

The $K(k, l)$-sets may be useful in construction of tactical configurations. If K_k is a $K(k, l)$-set and if for every $k_i \in K_k$, $C[k, l, \lambda; k_i]$ exists, then by Theorem 5 the existence of $C[k, l, \lambda; v]$ follows for every v.

We shall present in this paper a method constructing $K(k, l)$-sets for $l=2, 3$ and for every k. We shall also prove that if q is a power of a prime then tactical configurations $C[q+1, 3, 1; v]$ exist for infinite many values of v.

2. **Finite planes.** Let q be a prime-power (a power of a prime) and let G be a Galois field GF[q] of order q (see e.g. [**4**, pp. 242–288]). Further let F be a Galois field over G of order q^d obtained by adjoining to G a root of an irreducible equation.

$$t^d = m_0 + m_1 t + \cdots + m_{d-1} t^{d-1}, \qquad m_i \in G, \quad i = 0, 1, \ldots, d-1.$$

In the sequel the elements of G will be denoted by small Latin letters and those of F by Greek letters

$$\alpha = a_0 + a_1 t + \cdots + a_{d-1} t^{d-1}. \tag{2}$$

The zero element of F is $\phi = 0 + 0 \cdot t + \cdots + 0 \cdot t^{d-1}$.

We *extend* the field F to F^* by adjoining an element ∞ which has no representation in the form (2) and for which we define the field operations as follows

$$\begin{aligned} \phi \cdot \infty &= \phi, \\ (\alpha\infty + \beta)/(\gamma\infty + \delta) &= \alpha/\gamma \quad \text{for } \beta, \delta \neq \infty, \\ \alpha/\phi &= \infty \quad \text{for } \alpha \neq \phi. \end{aligned}$$

In a similar way as in fields of characteristic 0 and $d=2$ (see e.g. [**1**, pp. 22–28]) we introduce the linear transformations

$$\eta = T(\xi) = (\alpha\xi+\beta)/(\gamma\xi+\delta), \qquad \alpha, \beta, \gamma, \delta \neq \infty, \quad \alpha\delta-\beta\gamma \neq \phi.$$

The linear transformations are known to be one-one valued and to form a group.

The cross ratio

$$(\xi, \xi_2, \xi_3, \xi_4) = \frac{\xi-\xi_2}{\xi-\xi_4}\Big/\frac{\xi_3-\xi_2}{\xi_3-\xi_4}$$

is the image of ξ under the linear transformation which carries ξ_2, ξ_3 and ξ_4 into ϕ, 1 and ∞ respectively.

DEFINITION 4. A subset S of F^* is a *circle* if $(\xi_1, \xi_2, \xi_3, \xi_4) \in G$ for any four distinct elements $\xi_1, \xi_2, \xi_3, \xi_4$ of S and if no set properly containing S has this property.

As in fields of characteristic 0, a linear transformation transforms circles into circles, for any two circles there exists a linear transformation transforming one of them into the other and for any three distinct elements of F^* there exists exactly one circle containing them. Further, observing that the set $G \cup \{\infty\}$ is a circle, we deduce that every circle has exactly $q+1$ elements.

DEFINITION 5. The extended field F^* with the system of circles on it forms a *finite inversive geometry* IG(d, q).

Regarding the circles of a finite inversive geometry as blocks of a tactical configuration we obtain

THEOREM 6. *If q is a prime-power and d a positive integer, then there exists a finite inversive geometry* IG(d, q). *The* IG(d, q) *forms a tactical configuration* $C[q+1, 3, 1; q^d+1]$.

DEFINITION 6. A subset L of F is a *line* if and only if $L \cup \{\infty\}$ is a circle in the corresponding extended field F^*.

Clearly for any two distinct elements of F there exists exactly one line containing them and every line has exactly q elements.

DEFINITION 7. The field F with the system of lines on it forms a *finite affine geometry* AG(d, q).

Regarding the lines of a finite affine geometry as blocks of a tactical configuration we obtain the well-known (see e.g. [**6**, p. 172])

THEOREM 7. *If q is a prime-power and d a positive integer, then there exists a finite affine geometry* AG(d, q). *The* AG(d, q) *forms a tactical configuration* $C[q, 2, 1; q^d]$.

3. **Truncated finite planes.** Consider a finite affine plane AG$(2, q)$, where q is a prime-power. Arranging the elements of AG$(2, q)$ in a q by q lattice we may assume, without loss of generality, that the rows form one family of parallel lines and the columns another such family. Let k be an integer satisfying $2 \le k \le (q+1)/2$. Delete from AG$(2, q)$ some of its elements in such manner, that at least k columns remain intact and that other columns have either at most one or at least k elements left. In this *truncated plane*, each of the remaining (possibly truncated) lines has at least k and at most q elements. It is easily checked that for every v satisfying $kq \le v \le q^2$ it is

possible to truncate AG(2, q) in the above described way so that v elements be left. The truncated planes form C-designs $C[\{k, k+1, \ldots, q\}, 2, 1; v]$ for every v satisfying $kq \leq v \leq q^2$.

Tchebychef proved (see e.g. [5, p. 435]) that for $x \geq 8$ there is always a prime between x and $3x/2$. By checking for values $x<8$ we see that for $x \geq 2$ there exists always a prime-power q such that $x \leq q < 3x/2$. Therefore if for given $k \geq 2$, $2k-1 \leq q_1 < q_2$ are consecutive prime-powers, $kq_2 < q_1^2$ holds and consequently the intervals $[kq_1, q_1^2]$ and $[kq_2, q_2^2]$ overlap. From Theorem 5* follows

THEOREM 8. *Let k be an integer $k \geq 2$ and let q be the smallest prime-power satisfying $q \geq 2k-1$, then the integers n, $k \leq n \leq kq-1$, form a $K(k, 2)$-set.*

For specific values of k this result may be improved and smaller $K(k, 2)$-sets may be formed. The case of $k=2$ being trivial, we shall give an example in constructing such an improved $K(3, 2)$-set.

THEOREM 9. $\{3, 4, 5, 6, 8\}$ *is a* $K(3, 2)$-*set.*

PROOF. By Theorem 8 the set $\{3, 4, \ldots, 14\}$ is a $K(3, 2)$-set. Further, if $12 \leq v \leq 14$ consider AG(2, 4); for $v=12$ delete one vertical line, for $v=13$ delete one vertical line with exception of one element and for $v=14$ proceed as in the case $v=13$ and add an element ∞ adjoint to a family of parallel lines which are not the columns. If $v=11$ we prove the existence of $C[\{3, 5\}, 2, 1; 11]$ taking as elements (i, j), $i=0, 1$; $j=0, 1, 2, 3, 4$ and (∞) and as blocks $\{(0, j), (1, j+\mu), (1, j-\mu)\}$, $j=0, 1, 2, 3, 4$, $\mu=1, 2$; $\{(0, j), (1, j), (\infty)\}$, $j=0, 1, 2, 3, 4$; $\{(0, 0), (0, 1), (0, 2), (0, 3), (0, 4)\}$. If $v=10$ consider AG(2, 3) and add an element ∞ adjoint to a family of parallel lines. For $v=9$ use AG(2, 3). For $v=7$ we prove the existence of $C[3, 2, 1; 7]$ by taking as elements (i), $i=0, 1, \ldots, 6$ and as blocks $\{(i), (i+1), (i+3)\}$, $i=0, 1, \ldots, 6$.

We turn to the case $l=3$. Let k be an integer $k \geq 3$, and q be a prime-power such that

$$k+q \equiv 0 \pmod 2, \qquad k+q \equiv 1 \pmod 2$$

and

$$q \geq 2k-1, \qquad q \geq 2k-2. \tag{3}$$

Consider a finite inversive plane IG(2, q). For every v satisfying

$$q(q+k)/2 \leq v \leq q^2+1, \qquad q(q+k-1)/2 \leq v \leq q^2+1 \tag{4}$$

it is possible to delete some elements from IG(2, q) in such a way that at least

$$(q+k)/2, \qquad (q+k-1)/2$$

columns remain intact, other columns have either at most 2 or at least k elements left and the total number of remaining elements is v. Every circle has at most two elements in common with every column which is not included in it. Accordingly, if the element ∞ is the first among the deleted elements then each of the remaining (possibly truncated) circles has at least k and at most $q+1$ elements and we conclude that for every v satisfying (4) there exists a *truncated* finite *inversive plane* IG(2, q) which forms a C-design $C[\{k, k+1, \ldots, q+1\}, 3, 1; v]$.

Sylvester proved (see e.g. [5, p. 435]) that for every $x \geq 33$ there exists a prime between x and $8x/7$. By checking for values $x < 33$ it is easily found that if $20 < q_1 < q_2$ are two consecutive prime-powers, then $q_2/q_1 \leq 37/32$. Accordingly, for $q_1 > 20$ the intervals in (4) for $q = q_1$ and $q = q_2$ overlap and by Theorem 5* we obtain

THEOREM 10. *The set of integers n satisfying $k \leq n \leq q[(q+k)/2]-1$ ($[x]$ is the largest integer satisfying $[x] \leq x$) forms a $K(k, 3)$-set if $k \geq 11$ and q is the smallest prime-power satisfying* (3), *or if $3 \leq k \leq 10$ and $q = 23$.*

For specific values of k this result may be considerably improved and smaller $K(k, 3)$ may be formed. As an example we shall take $k=4$.

THEOREM 11. $\{4, 5, 6, 7, 9, 11, 13, 15, 19, 23, 27, 29, 31\}$ *is a $K(4, 3)$-set.*

PROOF. By Theorem 10, $\{4, 5, \ldots, 298\}$ is a $K(4, 3)$-set. The truncated plane IG(2, 19) takes care of values $209 \leq v \leq 298$, IG(2, 16)—of values $160 \leq v \leq 208$, IG(2, 13)—of values $104 \leq v \leq 159$, IG(2, 11)—of values $77 \leq v \leq 103$, IG(2, 9)—of values $54 \leq v \leq 76$, IG(2, 8)—of values $48 \leq v \leq 53$ and IG(2, 7)—of values $35 \leq v \leq 47$. For $v=21$ or 25 make use of truncated IG(2, 5) and for $v=17$ of IG(2, 4). For other values of v we shall prove the following lemmas.

LEMMA 1. *If $v \equiv 0 \pmod 2$, then $C[\{4, 6\}, 3, 1; v]$ exists.*

PROOF. The proof will be given by induction. For $v=4$ and 6 the lemma follows from Theorem 1. If $v > 6$ denote the elements by (i, j), $i=0, 1$; $j=0, 1, \ldots, t-1$, where $t=v/2$. By the assumption of the induction $C[\{4, 6\}, 3, 1; t]$ exists if $t \equiv 0 \pmod 2$, and $C[\{4, 6\}, 3, 1; t+1]$ exists if $t \equiv 1 \pmod 2$.

If $t \equiv 0 \pmod 2$ form the blocks $\{(0, j), (1, j), (0, j'), (1, j')\}$, $0 \leq j < j' \leq t-1$. Further let $\{(x_0), (x_1), (x_2), (x_3)\}$ and $\{(y_0), (y_1), (y_2), (y_3), (y_4), (y_5)\}$ be the blocks of $C[\{4, 6\}, 3, 1; t]$ on the elements $(j), j=0, 1, \ldots, t-1$. For every quadruple $\{(x_0), (x_1), (x_2), (x_3)\}$ construct the blocks

$$\{(i_0, x_0), (i_1, x_1), (i_2, x_2), (i_3, x_3)\}, \qquad \sum i_\alpha \equiv 0 \pmod 2;$$

and for every sixtuple $\{(y_0), (y_1), (y_2), (y_3), (y_4), (y_5)\}$ construct the blocks:

$$\begin{aligned} &\{(i, y_0), (i, y_1), (i, y_2), (i, y_3), (i, y_4), (i, y_5)\}, \quad i = 0, 1; \\ &\{(i, y_\alpha), (i, y_{\alpha+2}), (i+1, y_{\alpha+1}), (i+1, y_{\alpha+4})\}, \quad i = 0, 1; \alpha = 0, 1, \ldots, 5, \end{aligned}$$

(the indices have to be taken modulo 2 and 6 respectively);

$$\begin{aligned} &\{(0, y_\alpha), (0, y_{\alpha+2}), (1, y_{\alpha+3}), (1, y_{\alpha+5})\}, \qquad \alpha = 0, 1, \ldots, 5; \\ &\{(i, y_{2\beta}), (i, y_{2\beta+\varepsilon}), (i+1, y_{2\gamma}), (i+1, y_{2\gamma+\varepsilon})\}, \quad i = 0, 1; 0 \leq \beta < \gamma \leq 2; \varepsilon = \pm 1. \end{aligned}$$

If $t \equiv 1 \pmod 2$ form $C[\{4, 6\}, 3, 1; t+1]$ on the elements $(j), j=0, 1, \ldots, t-1, \infty$. The blocks will be of the form $\{(x_0), (x_1), (x_2), (x_3)\}$, $\{(y_0), (y_1), (y_2), (y_3), (y_4), (y_5)\}$, $\{(\infty), (s_0), (s_1), (s_2), (s_3), (s_4)\}$, $\{(\infty), (r_0), (r_1), (r_2)\}$. For the quadruples $\{(x_0), (x_1), (x_2), (x_3)\}$ and the sixtuples $\{(y_0), (y_1), (y_2), (y_3), (y_4), (y_5)\}$ form blocks of $C[\{4, 6\}, 3, 1; v]$ as above. For $\{(\infty), (r_0), (r_1), (r_2)\}$ form the block $\{(0, r_0), (0, r_1), (0, r_2), (1, r_0), (1, r_1), (1, r_2)\}$, and for $\{(\infty), (s_0), (s_1), (s_2), (s_3), (s_4)\}$ form the blocks

$\{(i, s_{\alpha+1}), (i, s_{\alpha+4}), (i+1, s_{\alpha+2}), (i+1, s_{\alpha+3})\}, \quad i = 0, 1; \alpha = 0, 1, 2, 3, 4;$
$\{(i, s_\alpha), (i, s_{\alpha+\mu}), (i, s_{\alpha-\mu}), (i+1, s_\alpha)\}, \quad i = 0, 1; \alpha = 0, 1, 2, 3, 4; \mu = 1, 2.$

LEMMA 2. *$C[\{4, 5, 9\}, 3, 1; 33]$ exists.*

PROOF. Denote the elements by (i, g_j), $i=0, 1, 2, 3; j=0, 1, \ldots, 7$ and (∞); g_j's are marks in the Galois field $GF[2^3]$, with $g_0=0$ and $g_j=g^{j-1}$ for $1 \leq j \leq 7$ where g is a primitive mark, namely a root of the (irreducible) equation $x^3=x+1$. The blocks are:

$$\{(i, g_0), (i, g_1), (i, g_2), (i, g_3), (i, g_4), (i, g_5), (i, g_6), (i, g_7), (\infty)\}, \quad i = 0, 1, 2, 3;$$
$$\{(0, g_j), (1, g_j), (2, g_j), (3, g_j), (\infty)\}, \quad j = 0, 1, \ldots, 7;$$
$$\{(0, g_j), (1, g_j+g^\alpha), (2, g_j+g^{\alpha+1}), (3, g_j+g^{\alpha+3}), (\infty)\}, \quad j = 0, 1, \ldots, 7; \alpha = 0, 1, \ldots, 6;$$
$$\{(0, g_{\alpha_0}), (1, g_{\alpha_1}), (2, g_{\alpha_2}), (3, g_{\alpha_3})\}, \quad \sum g_{\alpha_h} = 0, \qquad g_{\alpha_2} \neq g(g_{\alpha_0}+g_{\alpha_1})+g_{\alpha_0};$$
$$\{(i, g_{\alpha_0}), (i, g_{\alpha_1}), (i', g_{\alpha_2}), (i', g_{\alpha_3})\}, \quad 0 \leq i < i' \leq 3, \qquad g_{\alpha_0}+g_{\alpha_1} = g_{\alpha_2}+g_{\alpha_3} \neq 0.$$

REFERENCES

1. L. V. Ahlfors, *Complex analysis. An introduction to the theory of analytic functions of one complex variable*, McGraw-Hill, New York, 1953. MR **14**, 857.
2. R. C. Bose, E. T. Parker and S. Shrikhande, *Further results on the construction of mutually orthogonal Latin squares and the falsity of Euler's conjecture*, Canad. J. Math. **12** (1960), 189–203. MR **23** #A69.
3. R. C. Bose and S. Shrikhande, *On the construction of sets of mutually orthogonal Latin squares and the falsity of a conjecture of Euler*, Trans. Amer. Math. Soc. **95** (1960), 191–209. MR **22** #2557.
4. R. D. Carmichael, *Introduction to the theory of groups of finite order*, Dover, New York, 1956. MR **17**, 823.
5. L. E. Dickson, *History of the theory of numbers*, Vol. 1, Reprint, Chelsea, New York, 1952.
6. M. Hall, Jr., *Combinatorial theory*, Blaisdell, Waltham, Mass., 1967. MR **37** #80.
7. H. Hanani, *The existence and construction of balanced incomplete block designs*, Ann. Math. Statist. **32** (1961), 361–386. MR **29** #4161.
8. ———, *On some tactical configurations*, Canad. J. Math. **15** (1963), 702–722. MR **28** #1136.
9. E. Parker, *Construction of some sets of mutually orthogonal Latin squares*, Proc. Amer. Math. Soc. **10** (1959), 946–949. MR **22** #674.

TECHNION ISRAEL INSTITUTE OF TECHNOLOGY

THE CITY UNIVERSITY OF NEW YORK

HOMOGENEOUS 0-1 MATRICES

ALEXANDER HURWITZ[1,2]

(0-0). **Introduction.** In this paper we study combinatorial properties of certain matrices of zeros and ones, characterized by conditions on subsets of their rows and columns. The matrices constitute a generalization of the incidence matrices of structures well known in combinatorics, geometry, and statistics, such as tactical configurations, finite projective planes, and balanced incomplete block designs. See **[5]**, **[3]**, **[4]**, as well as other papers in this volume.

In our notation an incidence matrix of a balanced incomplete block design is a 2,1-homogeneous matrix where rows correspond to elements, columns correspond to blocks, and the equivalence between the standard parameters v, k, λ, b and r and our notation is $v=f_0'$, $k=f_1'$, $b=f_0$, $r=f_1$, and $\lambda=f_2$. A symmetric balanced incomplete block design corresponds to a 2,2-homogeneous matrix.

If we attempt to construct a 2,2-homogeneous matrix by an exhaustive search of a certain kind we find that during the process we construct a number of 2-homogeneous matrices. The process is characterized by inclusion graphs which are defined in this paper and whose properties are studied. For a list of known 2,2-homogeneous matrices, see **[6]**.

(1-1). **Definitions and notations.** If the matrices A and B have the same number of rows then define the *juxtaposition* AB of A and B to be the matrix resulting from adjoining the columns of B at the right of A. Further, denote the juxtaposition of i copies of a given matrix A by A^i. For example $A^3=AAA$ and $A^2B^0=AA$.

We consider matrices with entries in GF(2), the field consisting of 0 and 1. For any two vectors a and b of the same dimension over GF(2) define $a+b$ and ab to be the usual componentwise addition and multiplication over GF(2).

Define the *weight* $w(a)$ of a to be the number of ones in a. For example, if $a=(0, 1, 0, 1)$ and $b=(0, 0, 1, 1)$ then $w(ab)=1$ and $w(a+b)=2$.

Let A be an m by n matrix with rows $a_1, a_2, \ldots, a_m$. Define the *frequencies* $f_0(A), f_1(A), \ldots, f_m(A)$ of A as follows: let $f_1(A)=w(a_i)$ if $w(a_i)$ is the same for all i;

---

[1] Presented to the Society, June 20, 1970.

[2] The contents of this paper are taken from the author's dissertation **[6]** which was written under the direction of Professor T. S. Motzkin. The research for this work was supported in part by the Office of Naval Research under Contract NONR 233(76).

otherwise $f_1(A)$ is undefined. Similarly let $f_2(A)=w(a_i a_j)$ if $w(a_i a_j)$ is the same for all pairs i, j where $i \neq j$, and so forth for $f_3(A)$, $f_4(A), \ldots, f_m(A)$ using triples, quadruples, etc., of rows of A.

Define $f_0(A)=n$. Then for any m by n matrix B, both $f_0(B)$ and $f_m(B)$ are always defined.

Define $f_i'(A)=f_i(A^T)$, where A^T is A transposed; thus $f_0'(A)=m$. We write f_i for $f_i(A)$ and f_i' for $f_i'(A)$ when no ambiguity ensues.

Define A to be *j-homogeneous* if $f_0'(A) \geq j$ and $f_0, f_1, \ldots, f_j$ are defined (in **[6]** j-homogeneous is given a definition which applies to all matrices and does not use frequencies). Define A to be *j,k-homogeneous* if A is j-homogeneous and A^T is k-homogeneous. We abbreviate "j-homogeneous matrix (matrices)" by j-HM and "j,k-homogenous matrix (matrices)" by j,k-HM.

The following matrices occur frequently: the m by n zero matrix, $Z_{m,n}$; the m by n matrix of all ones, $J_{m,n}$; the m by m identity matrix, I_m; and the matrix $K_m=J_{m,m}-I_m$. When the values of the subscripts of the matrices are understood from the context, the subscripts will be dropped. For example I_4 and K_4 are 4,4-HM and I_4KK is a 4-HM.

From now on we will concern ourselves with 2-HM.

(1-2). *Inclusion graphs.* Construction of a 2,2-HM with $f_1=s$ and $f_2=t$ by an exhaustive search procedure, building up matrices row by row, can proceed as follows. Always working with rows of weight s, first write down a single row. Next adjoin a row—and possibly adjoin columns of all zeros to the first row—so that f_2 of the pair of rows is t. There are many ways of adjoining the second row, but they are all essentially the same. Hence define a matrix A to be *equivalent* to B if but for possible zero columns, B can be obtained from A by permuting rows and permuting columns. Clearly this is an equivalence relation; denote the equivalence class containing A by $\overline{A}$.

Note that if A and B are equivalent 2-HM then $f_1(A)=f_1(B)$ and $f_2(A)=f_2(B)$. We will extend notions and notations referring to matrices to the equivalence classes containing them. For example, $f_1(\overline{A})$ is the common value of $f_1(B)$ for all $B \in \overline{A}$.

The beginning of the exhaustive procedure described above for construction of a 2,2-HM with $f_1=3$ and $f_2=1$ is shown in Figure (1-3). There, periods are used for zeros to improve readability. The matrices are representatives of the equivalence classes involved and a line connects two matrices if by adjoining a suitable row to the first matrix the result is in the equivalence class of the second.

Note that the matrix in the lower right of Figure (1-3) is such that exactly one row, up to equivalence, can be adjoined to it. Note also that it is equivalent to $J_{4,1}II$.

We can formalize the exhaustive procedure as follows. Define $M(s, t)$ to be the set of 2-HM with $f_1=s$ and $f_2=t$ together with all matrices with one row having $f_1=s$, that is,

$$M(s, t) = \{A \mid f_1(A) = s, \text{ and if } f_0'(A) \geq 2 \text{ then } f_2(A) = t\}.$$

Define $\overline{M(s, t)}$ to be the set of equivalence classes of $M(s, t)$, that is,

$$\overline{M(s, t)} = \{\overline{A} \mid A \in M(s, t)\}.$$

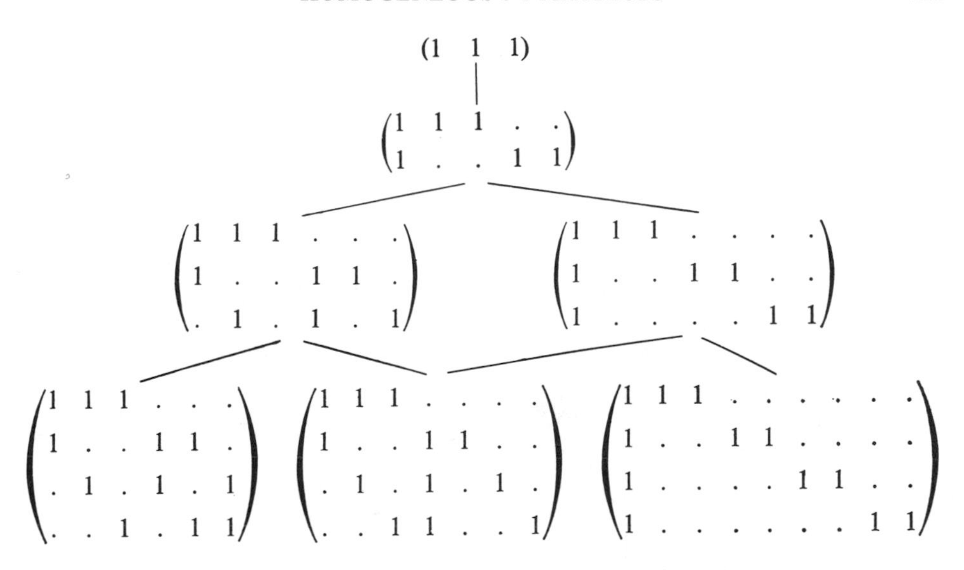

FIGURE (1-3). Search for 2,2-HM with $f_1=3, f_2=1$

Define the *inclusion graph* $G(s, t)$ to be a directed graph where the set of vertices is $\overline{M(s, t)}$ and where an edge goes from U to V if and only if given any $A \in V$ there is some row of A which when it is deleted the resulting matrix is a member of U. Define a vertex U of an inclusion graph $G(s, t)$ to be *maximal* if U is not an initial vertex of an edge of $G(s, t)$. Define the *initial vertex of an inclusion graph* to be the vertex containing a matrix with exactly one row.

Figure (1-4) shows the complete inclusion graph $G(3, 1)$ except of course for all of the infinite path which consists of matrices of the form JII. The direction of each edge is down the page and the number attached to each vertex is the number of columns of a representative without zero columns. This attached number together with f_0' of the vertex is enough to uniquely identify the vertex in the graph $G(3, 1)$. Note that the vertex labeled $\underline{7}$ is a maximal vertex. It contains a 2,2-HM which is an incidence matrix of the finite projective plane of order 2.

Corresponding to the previous set of definitions using equivalence we now make a set of definitions using a new equivalence relation. Define a matrix A to be *J-equivalent* to B if but for possible zero columns or columns of all ones (a J matrix) the matrix B can be obtained from A by permuting rows and permuting columns. Denote the J-equivalence class containing A by $\overline{A}^J$.

Note that J-equivalence preserves f_1-f_2 (but not f_1 and f_2 as does equivalence), that is, if A is J-equivalent to B then $f_1(A)-f_2(A)=f_1(B)-f_2(B)$.

Define $M(n)$ to be the set of 2-HM with $f_1-f_2=n$ together with all matrices with one row, that is,

$$M(n) = \{A \mid f_0'(A) \geq 2 \text{ and } f_1(A)-f_2(A) = n\} \cup \{A \mid f_0'(A) = 1\}.$$

Define $\overline{M(n)}^J$ to be the J-equivalence classes of $M(n)$.

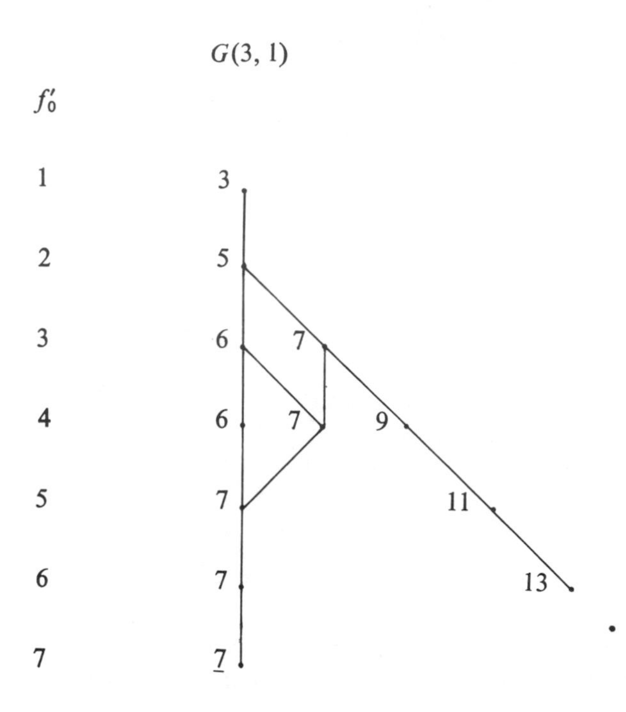

FIGURE (1-4)

By replacing equivalence by J-equivalence and $M(s, t)$ by $M(n)$ in the previous definitions we obtain definitions for the *J-inclusion graph* $G(n)$, *J-maximal* vertex and matrix, and the *initial vertex of* $G(n)$.

For example there is an edge of $G(2)$ from $\overline{K_4K_4^J}$ to $\overline{K_5K_5^J}$.

(2-1). **Bounds on the size of** 2-HM. The bounds in this section and the remarks in the next show that the interesting part of an inclusion graph $G(s, t)$ is a finite subgraph.

THEOREM (2-2). *Let A be a* 2-HM. *If there exist three rows of A, a, b, and c such that $w(abc)=x$ and $x<f_2$ then*

$$f_0'(A) \leq 1$$

$$+ \sum_{x_1+x_2+x_3+x_4=f_2} \binom{x}{x_1}\binom{f_2-x}{x_2}\binom{f_2-x}{x_3}\binom{f_1-2f_2+x}{x_4}\left[\frac{f_1-f_2}{f_2-x_1-\min(x_2, x_3)}\right], \tag{2-3}$$

where the values of x_1, x_2, x_3, and x_4 are integers chosen so that none of the binomial coefficients are zero and the brackets indicate the greatest integer value.

PROOF. Without loss of generality we can assume a, b, and c are as follows:

$$A = \begin{pmatrix} a \\ b \\ c \\ \vdots \\ d \\ \vdots \end{pmatrix} = \begin{pmatrix} a_1 & a_2 & a_3 & a_4 & a_5 & a_6 & a_7 & a_8 \\ b_1 & b_2 & b_3 & b_4 & b_5 & b_6 & b_7 & b_8 \\ c_1 & c_2 & c_3 & c_4 & c_5 & c_6 & c_7 & c_8 \\ \cdot & \cdot & \cdot & & & & & \\ d_1 & d_2 & d_3 & d_4 & d_5 & d_6 & d_7 & d_8 \\ \cdot & \cdot & \cdot & & & & & \end{pmatrix}$$

$$\begin{matrix} x & f_2-x & f_2-x & f_1-2f_2+x & \overbrace{\quad\quad}^{f_1-f_2} & & & \end{matrix}$$

$$= \begin{pmatrix} J & J & J & J & Z & Z & Z & Z \\ J & J & Z & Z & J & J & Z & Z \\ J & Z & J & Z & J & Z & J & Z \\ \cdots & & & & & & & \\ d_1 & d_2 & d_3 & d_4 & d_5 & d_6 & d_7 & d_8 \\ \cdots & & & & & & & \end{pmatrix}$$

where all entries are matrices with one row and with the number of columns indicated at the top of each column of matrices. The row d will be used to stand for an arbitrary row of A except a.

Let e_i be a matrix with one row and the same number of columns as d_i, for $i=1, 2, 3$, and 4. Consider the set of matrices

$$E(x_1, x_2, x_3, x_4) = \{e \mid e = e_1e_2e_3e_4,\ w(e_i) = x_i \text{ for } i = 1, 2, 3, 4\},$$

where x_1, x_2, x_3, and x_4 are nonnegative integers and $x_1+x_2+x_3+x_4=f_2$. Consider the set of all rows of A, except a, with a given initial segment e,

$$D(e) = \{d \mid d \text{ is a row of } A,\ d \neq a,\ d_i = e_i \text{ for } i = 1, 2, 3, \text{ and } 4\};$$

where $e \in E(x_1, x_2, x_3, x_4)$. Note that if $e \neq e'$ then $D(e)$ is disjoint from $D(e')$ and that $E(x_1, x_2, x_3, x_4)$ is disjoint from $E(x'_1, x'_2, x'_3, x'_4)$ unless $x_i = x'_i$ for $i=1, 2, 3$, and 4. Thus using # to denote cardinality

$$f'_0(A) = 1 + \sum_{x_1+x_2+x_3+x_4=f_2} \ \sum_{e \in E(x_1,x_2,x_3,x_4)} \#(D(e)). \tag{2-4}$$

We now show that for all $e \in E(x_1, x_2, x_3, x_4)$

$$\#(D(e)) \leq \left[\frac{f_1-f_2}{f_2-x_1-x_2}\right], \tag{2-5}$$

where the right-hand side is taken to be plus infinity if the denominator vanishes, which happens if and only if $b \in D(e)$. If $b \notin D(e)$ then every row $u \in D(e)$ must be such that $w(ub)=f_2$. By the definition of $D(e)$ and b we have $w(uba)=x_1+x_2$, therefore, the f_1-f_2 columns of b corresponding to b_5 and b_6 are partitioned into disjoint sets of cardinality $f_2-x_1-x_2$ by $D(e)$, one set for each row of $D(e)$, with possibly some columns left over. Hence (2-5) follows. By using c instead of b in the argument above we obtain

$$\#(D(e)) \leq \left[\frac{f_1-f_2}{f_2-x_1-x_3}\right]. \tag{2-6}$$

There are $\binom{n}{r}$ ordered n-tuples of zeros and ones with exactly r ones. Hence

$$\#(E(x_1, x_2, x_3, x_4)) \leq \binom{x}{x_1}\binom{f_2-x}{x_2}\binom{f_2-x}{x_3}\binom{f_1-2f_2+x}{x_4}. \tag{2-7}$$

Combining (2-5) and (2-6) we obtain

$$\#(D(e)) \leq \left[\frac{f_1-f_2}{f_2-x_1-\min(x_2, x_3)}\right]. \tag{2-8}$$

Thus the theorem follows by combining (2-7) and (2-8) with (2-4).

A weaker but simpler bound can be derived as follows. The right-hand side of (2-8), as it appears in the sum in (2-3), is less than or equal to f_1-f_2. Assume A has three rows satisfying the hypothesis of Theorem (2-2). Then by using the well-known identity

$$\sum_{y_1+y_2+y_3+y_4=k} \binom{n_1}{y_1}\binom{n_2}{y_2}\binom{n_3}{y_3}\binom{n_4}{y_4} = \binom{n_1+n_2+n_3+n_4}{k}$$

we have an upper bound for the right side of (2-3) and hence

$$f_0'(A) \leq 1+\binom{f_1}{f_2}(f_1-f_2). \tag{2-9}$$

This bound is attained in the following two cases. First, the 2,2-HM corresponding to a finite projective plane of order n. Here $f_1=n+1$, $f_2=1$, and $f_0'=n^2+n+1$. Second, the matrix K_n. Here $f_0'=n$, f_1+1, $f_1=f_2+1$. Thus *both kinds of matrices are maximal.*

We note (see e.g., [**6**]), that if $n\geq 2$, then a 2,2-HM with $f_0'=n^2+n+1$, $f_1=n+1$ and $f_2=1$ is an incidence matrix of a finite projective plane of order n.

(3-1). **Inclusion graphs and mappings.** In this section we describe some of the global features of inclusion graphs and some mappings between inclusion graphs which preserve edges.

(3-2). *The infinite path.* The significance of Theorem (2-2) is that the characteristic part of an inclusion graph is finite; in fact, given any s and t, $s\geq t$, each 2-HM A with $f_1=s$, $f_2=t$ and a sufficiently large number of rows satisfies $f_m(A)=f_2(A)$, where $m=f_0'(A)$, and thus is equivalent to a matrix of the form

$$J_x I^y \tag{3-3}$$

where $x=f_2$ and $y=f_1-f_2$. Further, given a matrix of the form (3-3) there is exactly one way, up to equivalence, of adjoining a row to get a matrix of the form (3-3). Hence we have the following theorem:

THEOREM (3-4). *Every inclusion graph $G(s, t)$ has exactly one infinite path starting from the initial vertex, and the set of vertices of this path is equal to the initial vertex together with the set of all vertices V with $f_2(V)=f_m(V)$, where $m=f_0'(V)\geq 2$.*

The following theorem shows that some inclusion graphs are contained in other inclusion graphs as subgraphs.

THEOREM (3-5). *For every inclusion graph $G(s, t)$ and every $x\geq 0$ there is a one-to-one mapping T from the set of vertices of $G(s, t)$ into the set of vertices of $G(s+t, t+x)$ which preserves edges.*

PROOF. Define a mapping T from the vertices of $G(s, t)$ to the vertices of $G(s+x, t+x)$ as follows: for every vertex V of $G(s, t)$ choose a representative $A \in V$ and define $T(V)=\overline{AJ_x}$ which is a vertex of $G(s+x, t+x)$. Clearly the mapping T is independent of the choice of representatives and is one-to-one because given any 2-HM B and C, B is equivalent to C if and only if BJ_x is equivalent to CJ_x. The mapping T preserves edges because if an edge in $G(s, t)$ leads from U to V and hence there is an $A \in U$ and a $B \in V$ such that removing a row of B yields A, then removing the corresponding row of BJ_x will yield AJ_x and thus an edge goes from $\overline{AJ_x}=T(U)$ to $\overline{BJ_x}=T(V)$. Thus the theorem is proved.

By a similar argument to Theorem (3-5) only replacing J_x by I^r in the mapping T we obtain the mapping $T'(V)=\overline{AI^r}$ and the theorem:

THEOREM (3-6). *For every inclusion graph $G(s, t)$ and every $r\geq 0$ there is a one-to-one mapping T' from the set of vertices of $G(s, t)$ into the set of vertices of $G(s+r, t)$ which preserves edges.*

By combining Theorem (3-5) and Theorem (3-6) we obtain:

THEOREM (3-7). *Every inclusion graph $G(s, t)$ contains subgraphs which are isomorphic to the inclusion graphs $G(u, v)$ for all u and v such that $s-u\geq t-v\geq 0$.*

Another kind of mapping can be used to map an infinite number of inclusion graphs into a given inclusion graph. We first define the *pivot transformation* as follows.

Given a 2-HM A with rows $a_1, a_2, \ldots, a_m$, $m\geq 2$, choose a row of A, say a_i. We call a_i the *pivot*. Form a new matrix B, the result of a pivot transformation of A, such that the rows b_j of B are $b_j=a_j+a_i$, $i\neq j$, $j=1, 2, \ldots, m$. The matrix B has $f_0'(A)-1$ rows. We now show B to be a 2-HM.

Let b and c be any rows of A, and b' and c' be the corresponding rows of B under this transformation. Without loss of generality we can diagram a_i, b, and c as below (see Figure (3-8)) where empty boxes indicate all zeros and boxes with letters in them indicate all ones, the letters indicating the number of ones.

a_i		f_1						
b	e	d			f_2-e	g		
c	e		d_1		f_2-e		g_1	

FIGURE (3-8)

Let h^- denote the complement of the row h. Define the following integers which count parts of the rows under discussion:

$$w(a_ibc) = e, \qquad w(a_ibc^-) = d, \quad w(a_ib^-c) = d_1,$$

$$w(a_i^-bc^-) = g, \qquad w(a_i^-b^-c) = g_1.$$

Note that $w(a_i^-bc)=f_2-e$. All of the following f_i refer to A.

First we note that $e+d=f_2=e+d_1$, and hence $d=d_1$. Next we see that

$$w(b) = e+d+(f_2-e)+g = w(c) = e+d+(f_2-e)+g_1 = f_1.$$

Hence $g=g_1$. Since $w(b')=f_1-e-d+(f_2-e)+g, f_2-e+g = f_1-f_2$, and $e+d=f_2$, we have that $w(b')=2(f_1-f_2)$ and thus $f_1(B)$ exists. Further

$$w(b'c') = f_2-e+(f_1-e-d-d_1) = f_1-f_2;$$

hence $f_2(B)$ exists and B is a 2-HM.

Using a pivot transformation we can prove the following theorem.

THEOREM (3-9). *For every inclusion graph $G(s, t)$ there is a mapping T from the set of vertices of $G(s, t)$ minus the initial vertex into the set of vertices of $G(2(s-t), s-t)$ which preserves edges.*

PROOF. Choose a system of representatives, $A_1, A_2, \ldots$ of the vertices of $G(s, t)$ so that if there is an edge from $\bar{A}_i$ to $\bar{A}_j$ then A_i is equal to the first $f_0'(A_i)$ rows of A_j. Let A_1 be the representative of the initial vertex. The initial vertex is connected by a path to every vertex. Hence every representative has as its first row the representative of the initial vertex, A_1. Apply the pivot transformation to every $A_i \neq A_1$ using A_1 as the pivot, and call the result B_i. Define T so that

$$T(\bar{A}_i) = \bar{B}_i$$

for every $A_i \neq A_1$. The mapping T has all the required properties and thus the theorem is proved.

We now give an application of Theorem (3-9). The following theorem describes the form of all 2-HM with a given value of f_1-f_2 and a sufficiently large number of rows, the number depending only on the value of f_1-f_2.

THEOREM (3-10). *Let U be a maximal vertex of $G(2t, t)$ such that for any other maximal vertex V, $f_0'(V) \leq f_0'(U)$; then any 2-HM A such that $f_1(A) - f_2(A) = t$ and $f_0'(A) \geq f_0'(U) + 2$ is equivalent to $J_y K^x I^z$, for some x, y, and z.*

PROOF. The vertex $\overline{A}$ is a member of $G(f_1(A), f_2(A))$. Since

$$f_0'(A) \geq f_0'(U) + 2,$$

the image of $\overline{A}$ in $G(2t, t)$ under a mapping T given in Theorem (3-9) must be on the infinite path. Hence

$$T(A) = \overline{J_t I_m^t}$$

where $m = f_0'(A) - 1$.

By the definition of T, $\overline{J_t I^t}$ is the result of applying the pivot transformation to a 2-HM which is a member of $\overline{A}$. Let A_i be the representative of $\overline{A}$ used in the mapping T and let B_i be the result of the pivot transformation applied to A_i. Thus $A_i \in \overline{A}$ and $B_i \in \overline{J_t I^t}$. Call the pivot row p. Thus we can write

$$A_i = \begin{pmatrix} p \\ C \end{pmatrix}.$$

Let $m = f_0'(C)$. The matrix B_i has columns with at most three different weights, that is, weights equal to zero, one, and $f_0'(B_i) = m$. The columns where p has a zero are the same in B_i as in C. The columns of C where p has a one are the complements of the corresponding columns in B_i. Therefore, C has columns with at most four different weights, namely, zero, one, m, $m-1$.

We will now show that the columns of C which have weight $m-1$ can be rearranged as a juxtaposition of $t \geq 0$ copies of the K matrix, that is K_m^t.

Let x_j, $j = 1, 2, \ldots, m$, be the number of columns of C such that the jth row contains the only zero in that column. We can assume C has at least three rows, otherwise we can assume without loss of generality that $t = 0$. Since C is a 2-HM, comparing the first and jth rows we have

$$f_2(C) = f_m(C) + \sum_{k=1}^{m} x_k - x_1 - x_j \tag{3-11}$$

for $j = 2, 3, \ldots, m$, that is, only columns with weight m or $m-1$ can affect the value of $f_2(C)$. Thus $x_2 = x_3 = \cdots = x_m$. By replacing x_1 by x_2 in (3-11) we obtain $x_1 = x_2 = \cdots = x_m$. Thus C is equivalent to $K^x D$ where x is the common value of the x_j's and D has no columns of weight $m-1$. Since C is a 2-HM we have that

$$f_1(C) = f_1(K^x) + f_1(D)$$

and thus D is equivalent to $J_v I^z$ and C is equivalent to $J_v K^x I^z$ where $f_1(C) = x(m-1) + v + z$. Because ones must appear in the row p where a column of weight $m-1$ appears in B_i, A_i is equivalent to

$$E = \begin{pmatrix} J & J & Z & J & J & & J & Z & Z & & Z \\ Z_q & J_y & J_r & K & K & \ldots & K & I & I & \ldots & I \end{pmatrix}$$

where the top row of matrices all have one row and corresponds to p, and the bottom row of matrices corresponds to C which has x copies of the K matrix, has z copies of the I matrix, and has $v=y+r$.

There remains to be shown that E, and hence A, is equivalent to $J_yK^xI^z$. Since E is a 2-HM it suffices to show that E is equivalent to $K^x_{m+1}F$ where F has columns of weight either $m+1$, one, or zero. Let a, b, and c be distinct rows of E, and let a be the first row of E. Then, $w(ab)=y+x(m-1)=f_2(E)=w(bc)=y+r+x(m-2)$ and hence $x=r$. The matrix, which forms the center part of E,

$$\begin{pmatrix} Z & J & J & \cdots & J \\ J_r & K & K & & K \end{pmatrix},$$

where there are x copies of the K matrix, is thus equivalent to K^x_{m+1}. Hence A is equivalent to $J_yK^xI^z$ and the theorem is proved.

A corollary of the preceding theorem describes J-inclusion graphs.

COROLLARY (3-12). *Every J-inclusion graph $G(t)$ has exactly $t+1$ infinite paths starting from the initial vertex, and any two such paths have just the initial vertex and the vertex adjacent to the initial vertex in common.*

PROOF. First we show that there are $t+1$ infinite paths in $G(t)$. There is exactly one vertex V, the initial vertex, such that $f_0'(V)=1$ and exactly one vertex W such that $f_0'(W)=2$. In fact, W contains I_2^t.

Two matrices of the form $K_r^xI_r^z$ and $K_r^uI_r^v$ are J-equivalent if and only if $x=u$ and $z=v$. Hence there are $t+1$ paths π_x, $x=0, 1, 2, \ldots, t$, where π_x consists of V, W, and vertices containing $K_r^xI_r^z$, $x+z=t$, for $r=3, 4, 5, \ldots$. If $x\neq y$ then π_x and π_y have only V and W in common.

We now show that every infinite path starting from V is equal to π_x for some $x=0, 1, \ldots,$ or t.

No matter which row of a matrix J-equivalent to $K_m^uI_m^v$, $m\geq 4$, is removed the result is J-equivalent to $K_{m-1}^uI_{m-1}^v$. Therefore the only path from V to a vertex containing $B=K_m^uI_m^v$, $m\geq 4$ and $u+v=t$, is the part of π_u from V to $\bar{B}^J$.

Let π be any infinite path starting from V. Suppose π is different from π_x for every $x=0, 1, \ldots, t$. Then π must have vertices $P_0, P_1, \ldots, P_t$ such that P_x is not on π_x for $x=0, 1, \ldots, t$. Let Q be a vertex on π such that $f_0'(Q)>f_0'(P_x)$ for $x=0, 1, \ldots, t$ and $f_0'(Q)\geq f_0'(U)+2$ where U is a maximal vertex of $G(2t, t)$ as described in Theorem (3-10). Then Q must contain $J_uK^vI^w$ for some u, v, w, with $v+w=t$. Since $J_uK^vI^w$ is J-equivalent to K^vI^w, Q also contains K^vI^w. The path from V to Q along π contains P_v and by the remarks above is identical to the part of the path from V to the vertex of π_v containing K^vI^w and hence P_v is on π_v. This contradicts the definition of P_v and thus the corollary is proved.

Note there is a mapping from the graph $G(s, t)$, where $s-t=n$, into the graph $G(n)$ which preserves edges, namely, the mapping which carries a vertex of $G(s, t)$ which contains a 2-HM A to the vertex $\bar{A}^J$. Hence a J-maximal 2-HM is maximal.

(4-0). **Properties of some** 2-HM. A well-known result states, in our notation, that given a 2,1-HM with $f_1 > f_2$ it has $f_0 \geq f_0'$ (Fisher's inequality). A proof using matrices appears in [**9**, p. 99]. The same proof works for 2-HM.

THEOREM (4-1). *A* 2-HM *with* $f_1 > f_2$ *has* $f_0 \geq f_0'$.

COROLLARY (4-2). *A* 2,2-HM *is a square matrix if* $f_1 > f_2$.

Another well-known result stated in our notation is the theorem:

THEOREM (4-3). *Every* 2,1-HM *with* $f_0' \geq 2$, *has frequencies such that*

$$f_0' f_1 = f_0 f_1', \tag{4-4}$$

$$f_2(f_0' - 1) = f_1(f_1' - 1). \tag{4-5}$$

COROLLARY (4-6). *If A is a* 2,2-HM *with* $f_1 > f_2$ *then* $f_i(A) = f_i'(A)$, $i = 0, 1, 2$.

The following theorem shows that 2,2-HM are not contained in any vertex which is the image of the kind of mappings described in Theorem (3-5) and Theorem (3-6), that is, mappings which use juxtaposition transformations.

THEOREM (4-7). *A* 2,2-HM *which has* $f_1 > f_2$ *is not the juxtaposition of any two* 2-HM.

PROOF. Let C be a 2,2-HM where $f_1 > f_2$ and $C = AB$ where A and B are each 2-HM. Then $f_1(A)$ and $f_2(A)$ exist because A is 2-HM and $f_1'(A)$ and $f_2'(A)$ exist because A is part of C which is a 2,2-HM. Hence A is a 2,2-HM. But if A is a 2,2-HM then by Corollary (4-6) A is a square matrix, which is impossible because C is a square matrix. Thus the theorem is proved.

(5-0). **Maximality of** 2,1-HM **and** 2,2-HM.

THEOREM (5-1). *If A is a* 2,1-HM *and*

$$f_1(A) \equiv f_2(A) \qquad (\operatorname{mod} f_1'(A)) \tag{5-2}$$

does not hold then A is maximal, where if $f_1' = 0$ *we take the congruence to be an equation.*

PROOF. Assume A is a 2,1-HM and is not maximal. We will show (5-2) holds. The matrix A is equivalent to a J or Z matrix if and only if $f_1 = f_2$. The J and Z matrices are not maximal. Assume $f_1 > f_2$. Hence $f_1' > 0$. Since A is not maximal we can adjoin a row to A and obtain a 2-HM B which without loss of generality we may write as

$$B = \begin{pmatrix} A & Z \\ c & c_1 \end{pmatrix}.$$

By definition of a 2-HM and counting the total number of ones in the columns of A where c has a one we obtain

$$w(c) = f_0'(A) f_2(A) / f_1'(A). \tag{5-3}$$

The right-hand side of (5-3) is an integer if and only if (5-2) holds, as we now show. Solving (4-5), if $f_2 > 0$, for f_0' and substituting for f_0' in the right-hand side of (5-3), we obtain

$$(f_1(f_1'-1)+f_2)/f_1'$$

which is an integer if and only if $f_1 \equiv f_2 \pmod{f_1'}$. Finally, if $f_2 = 0$ the right-hand side of (5-3) is the integer 0 and (5-2) holds because by (4-5) $f_1' = 1$.

COROLLARY (5-4). *If A is* 2,2-HM *with $f_1 > f_2 \geq 1$ then A is maximal.*

PROOF. By Corollary (4-6) we have $f_1 = f_1' > 1$. Hence (5-2) does not hold and the corollary follows.

CONJECTURE. The converse of Theorem (5-1) holds, that is, if (5-2) holds then A is not maximal.

(6-0). **Maximality of juxtaposed matrices.** Let A and B be 2-HM and $C = AB$. If A and B are not maximal then C is not maximal. If A is maximal then C may or may not be maximal, and both cases occur.

The several classes of juxtaposed 2,1-HM and 2,2-HM listed in this section are all determined in essentially the same way. Proofs are given in **[6]** and will be omitted here except for the following theorem which is representative.

THEOREM (6-1). *If D and E are* 2,1-HM *then the juxtaposition, D^tE^u, is maximal if there is no solution in integers x and y of the following system:*

$$x, y \geq 0, \tag{6-2}$$

$$tf_0(D) \geq x, \tag{6-3}$$

$$uf_0(E) \geq y, \tag{6-4}$$

$$tf_1(D) + uf_1(E) \geq x + y, \tag{6-5}$$

$$(tf_2(D) + uf_2(E))f = xf_1'(D) + yf_1'(E), \tag{6-6}$$

where $f = f_0'(D) = f_0'(E)$.

PROOF. Assume D^tE^u is not maximal. We will find integers x and y which satisfy the above relations. We can find a row c such that c can be adjoined to D^tE^uZ to form a 2-HM B,

$$B = \begin{pmatrix} D & D & \cdots & D & E & E & \cdots & E & Z \\ c_1 & c_2 & & c_t & c_{t+1} & c_{t+2} & & c_{t+u} & c_{t+u+1} \end{pmatrix}.$$

Define

$$x = \sum_{j=1}^{t} w(c_j), \quad \text{and} \quad y = \sum_{j=t+1}^{t+u} w(c_j).$$

Clearly x and y satisfy (6-2) to (6-5) by the definitions of B, x and y. The left side of (6-6) counts the total number of ones in the columns of D^tE^u where c has a one by using the fact that B is a 2-HM, and the right side of (6-6) counts the number of ones by using the fact that D and E are 2,1-HM, hence (6-6) holds.

THEOREM (6-7). *Let $A=K^tI^u$ and $f_0'(A)=f\geq 3$. The 2-HM A is maximal if and only if $f-1\nmid t$ and $u<\max(1, r-s)$ where $t=s(f-1)-r$ and $0\leq r\leq f-2$.*

If K^tI^u is not maximal a new row can be adjoined as follows. Using the notation of Theorem (6-1) choose the row c as:

$$c = J_xZ_{tf-x}J_yZ_{uf-y}J_{t(f-1)+u-x-y}.$$

COROLLARY (6-8). *If K^tI^u is maximal then K^{t+u} is maximal.*

THEOREM (6-9). *If P_n is an incidence matrix of a finite projective plane of order n then P_nI^u is maximal if and only if $u\leq n^2-n-1$.*

THEOREM (6-10). *If A is a 2,2-HM with $f_1-2\geq f_2\geq 3$ then AI is maximal.*

THEOREM (6-11). *If A is a 2,2-HM with $f_1\geq f_2+2$ then AJ_t is maximal for all t.*

(7-0). **Nonjuxtaposed** 2-HM. The juxtaposition of any two 2-HM with the same number of rows is again a 2-HM. Therefore given an integer $r\geq 2$ there are an infinite number of 2-HM with r rows. But, as we shall see in this section, every 2-HM with r rows is "column-equivalent" to the juxtaposition of matrices chosen from a finite set of 2-HM with r rows. Two matrices are *column-equivalent* if by a permutation of columns of the first matrix we can obtain the second.

The noncolumn-equivalence condition in the following theorem is necessary. Consider the 2-HM of the form

$$A_u = \begin{pmatrix} J_{1,u} & Z_{1,u} \\ Z_{1,u} & J_{1,u} \end{pmatrix} \qquad u = 1, 2, 3, \ldots.$$

Clearly there are an infinite number of A_u, each A_u has two rows and each is a nonjuxtaposed 2-HM. However, A_u is column-equivalent to I_2^u.

THEOREM (7-1). *For any integer $r\geq 2$ there are only a finite number of 2-HM with exactly r rows which are not column-equivalent to a juxtaposition of two or more 2-HM.*

PROOF. Let A be a 0-1 matrix, and $M=\{1, 2, \ldots, r\}$. For all $S\subseteq M$ define the variable x_S to be the number of columns of A each of which has a one in row i if and only if $i\in S$. For example if

$$A = \begin{pmatrix} 1 & 0 & 0 & 1 & 1 & 0 & 1 & 1 & 1 & 0 & 0 \\ 1 & 0 & 1 & 0 & 1 & 1 & 0 & 1 & 0 & 1 & 0 \\ 1 & 0 & 1 & 1 & 0 & 1 & 1 & 0 & 0 & 0 & 1 \end{pmatrix}$$

then $x_M=1$, $x_\varnothing=1$, $x_{\{1\}}=x_{\{2\}}=x_{\{3\}}=1$ and $x_{\{2,3\}}=x_{\{1,3\}}=x_{\{1,2\}}=2$.

Now, let A be a 2-HM with r rows. The following system of equations and inequalities must be satisfied:

$$(7\text{-}2)\qquad \begin{cases} \text{(i) } f_1 = \sum_{i\in S\subseteq M} x_S & \text{for all } i \in M, \\ \text{(ij) } f_2 = \sum_{\{i,j\}\subseteq S\subseteq M} x_S & \text{for all } \{i,j\} \subseteq M,\ i \neq j, \\ (*)\ f_1, f_2, x_S \geq 0 & \text{for all } S \subseteq M. \end{cases}$$

Above, (i) represents r equations, and (ij) represents $r(r+1)/2$ equations.

Conversely, for any set of integers f_1, f_2, and x_S, $S \subseteq M$, which satisfy (7-2) there exists exactly one 2-HM up to a permutation of columns.

In solving these systems we can work entirely over the field of rational numbers.

All the rational solutions of a system such as (7-2) form a pointed finite cone with its vertex at the origin. The set of integer solutions such that no solution is a nonnegative integral combination of the others we call the set of *indecomposable solutions*. There is a one-to-one correspondence between indecomposable solutions and column-equivalence classes which do not contain juxtaposed 2-HM. The smallest nonzero vector with integer components along any edge of the finite cone is clearly an indecomposable solution. We call these *indecomposable edge solutions*. By the definition of a finite cone every vector in the cone is a nonnegative rational combination of vectors chosen from the cone's edges. Thus in particular every vector in the cone is a nonnegative rational combination of the indecomposable edge solutions. Therefore, the indecomposable solutions which are not on an edge, if any, are nonnegative rational combinations of the indecomposable edge solutions and the coefficients of these combinations must be less than one, otherwise an integral multiple of an indecomposable edge solution could be subtracted. Thus we have that if $\{b_k\}$ is the set of indecomposable edge solutions and c is any indecomposable solution then we can write

$$c = \sum y_k b_k$$

where the sum runs over all indecomposable edge solutions and y_k are rational numbers such that $0 \leq y_k \leq 1$. Hence the length of c is

$$\|c\| = \left\| \sum y_k b_k \right\| \leq \sum |y_k|\, \|b_k\| \leq \sum \|b_k\|.$$

Consequently every indecomposable solution has length less than or equal to the sum of the lengths of the b_k. Thus the number of distinct indecomposable solutions is finite because they have nonnegative integer components with a given bound and the theorem is proved.

(8-0). **Related work.** The only places this author has found something related to 2,0-HM are in [7], in [8], and in [1] in a corollary to a theorem called "the first fundamental theorem of the method of differences." In [2], "adjoinable" is related to maximality and paths through inclusion graphs. See also [4, Chapter 16].

Bibliography

1. R. C. Bose, *On the construction of balanced incomplete block designs*, Ann. Eugenics **9** (1939), 353–399. MR **1**, 199.

2. W. S. Connor, Jr., *On the structure of balanced incomplete block designs*, Ann. Math. Statist. **23** (1952), 57–71; correction: **24** (1953), 135. MR **13**, 617.

3. M. Hall, Jr., "Block designs," in *Applied combinatorial mathematics*, E. F. Beckenbach, (editor), Wiley, New York, 1964, pp. 369–405. MR **30** #4687.

4. ———, *Combinatorial theory*, Blaisdell, Waltham, Mass., 1967. MR **37** #80.

5. H. Hanani, *On quadruple systems*, Canad. J. Math. **12** (1960), 145–157. MR **22** #2558.

6. A. Hurwitz, *Matrices of zeros and ones all of whose pairs of rows are isomorphic*, Dissertation, University of California, Los Angeles, 1965, University Microfilms Inc., Order No. 66-231, 218 pages.

7. P. W. M. John, *Balanced designs with unequal members of replicates*, Ann. Math. Statist **35** (1964), 897–899. MR **28** #5539.

8. V. R. Rao, *A note on balanced designs*, Ann. Math. Statist. **29** (1958), 290–294. MR **20** #401.

9. H. J. Ryser, *Combinatorial mathematics*, Carus Math. Monograph, no. 14, Math. Assoc. of America and Wiley, New York, 1963. MR **27** #51.

IBM Corporation, Los Angeles

THE GREEDY ALGORITHM FOR FINITARY AND COFINITARY MATROIDS

VICTOR KLEE[1]

Introduction. Kruskal's procedure [**15**] for constructing the shortest spanning tree of a finite graph was aptly called the "greedy algorithm" by Edmonds [**9**], who extended it to an arbitrary finite matroid. The extension was made independently by Gale [**11**] and Welsh [**24**]. Here we show that the algorithm applies to two important classes of infinite matroids—namely, the finitary matroids and their duals, the cofinitary matroids. A corollary is that in a finitary or cofinitary matroid, all bases are of the same cardinality. In the finitary case, equipollence of bases was first proved by Rado [**18**] and later by several other authors (Robertson and Weston [**22**], Kertesz [**14**], Bleicher and Preston [**2**], Hughes [**12**], Dlab [**6**]). In the cofinitary case, it appears to be new, though I have heard that it has also been proved recently by D. Bean and D. Higgs.[2]

In addition to the result on the greedy algorithm, we establish some basic though elementary results on the axiomatics of not-necessarily-finite matroids and their duals. These should be compared with the recent work of Brualdi and Scrimger [**4**], and they may also overlap with as-yet-unpublished work of Bean, Higgs, L. M. Brown, P. J. Chase, and G.-C. d'Ambly.[2] The axioms, in terms of closure-like operators, are similar to those of Schmidt [**23**], Marczewski [**16**], Crapo and Rota [**5**], and others. As compared to formulations in terms of other notions (circuits, bases, etc.), the operator approach has the advantage of greater generality and a simpler treatment of duality, though for many applications it is not the most natural approach. The present treatment is self-contained and much of it is expository, consisting of a rearrangement of known material in a more general setting.

The next section contains a statement of the main result, preceded by those definitions required for understanding it. Then comes a section on general matroid axiomatics, followed by a section characterizing certain classes of matroids in terms of their circuits. These sections go considerably beyond what is required for the greedy algorithm. The final section characterizes finitary and cofinitary matroids

[1] Research partially supported by the Office of Naval Research and partially by the Boeing Scientific Research Laboratories.

[2] See the references to Higgs [**26**], [**27**] added in proof.

in terms of their bases and ends with a result (Theorem 9) that describes a procedure (the greedy algorithm) for constructing a certain "best" basis with respect to an arbitrary well-ordering of a finitary or cofinitary matroid.

I am indebted to J. Edmonds and D. R. Fulkerson for several helpful conversations concerning matroids. I apologize for my terminology to H. Crapo and G.-C. Rota, who feel that the term *matroid* is "ineffably cacophonic" **[5]** and would prefer *pregeometry*.

Definitions and statement of main result. Our considerations are relative to a fixed set M. The class of all subsets of M is denoted by 2^M and $'$ denotes complementation. A set $X \subset M$ is *cofinite* provided that X' is finite. A function f on 2^M is *enlarging* provided that $X \subset M \Rightarrow X \subset fX$ and *isotonic* provided that $X \subset Y \subset M \Rightarrow fX \subset fY$. For any enlarging f on 2^M to 2^M there is a unique enlarging g on 2^M to 2^M such that

(D) *For any partition* $M = p \cup X \cup Y$, *either* $p \in fX$ *or* $p \in gY$.

Here $\{p\}$, X, and Y are pairwise disjoint sets covering M, and the "either . . . or" is exclusive. The function g is called the *dual* of f and the system (M, g) is the *dual* of the system (M, f). Plainly g is the dual of f if and only if f is the dual of g.

As the term is used here, an *operator* on M is a function f on 2^M to 2^M such that f is enlarging and isotonic. Note that *the dual of an operator is an operator*. Indeed, if f is enlarging and its dual g is not isotonic there exist $X \subset Y \subset M$ and $p \in gX \sim gY$. But then $p \notin Y$ and with $W = Y \sim X$ there is a partition $M = p \cup W \cup X \cup Z$ for suitably defined Z. From $p \in g\, X$ and $p \notin g\, W \cup X$ it follows by (D) that $p \notin f\, W \cup Z$ and $p \in f\, Z$, whence f is not isotonic. (We sometimes write $p \in f\, Z$ for $p \in fZ$, and similarly with $\notin$ or $\subset$ in place of $\in$. This reduces the need for parentheses. For example, $p \notin f\, W \cup Z$ means $p \notin f(W \cup Z)$.)

An operator f is called, respectively, *finitary*, *cofinitary*, *weakly idempotent*, or *weakly exchanging* provided that the following conditions are satisfied for all $p, x \in M$ and $Y \subset M$:

(C_F) *If* $p \in fY$ *there is a finite* $U \subset Y$ *for which* $p \in fU$.

(H_F) *If* $p \notin fY$ *there is a cofinite* $V \supset Y$ *for which* $p \notin fV$.

(wI) *If* $x \in fY$ *then* $f(x \cup Y) \subset fY$.

(wE) *If* $p \in fY$ *and* $p \notin f\, Y \sim x$ *then* $x \in f\, p \cup (Y \sim x)$.

A *finitary* (resp. *cofinitary*) *matroid* is a system (M, f) such that f is an operator which is finitary (resp. cofinitary), weakly idempotent, and weakly exchanging.

Consider, again, an operator f on M. For $X \subset M$ the points of fX are said to *depend* on X. The set X is *dependent* provided that some point of X depends on the remainder of X; otherwise X is *independent*. The set X is *spanning* provided that $fX = M$; otherwise X is *nonspanning*. Any subset of an independent set is independent and any superset of a spanning set is spanning. The minimal dependent sets are called *circuits* by analogy with graph theory and the maximal nonspanning sets are called *hyperplanes* by analogy with affine geometry. The independent spanning sets are called *bases*; they are simultaneously maximal independent sets and minimal spanning sets.

When the notions of the preceding paragraph are discussed for more than one operator, prefixes are used. If f and g are dual operators on M, the f-independent sets are precisely the complements of the g-spanning sets. Accordingly, the f-circuits are precisely the complements of the g-hyperplanes, the f-bases are precisely the complements of the g-bases, etc.

Our main theorem is the following.

SPECIAL BASES IN FINITARY AND COFINITARY MATROIDS. *Suppose that (M,f) is a finitary matroid, (M, g) is the dual cofinitary matroid, and $\prec$ is an antireflexive well-ordering of M. For each $m \in M$ let $P_m = \{x : x \prec m\}$ and $S_m = \{y : m \prec y\}$. Let*

$$A = \{m \in M : m \notin fP_m\} \quad \textit{and} \quad B = \{m \in M : m \notin gS_m\}.$$

Then A and B are complementary in M, A being an f-basis and B a g-basis. For any f-basis U there is a biunique mapping α of U into A such that $\alpha(u) = u$ for all $u \in U \cap A$ while $\alpha(u) \prec u$ for all $u \in U \sim A$. For any g-basis V there is a biunique mapping β of B into V such that $\beta(b) = b$ for all $b \in B \cap V$ while $\beta(b) \prec b$ for all $b \in B \sim V$.

Thus for an arbitrary well-ordering of M there is an f-basis A that is *first* in a very strong sense and a g-basis B that is *last* in an equally strong sense. (Note that α maps U into A, while β maps B into V.) Implicit in the above theorem is a simple inductive procedure for constructing A and B, and that procedure is here called the *greedy algorithm*. A useful property of A, established in the proof given later, is that

$$A = \{m \in M : m \notin f P_m \cap A\}.$$

Thus to identify a point m of M as belonging to A it is not necessary to show that it does not depend on the set P_m of all its predecessors. It suffices instead, when A is being constructed inductively, to show that m does not depend on those of its predecessors that have already been assigned to A.

Axiomatics and preliminary results. In addition to supplying some specific results that are needed in subsequent sections, the present section is concerned in a more general way with the axiomatics of not-necessarily-finite matroids and their duals.

In conjunction with certain other conditions on a matroid, conditions (C_F) and (H_F) say respectively that circuits are finite and hyperplanes are cofinite. The following two conditions, which are dual to each other, also have important implications for the relationships of the operator f to its circuits and hyperplanes respectively:

(C) *If $p \in fY$ there is a minimal $U \subset Y$ for which $p \in fU$ and U is independent.*
(H) *If $p \notin fY$ there is a maximal $V \supset Y$ for which $p \notin fV$ and $p \cup V$ is spanning.*

In addition to conditions (wI) and (wE), we consider the following weaker or stronger conditions, which are dual in pairs:

(vwI) *If X is finite, Y is independent, and $X \subset fY$, then $f(X \cup Y) \subset fY$.*
(vwE) *If X is finite, $p \cup Y$ is spanning, $p \in f\, Y$, and $p \notin f\, Y \sim X$, then*

$$x \in f p \cup (Y \sim x) \quad \textit{for some } x \in X.$$

(I) *If* $X \subset fY$ *then* $f(X \cup Y) \subset fY$.

(E) *If* $p \in f\, Y$ *and* $p \notin f\, Y \sim X$ *then* $x \in f\, p \cup (Y \sim x)$ *for some* $x \in X$.

As f is isotonic, the final "$\subset$" of (vwI), (wI), and (I) may be replaced by "$=$" and (I) amounts to saying $f(fY) = fY$.

When (K) is any of the conditions or a conjunction of some of the conditions listed above, an operator satisfying (K) is called a *K-operator*.

1. Relations among conditions on operators. *The above conditions on operators are related as follows*:

$$\begin{array}{ll} \mathrm{I} \Rightarrow \mathrm{wI} \Rightarrow \mathrm{vwI}, & \mathrm{E} \Rightarrow \mathrm{wE} \Rightarrow \mathrm{vwE}, \\ \mathrm{wI} \wedge \mathrm{H} \Rightarrow \mathrm{vwI}, & \mathrm{wE} \wedge \mathrm{C} \Rightarrow \mathrm{E}, \\ \mathrm{vwI} \wedge \mathrm{C_F} \Rightarrow \mathrm{I} \wedge \mathrm{C}, & \mathrm{vwE} \wedge \mathrm{H_F} \Rightarrow \mathrm{E} \wedge \mathrm{H}, \\ \mathrm{wI} \wedge \mathrm{H_F} \Rightarrow \mathrm{C}, & \mathrm{wE} \wedge \mathrm{C_F} \Rightarrow \mathrm{H}. \end{array}$$

Proof. For the first row, note that (wI) and (wE) are respectively equivalent to the modifications of (vwI) and (vwE) in which the phrases "Y is independent" and "$p \cup Y$ is spanning" are missing. We remark also that (wI) and (wE) can be stated as follows:

(wI) $$p \in f\, x \cup Y \wedge p \notin f\, Y \Rightarrow x \notin f\, Y,$$

(wE) $$p \in f\, x \cup Y \wedge p \notin f\, Y \Rightarrow x \in f\, p \cup Y.$$

To take care of the second row we show $\mathrm{wE} \wedge \mathrm{C} \Rightarrow \mathrm{E}$; the other implication follows by duality. Let f be a wEC-operator on M and consider p, Y, and X such that $p \in f\, Y$ but $p \notin f\, Y \sim X$. By (C) there is a minimal $U \subset Y$ for which $p \in f\, U$. Independence of U is not needed here. If U misses X then $p \in f\, Y \sim X$ by isotonicity, and as that is contrary to hypothesis there exists $x \in U \cap X$. Then $p \notin f\, U \sim x$ by minimality, and

$$x \in f(p \cup (U \sim x)) \subset f(p \cup (Y \sim X))$$

by (wE) and isotonicity.

To settle the third row we show $\mathrm{vwI} \wedge \mathrm{C_F} \Rightarrow \mathrm{I} \wedge \mathrm{C}$. Let f be a vwI$\mathrm{C_F}$-operator on M and consider p and Y with $p \in f\, Y$. By ($\mathrm{C_F}$) there is a finite $U \subset Y$ for which $p \in f\, U$ and of course U contains a minimal W for which $p \in f\, W$. We claim W is independent. If it is not, there is a partition $W = w_1 \cup W_1$ with $w_1 \in f\, W_1$. If W_1 is dependent there is a partition $W_1 = w_2 \cup W_2$ with $w_2 \in f\, W_2$. As the empty set is independent, we arrive eventually at a partition

$$W = w_1 \cup w_2 \cup \cdots \cup w_k \cup W_k,$$

where W_k is independent and

$$w_k \in f\, W_k,\ w_{k-1} \in f\, w_k \cup W_k, \ldots, w_1 \in f\, w_2 \cup \cdots \cup w_k \cup W_k.$$

By repeated application of (vwI) it then follows that $w_{k-1} \in f\, W_k, \ldots, w_1 \in f\, W_k$, and $p \in f\, W_k$, contradicting the minimality of W. Thus W is independent and $\mathrm{vwI} \wedge \mathrm{C_F} \Rightarrow \mathrm{C}$.

Now again let f be a vwIC$_F$-operator on M, and suppose that $X \subset fY \subset M$ and $p \in f X \cup Y$. By the preceding paragraph there is a finite independent $U \subset X \cup Y$ such that $p \in f U$. Among all such U's, choose one for which $U \cap X$ is minimal. We want $U \cap X = \varnothing$ in order to conclude $p \in f Y$ and establish (I). Suppose there exists $u \in U \cap X$, whence with $u \in f Y$ there is (by the preceding paragraph) a finite independent $V \subset Y$ with $u \in f V$. With $U_1 = (U \sim u) \cup V$, it follows from (vwI) that $fU \subset fU_1$. As $U_1 \cap X \subsetneq U \cap X$, the minimality of $U \cap X$ is contradicted and we conclude vwI $\wedge$ C$_F \Rightarrow$ I.

To take care of the fourth row we show wE $\wedge$ C$_F \Rightarrow$ H. Suppose that f is a C$_F$-operator on M, $p \in M$, $Y \subset M$, and $p \notin f Y$. Let **Z** denote the class of all subsets Z of M such that $p \notin f Y \cup Z$. By (C$_F$), membership in **Z** is a property of finite character and hence by Tukey's Lemma there is a maximal member Z_0 of **Z**. Let $V = Y \cup Z_0$. To establish (H) we want to show $p \cup V$ is spanning. Consider an arbitrary point $m \in M$. Then $p \in f m \cup V$ by the maximality of Z, though $p \notin f V$. It follows from (wE) that $m \in f p \cup V$. □

We add one remark to the implications of Theorem 1. Suppose that p depends on U but not on any proper subset of U. If f satisfies (wI) then plainly U is independent. Hence in the presence of (wI) condition (C) may be replaced by its weaker form in which the final clause "and U is independent" is omitted. An example below shows that (vwI) is insufficient for this purpose. Of course the dual statements apply to (vwE), (wE), and (H).

We earlier defined the terms *finitary matroid* and *cofinitary matroid*, but not the term *matroid* as such. We now define a *matroid* as a system (M, f) such that the operator f is very weakly idempotent (vwI) and very weakly exchanging (vwE). This seems to be about the minimum requirement on f under which any interesting theory can be developed, and it is equivalent to the usual notion of matroid in the finite case. In most instances, additional conditions must be imposed, but as the conditions imposed on a matroid imply different conditions on its dual, it seems advisable to define the term *matroid* in the weak way just described and then add additional conditions as necessary. Additional conditions are indicated by means of prefixes. For example, finitary matroids are defined as wIwEC$_F$-matroids, but in view of Theorem 1 they could be described equivalently as wEC$_F$-matroids or IECHC$_F$-matroids.

Let us pause to consider two instructive examples of dual operators. For the first example, let M be a topological space. For each $X \subset M$ let fX be the closure of X and let gX consist of X together with all points $m \in M$ such that m is interior to $m \cup X$. Then (M, f) is an IwE-matroid and (M, g) is a wIE-matroid. The following conditions on M are equivalent: f is exchanging (satisfies (E)); g is idempotent (satisfies (I)); every closed subset of M is discrete. These conditions are satisfied by the discrete topology on M, as well as by the topology whose closed sets are precisely the finite subsets of M along with M itself.

For the second example, let M be a partially ordered set and for each $m \in M$ let P_m denote the set of all predecessors of M. For each $X \subset M$ let fX consist of X together with all points m such that $P_m \subset X$ and let gX consist of X together with all points m such that P_m intersects X. The dual operators f and g satisfy conditions (E) and (I) respectively. If $M=\{r, s, t\}$ with $P_r=P_s=\varnothing$ and $P_t=\{r, s\}$, then f does not satisfy condition (vwI) even though it does satisfy the special case of (vwI) in which X consists of a single point. Compare this with the behavior of condition (wI). If, on the other hand, P_m is infinite for all $m \in M$, then f satisfies (vwI) and hence (M, f) and (M, g) are dual matroids. Whenever $p \in fY$ there is a minimal $U \subset Y$ for which $p \in fU$, but U is never independent. Compare with the paragraph following the proof of Theorem 1.

The next two results are used in the treatment of the greedy algorithm. As they are dual to each other, it suffices to prove the first.

2. PROPERTIES OF BASES AND CIRCUITS. *Relative to any* wE-*operator, the bases are identical with the maximal independent sets. Any member of a circuit depends on the remaining members of the circuit. If p, Y, and U are as in* (C) *then $p \cup U$ is a circuit.*

2*. PROPERTIES OF BASES AND HYPERPLANES. *Relative to any* wI-*operator, the bases are identical with the minimal spanning sets. The union of any hyperplane and any point of its complement is spanning. If p, Y, and V are as in* (H) *then V is a hyperplane.*

PROOF. Suppose that f is a wE-operator. Consider an arbitrary maximal independent set B and point $p \in B'$. Then $p \in fB$ or there is a point $b \in B$ with $b \in f(B \sim b) \cup p$. In the second instance $p \in fB$ by (wE). It follows that B is spanning and hence is a basis.

Consider next an arbitrary circuit C and point $p \in C$. As C is dependent, there exists $c \in C$ such that $c \in fC \sim c$. If $c \neq p$ it follows by the minimality of C that $c \notin fC \sim \{p, c\}$, whence $p \in fC \sim p$ by (wE).

Suppose, finally, that p, Y, and U are as in condition (C). If $p \cup U$ is not a circuit there exists $w \in U$ such that $p \cup (U \sim w)$ is dependent, whence $u \in f\, p \cup (U \sim \{u, w\})$ for some $u \in U \sim w$. As $u \notin f\, U \sim \{u, w\}$ it follows from (wE) that $p \in f\, U \sim w$, contradicting the minimality of U. ☐

Following Edmonds and Fulkerson [**10**], a class $\mathbf{S}$ of subsets of M is called a *clutter* on M provided that no member of $\mathbf{S}$ contains another member of $\mathbf{S}$. The *complementary clutter* is the class $\mathbf{S}' = \{S' : S \in \mathbf{S}\}$. With any clutter $\mathbf{S}$ on M we associate four operators $\gamma_{\mathbf{S}}$, $\eta_{\mathbf{S}}$, $\beta_{\mathbf{S}}$, and $\bar{\beta}_{\mathbf{S}}$, as defined below. For all $X \subset M$ and $p \in M$:

$p \in \gamma_{\mathbf{S}} X \Leftrightarrow p \in X$ *or there exists* $S \in \mathbf{S}$ *with* $p \in S \subset p \cup X$;

$p \in \eta_{\mathbf{S}} X \Leftrightarrow p \in S$ *whenever* $S \in \mathbf{S}$ *and* $X \subset S$;

$p \in \beta_{\mathbf{S}} X \Leftrightarrow p \in X$, *or there exists* $W \subset X$ *such that* W *is contained in a member of* $\mathbf{S}$ *but* $p \cup W$ *is not contained in any member of* $\mathbf{S}$;

$p \in \bar{\beta}_{\mathbf{S}} X \Leftrightarrow p \in X$, *or every set* W *containing* $p \cup X$ *and containing a member of* S *is such that* $W \sim p$ *also contains a member of* $\mathbf{S}$.

Proofs of the following statements are routine and are left to the reader.

3. Properties of certain operators. *For any clutter* $\mathbf{S}$ *on* M, *the operators* $\gamma_{\mathbf{S}}$ *and* $\eta_{\mathbf{S}'}$ *are dual to each other, as are the operators* $\beta_{\mathbf{S}}$ *and* $\bar{\beta}_{\mathbf{S}'}$. $\gamma_{\mathbf{S}}$ *is an* EC-*operator whose circuits are precisely the members of* $\mathbf{S}$ *and* $\eta_{\mathbf{S}}$ *is an* IH-*operator whose hyperplanes are precisely the members of* $\mathbf{S}$. *If* $\mathbf{S}$ *is nonempty then* $\beta_{\mathbf{S}}$ *is an operator whose maximal independent sets are precisely the members of* $\mathbf{S}$ *and* $\bar{\beta}_{\mathbf{S}}$ *is an operator whose minimal spanning sets are precisely the members of* $\mathbf{S}$.

Matroids in terms of circuits. In connection with the greedy algorithm we require a well-known intersection property of the circuits of a finitary matroid. However, we go beyond that and characterize certain classes of not-necessarily-finitary matroids in terms of their circuits. In the case of finite matroids, Theorem 5 below is essentially Whitney's characterization of matroids in terms of circuits [**25**], while Theorem 4 is a related characterization of A. Lehman and Edmonds [**8**]. For extensions to finitary matroids see Robertson and Weston [**22**], Rado [**19**], and Asche [**1**]. Note that our circuits are not required to be finite. However, Lehman's observation [**8**] that $(\gamma) \Rightarrow (\gamma^*)$ in the case of finite circuits follows from parts of our Theorems 1–2 and 4–5. At the end of this section is an example showing that (γ) does not imply (γ^*) in general.

Recall that wEC-matroids are identical with EC-matroids. We use the shorter term.

4. EC-matroids in terms of circuits. *The circuits of any* EC-*matroid form a clutter satisfying the following conditions*:

(γ) *If* C_1 *and* C_2 *are distinct circuits and* $q \in C_1 \cap C_2$ *there is a circuit* C_3 *with* $C_3 \subset (C_1 \cup C_2) \sim q$.

For any clutter $\mathbf{C}$ *on* M *satisfying this condition,* $\gamma_{\mathbf{C}}$ *is the only operator on* M *that turns* M *into an* EC-*matroid whose circuits are precisely the members of* $\mathbf{C}$.

5. wIEC-matroids in terms of circuits. *The circuits of any* wIEC-*matroid form a clutter satisfying the following condition*:

(γ^*) *If* C_1 *and* C_2 *are circuits with* $p \in C_1 \sim C_2$ *and* $q \in C_1 \cap C_2$ *there is a circuit* C_3 *with* $p \in C_3 \subset (C_1 \cup C_2) \sim q$.

For any clutter $\mathbf{C}$ *on* M *satisfying this condition,* $\gamma_{\mathbf{C}}$ *is the only operator on* M *that turns* M *into a* wIEC-*matroid whose circuits are precisely the members of* $\mathbf{C}$.

Proof. Recall that the conditions (vwI) and (vwE) are embodied in our definition of matroid. Hence the operator f of Theorem 4 is assumed to satisfy conditions (vwI), (wE), and (C), while in Theorem 5 (vwI) is replaced by the stronger condition (wI). The full idempotency condition (I) is not required in either case.

Let C_1 and C_2 be distinct circuits of an EC-matroid (M, f), with $p \in C_1 \sim C_2$ and $q \in C_1 \cap C_2$. From isotonicity and the second part of Theorem 2 it follows that

$$p \in f(C_1 \sim p) \subset f(q \cup ((C_1 \cup C_2) \sim \{p, q\}))$$

and

$$q \in f(C_2 \sim q) \subset f((C_1 \cup C_2) \sim \{p, q\}).$$

If f satisfies (vwI) and the set $(C_1 \cup C_2) \sim \{p, q\}$ is independent, and also if f satisfies (wI), then $p \in f(C_1 \cup C_2) \sim \{p, q\}$, and from condition (C) and the third part of Theorem 2 there follows the existence of a circuit C_3 as in (γ^*). If, on the other hand, the set $(C_1 \cup C_2) \sim \{p, q\}$ is dependent, then (C) and Theorem 2 guarantee the existence of a circuit C_3 satisfying (γ).

Now let $\mathbf{C}$ be a clutter on M satisfying (γ). As was noted earlier, $\gamma_{\mathbf{C}}$ is an EC-operator whose circuits are precisely the members of $\mathbf{C}$. We want to show that $\gamma_{\mathbf{C}}$ satisfies (vwI), and satisfies (wI) when $\mathbf{C}$ satisfies (γ^*). Consider p, X, and Y such that X is finite, $X \subset \gamma_{\mathbf{C}} Y$, and $p \in \gamma_{\mathbf{C}} X \cup Y$. By the definition of $\gamma_{\mathbf{C}}$ there exists $C \in \mathbf{C}$ such that $p \in C \subset p \cup X \cup Y$, and among all such circuits C there is one, C_1, for which $C_1 \cap X$ is minimal. If $C_1 \cap X = \varnothing$ then $p \in \gamma_{\mathbf{C}} Y$, the desired conclusion. Suppose, on the other hand, that there exists $x \in C_1 \cap X$, and choose $C_2 \in \mathbf{C}$ such that $x \in C_2 \subset x \cup Y$. With the aid of (γ^*) we can eliminate x from $C_1 \cup C_2$ to produce $C_3 \in \mathbf{C}$ with

$$p \in C_3 \subset (p \cup C_1 \cup C_2) \sim x,$$

thus contradicting the minimality of $C_1 \cap X$. Using (γ) rather than (γ^*), we can be sure only that

$$C_3 \subset (p \cup C_1 \cup C_2) \sim x,$$

but if Y is independent this implies $p \in \gamma_{\mathbf{C}} Y$ and (vwI) follows.

Finally, consider an arbitrary operator f on M such that (M, f) is an EC-matroid whose circuits are precisely the members of $\mathbf{C}$. Then plainly $fY \supset \gamma_{\mathbf{C}} Y$ for all $Y \subset M$, and the reverse inclusion follows with the aid of Theorem 2. □

Note that any finitary matroid and its dual are both EC-matroids, or, equivalently, both IH-matroids. The following should be compared with the self-dual description of matroids given by Minty [17].

6. Dual EC-matroids in terms of circuits. *If (M, f) and (M, g) are dual EC-matroids their circuits form clutters related as follows*:

(γ^δ) *For any partition $M = p \cup X \cup Y$, either there is an f-circuit C with $p \in C \subset p \cup X$ or there is a g-circuit D with $p \in D \subset p \cup Y$.*

If $\mathbf{C}$ and $\mathbf{D}$ are two clutters on M related by (γ^δ), then $\gamma_{\mathbf{C}}$ and $\gamma_{\mathbf{D}}$ are the only operators on M that turn M into two EC-matroids whose circuits are precisely the members of $\mathbf{C}$ and $\mathbf{D}$ respectively.

Proof. If (M, f) and (M, g) are dual EC-matroids, (γ^δ) follows from the relevant definitions in conjunction with the third part of Theorem 2. For the converse it suffices, in view of Theorem 5, to show that if the clutters $\mathbf{C}$ and $\mathbf{D}$ are related by (γ^δ) then they both satisfy (γ^*). Let $C_1, C_2 \in \mathbf{C}$ with $p \in C_1 \sim C_2$ and $q \in C_1 \cap C_2$. Let $X = (C_1 \cup C_2) \sim \{p, q\}$ and $Y = (X \cup p)'$, so that $M = p \cup X \cup Y$ is a partition of M. If there is no member C_3 of $\mathbf{C}$ with $p \in C_3 \subset p \cup X$ then by (γ^δ) there is a member D of $\mathbf{D}$ with $p \in D \subset p \cup Y$. But then $C_1 \cap D = \{p\}$ and the exclusiveness of the "either . . . or" in (γ^δ) is contradicted. Hence $\mathbf{C}$ satisfies (γ^*) and of course the same reasoning applies to $\mathbf{D}$. □

By dualization, Theorems 4 and 5 yield characterizations of IH-matroids and wEIH-matroids in terms of hyperplanes, and Theorem 6 yields a characterization of dual IH-matroids (equivalently, dual EC-matroids) in terms of hyperplanes. The roles of the operators $\gamma_{\mathbf{C}}$ and $\gamma_{\mathbf{D}}$ are played by operators $\eta_{\mathbf{H}}$ and $\eta_{\mathbf{J}}$, where **H** and **J** are classes of hyperplanes. The duals of conditions (γ), (γ^*), and (γ^δ) are respectively as follows:

(η) *If H_1 and H_2 are distinct hyperplanes and $q \in (H_1 \cup H_2)'$ there is a hyperplane H_3 with $H_3 \supset (H_1 \cap H_2) \cup q$.*

(η^*) *If H_1 and H_2 are hyperplanes with $p \in H_1 \sim H_2$ and $q \in (H_1 \cup H_2)'$ there is a hyperplane H_3 with $p \notin H_3 \supset (H_1 \cap H_2) \cup q$.*

(η^δ) *For any partition $M = p \cup X \cup Y$, either there is an f-hyperplane H with $p \notin H \supset X$ or there is a g-hyperplane J with $p \notin J \supset Y$.*

We close this section with some examples of EC-matroids whose circuits may be infinite. Let S be a set, A a field, and A^S the set of all functions on S to A. A subset Φ of A^S is called *point-finite* provided that for each $s \in S$ the set $\{\phi \in \Phi : \phi(s) \neq 0\}$ is finite; we may then define the function $\sum \Phi \in A^S$ by the condition that

$$\left(\sum \Phi\right)(s) = \sum_{\phi \in \Phi} \phi(s) \quad \text{for all } s \in S.$$

Now let M be an arbitrary subset of A^S, and for each subset X of M let fX consist of X itself together with all functions of the form $p = \sum \Phi$, Φ being an arbitrary point-finite subset of X. It is easily verified that f is a wIwE-operator and hence the system (M, f) is a matroid. To see that f satisfies condition (C), consider $p \in M$ and $X \subset M$ with $p \in (fX) \sim X$—say $p = \sum \Phi$ as described. Let **Z** denote the family of all subsets Ψ of Φ such that $\sum \Psi = 0$. From the point-finiteness of Φ it follows that **Z** includes the union of any linearly ordered subfamily of **Z**, whence by Zorn's lemma **Z** has a maximal member Ψ_0. With $U = \Phi \sim \Psi_0 \subset X$, we have $p \in fU$ and U is minimal with respect to this property.

For an interesting specialization of the above example, let A be the field GF(2) and let M be such that each of its members has the value 1 at two points of S and the value 0 elsewhere. Then M may be regarded, in the usual way, as the arc-set of an undirected graph whose node-set is S. With f as in the preceding paragraph, the f-circuits are the graph's ordinary circuits together with its doubly infinite (simple) paths.

For an example that is related in an interesting way to the example of the preceding paragraph and to the usual circuit matroid of a graph, let S be the node-set and M the arc-set of an undirected graph without loops or slings. Let a subset C of M be called a *circuit* provided that C together with the nodes of its members forms a graph G satisfying the following four conditions:

(i) G is connected;

(ii) each of G's nodes is of G-valence 2, 3, or 4;

(iii) if G is finite, it has precisely one node of valence 4 and none of valence 3, or precisely two of valence 3 and none of valence 4;

(iv) if G is infinite, it has no node of valence 4, at most one of valence 3, and contains at most two node-disjoint singly infinite paths.

These conditions amount to saying that G is topologically equivalent to one of the five graphs shown below, where an arrow indicates a singly infinite path.

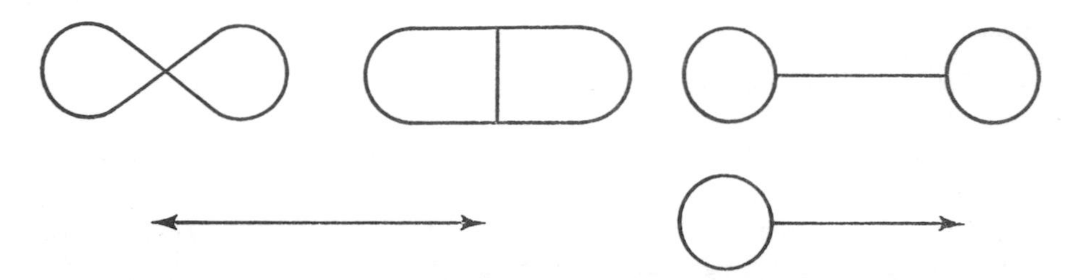

It is a bit tedious but not difficult to verify that these circuits satisfy condition (γ^*).

Finally, we describe, as promised earlier, a clutter satisfying condition (γ) but not (γ^*). Let $\mathbf{C}$ be any clutter of sets (called circuits) and C_0 any member of $\mathbf{C}$ such that the following strengthened form of (γ) is satisfied:

$(\gamma^\S)$ *If C_1 and C_2 are distinct circuits and $q \in C_1 \cap C_2$ there is a circuit $C_3 \neq C_0$ with $C_3 \subset (C_1 \cup C_2) \sim p$.*

Now let p be a point not in any member of $\mathbf{C}$ and let $\mathbf{C}^\S$ consist of the set $C_0 \cup p$ together with all members of $\mathbf{C}$ other than C_0. Then $\mathbf{C}^\S$ satisfies (γ), but plainly does not satisfy (γ^*) if C_0 intersects another member of $\mathbf{C}$. For a specific clutter $\mathbf{C}$ to which this construction is applicable, let $\mathbf{C}$ be the class of all subsets of the real line expressible as the union of a finite number of pairwise disjoint nondegenerate closed intervals whose lengths add up to 1.

As $(\gamma) \Rightarrow (\gamma^*)$ when all circuits are finite, the following is a consequence of the preceding construction: *If $\mathbf{C}$ is the circuit clutter of a finitary matroid and C_0 is any member of $\mathbf{C}$ that intersects another member of $\mathbf{C}$, then there exist members C_1 and C_2 of $\mathbf{C}$ and a point $q \in C_1 \cap C_2$ such that C_0 is the only member of $\mathbf{C}$ contained in $(C_1 \cup C_2) \sim q$.*

Bases in finitary and cofinitary matroids. For any class $\mathbf{S}$ of subsets of a given set M, let $\mathbf{S}^{\not\subset}$ denote the class of all sets $X \subset M$ such that X is maximal with respect to the condition

$$S \in \mathbf{S} \Rightarrow S \not\subset X,$$

and let $\mathbf{S}^{\not\supset}$ denote the class of all sets $X \subset M$ such that X is minimal with respect to the condition

$$S \in \mathbf{S} \Rightarrow S \not\supset X.$$

(For closely related notions see Edmonds and Fulkerson **[10]**.)

7. BASES AND CIRCUITS RELATED BY THE OPERATIONS $^{\not\subset}$ AND $^{\not\supset}$. *Let $\mathbf{S}$ be a clutter on M. If every subset of M not containing any member of $\mathbf{S}$ is contained in a member*

of $\mathbf{S}^{\not\subset}$ *then* $(\mathbf{S}^{\not\subset})^{\not\supset}=\mathbf{S}$. *If every subset of* M *not contained in any member of* $\mathbf{S}$ *contains a member of* $\mathbf{S}^{\not\supset}$ *then* $(\mathbf{S}^{\not\supset})^{\not\subset}=\mathbf{S}$. *Now consider the following conditions on clutters* $\mathbf{B}$ *and* $\mathbf{C}$ *on* M:

(β) *If* $B_1, B_2 \in \mathbf{B}$ *and* $b_1 \in B_1 \sim B_2$ *there exists* $b_2 \in B_2 \sim B_1$ *such that* $b_1 \cup (B_2 \sim B_2) \in \mathbf{B}$;

(γ) *If* $C_1, C_2 \in \mathbf{C}$ *with* $C_1 \neq C_2$ *and* $q \in C_1 \cap C_2$ *there exists* $C_3 \in \mathbf{C}$ *with* $C_3 \subset (C_1 \cup C_2) \sim q$.

If $\mathbf{C}$ *satisfies* (γ) *and* $\mathbf{B}=\mathbf{C}^{\not\subset}$ *then* $\mathbf{B}$ *satisfies* (β). *If* $\mathbf{B}$ *satisfies* (β), $\mathbf{C}=\mathbf{B}^{\not\supset}$, *and every subset of* M *not contained in a member of* $\mathbf{B}$ *contains a member of* $\mathbf{B}^{\not\supset}$, *then* $\mathbf{C}$ *satisfies* (γ).

Proof. As the first two statements are equivalent under complementation, we consider only the first. Suppose that $\mathbf{S}$ is as in the first hypothesis, and consider an arbitrary member S of $\mathbf{S}$. Plainly S is not contained in any member of $\mathbf{S}^{\not\subset}$, but by hypothesis every proper subset of S is contained in a member of $\mathbf{S}^{\not\subset}$. Hence $S \in (\mathbf{S}^{\not\subset})^{\not\supset}$. Now consider, on the other hand, an arbitrary member T of $(\mathbf{S}^{\not\subset})^{\not\supset}$. As T is not contained in any member of $\mathbf{S}^{\not\subset}$, T must contain a member S of $\mathbf{S}$. But of course S is not contained in any member of $\mathbf{S}^{\not\subset}$ and hence from T's minimality it follows that $T=S$.

Now suppose that $\mathbf{C}$ satisfies (γ), let $\mathbf{B}=\mathbf{C}^{\not\subset}$, and consider $B_1, B_2 \in \mathbf{B}$ and $b_1 \in B_1 \sim B_2$. By the maximality of B_2 there exists $C_1 \in \mathbf{C}$ such that $b_1 \in C_1 \subset b_1 \cup B_2$. Plainly $C_1 \not\subset B_1$ and hence there exists $b_2 \in (B_2 \sim B_1) \cap C_1$. Let $B_3 = b_1 \cup (B_2 \sim b_2)$. If B_3 contains a member C_2 of $\mathbf{C}$ then plainly $b_1 \in C_2$ and from (γ) there follows the existence of $C_3 \in \mathbf{C}$ such that

$$C_3 \subset (C_1 \cup C_2) \sim b_1 \subset B_2,$$

an impossibility. To show $B_3 \in \mathbf{B}$ it remains to show that B_3 is maximal relative to not containing any member of $\mathbf{C}$. Consider, then, an arbitrary point $m \in B_3'$. If $m=b_2$ then $m \cup B_3 \supset C_2$. If $m \neq b_2$ then by the maximality of B_2 there exists $C_4 \in \mathbf{C}$ with $m \in C_4 \subset m \cup B_2$. If $b_2 \notin C_4$ then $C_4 \subset m \cup B_3$, while if $b_2 \in C_4$ there follows from (γ) the existence of $C_5 \in \mathbf{C}$ such that

$$C_5 \subset (C_1 \cup C_4) \sim b_2 \subset m \cup B_3.$$

We conclude, therefore, that $\mathbf{B}$ satisfies (β).

Now suppose, on the other hand, that $\mathbf{B}$ satisfies (β), and consider $C_1, C_2 \in \mathbf{C}$ with $C_1 \neq C_2$ and $q \in C_1 \cap C_2$. Suppose there exists $B_2 \in \mathbf{B}$ such that $(C_1 \cup C_2) \sim q \subset B_2$. We will show that $C_1 \sim q \subset C_2$, whence $C_1 \subset C_2$ and the contradiction completes the proof. Consider an arbitrary point $x \in C_1 \sim q$. By the minimality of C_1 there exists $B_1^x \in \mathbf{B}$ with $C_1 \sim x \subset B_1^x$. As $q \in B_1^x \sim B_2$, (β) guarantees the existence of $b_2^x \in B_2 \sim B_1^x$ such that

$$q \cup (B_2 \sim b_2^x) = B_3^x \in \mathbf{B}.$$

Then $C_1 \subset B_3^x$ if $b_2^x \notin C_1$ and $C_2 \subset B_3^x$ if $b_2^x \notin C_2$. As both are impossible, it follows that $b_2^x \in C_1 \cap C_2$. But $C_1 \sim x \subset B_1^x$ and $b_2^x \notin B_1^x$, so it follows that $b_2^x = x$. As x was an arbitrary point of $C_1 \sim q$, we conclude that $C_1 \sim q \subset C_2$. □

The following result is due essentially to Dlab [6], but is proved differently here. See Whitney [25] for the finite case.

8. FINITARY MATROIDS IN TERMS OF BASES. *The bases of any finitary matroid form a clutter satisfying the following two conditions:*

(β) *If B_1, B_2 are bases and $b_1 \in B_1 \sim B_2$, there exists $b_2 \in B_2 \sim B_1$ such that $b_1 \cup (B_2 \sim b_2)$ is a basis;*

(β_ϕ) *If a set is not contained in any basis then it admits a finite subset not contained in any basis.*

For any clutter **B** *on M satisfying these conditions, $\beta_{\mathbf{B}}$ is the only operator on M that turns M into a finitary matroid whose bases are precisely the members of* **B**.

PROOF. Let (M, f) be a finitary matroid whose sets of bases and circuits are respectively **B** and **C**. Then $\mathbf{B} = \mathbf{C}^{\ddagger}$. The clutter **C** satisfies condition (γ) by Theorem 4, whence **B** satisfies (β) by Theorem 7. As the members of **C** are finite, condition (β_ϕ) is readily verified.

Now consider an arbitrary clutter **B** on M satisfying conditions (β) and (β_ϕ), and let $\mathbf{C} = \mathbf{B}^{\ddagger}$. It follows from ($\beta_\phi$) that all members of **C** are finite, from Theorem 7 that $\mathbf{B} = \mathbf{C}^{\ddagger}$ and **C** satisfies condition (γ), and then from Theorem 4 that $\gamma_{\mathbf{C}}$ is the unique operator on M that makes M into a CE-matroid whose circuits are precisely the members of **C**. As the circuits are finite, $(M, \gamma_{\mathbf{C}})$ is in fact a finitary matroid whose bases are precisely the members of **B**. But it is clear from a review of the relevant definitions that $\gamma_{\mathbf{C}} = \beta_{\mathbf{B}}$. To complete the proof, note that if f is any operator on M turning M into a finitary matroid whose circuits are precisely the members of **B**, then the f-circuits are precisely the members of **C** and hence $f = \gamma_{\mathbf{C}}$ by Theorem 4. □

Dlab [6] has given many other characterizations of the bases of finitary matroids. In particular, he has shown that they are characterized by condition (β_ϕ) in conjunction with condition ($\bar{\beta}$) below. However, the following result is merely the dual of Theorem 8.

8*. COFINITARY MATROIDS IN TERMS OF BASES. *The bases of any cofinitary matroid form a clutter satisfying the following two conditions:*

($\bar{\beta}$) *If B_1 and B_2 are bases and $b_1 \in B_1 \sim B_2$ there exists $b_2 \in B_2 \sim B_1$ such that $B_2 \cup (B_1 \sim B_1)$ is a basis;*

(β_ϕ^*) *If a set does not contain any basis then it admits a cofinite superset not containing any basis.*

For any clutter **B** *on M satisfying these conditions, $\bar{\beta}_{\mathbf{B}}$ is the only operator on M that turns M into a cofinitary matroid whose bases are precisely the members of* **B**.

Let us now restate and then prove the main results. Note that their proofs use only a small fraction of the facts that have been assembled up to this point.

9. SPECIAL BASES IN FINITARY AND COFINITARY MATROIDS. *Suppose that* (M, f) *is a finitary matroid,* (M, g) *is the dual cofinitary matroid, and* $\prec$ *is an antireflexive well-ordering of* M. *For each* $m \in M$ *let* $P_m = \{x : x \prec m\}$ *and* $S_m = \{y : m \prec y\}$. *Let*

$$A = \{m \in M : m \notin fP_m\} \quad \text{and} \quad B = \{m \in M : m \notin gS_m\}.$$

Then A *and* B *are complementary in* M, A *being an* f-*basis and* B *a* g-*basis. For any* f-*basis* U *there is a biunique mapping* α *of* U *into* A *such that* $\alpha(u) = u$ *for all* $u \in U \cap A$ *while* $\alpha(u) \prec u$ *for all* $u \in U \sim A$. *For any* g-*basis* V *there is a biunique mapping* β *of* B *into* V *such that* $\beta(b) = b$ *for all* $b \in B \cap V$ *while* $\beta(b) \prec b$ *for all* $b \in B \sim V$.

10. EXISTENCE AND EQUIPOLLENCE OF BASES. *Any finitary or cofinitary matroid admits a basis. All of its bases are of the same cardinality and all of its basis-complements are of the same cardinality.*

PROOF. We begin by deriving Theorem 10 from Theorem 9. As the bases of a matroid are the base-complements of its dual, it suffices to consider bases in finitary and cofinitary matroids. Existence of bases follows from Theorem 9 in conjunction with the well-ordering theorem, or it can be derived directly from the maximality principle. Now note that for any basis C of a matroid (M, h), there are antireflexive well-orderings $\prec_1$ and $\prec_2$ of M such that $C = \{m : m \notin hP_m\}$ relative to $\prec_1$ and $C = \{m : m \notin hS_m\}$ relative to $\prec_2$. If (M, h) is finitary or cofinitary the equipollence of bases therefore follows from the Schröder-Bernstein theorem in conjunction with the existence of the mappings α and β described in Theorem 9.

Turning now to Theorem 9, we show that with A as described,

$$(\dagger) \qquad m \in A' \Rightarrow m \in f A \cap P_m.$$

Suppose, to the contrary, that the set $\{m \in A' : m \notin f A \cap P_m\}$ is nonempty and let x be its first member. Plainly $A \cap P_x \subset f(A \cap P_x)$, while for $y \in P_x \sim A$ the firstness of x implies $y \in f(A \cap P_y) \subset f(A \cap P_x)$. Thus $P_x \subset f(A \cap P_x)$ and with $x \in A'$ we have

$$x \in fP_x \subset ff(A \cap P_x) = f(A \cap P_x),$$

a contradiction proving (†). From (†) it is clear that $fA = M$. If A is f-dependent then A contains a (necessarily finite) f-circuit C and for the last member l of C we have $l \in f(C \sim l) \subset fP_l$, an impossibility with $l \in A$. It follows that A is an f-basis for M and its complement B is a g-basis for M.

To produce the mapping α as described, we could use the known equipollence of bases for finite matroids in conjunction with the P. Hall and M. Hall theorem on existence of systems of distinct representatives. (See Rado **[21]** for an especially simple proof of the latter.) However, we shall instead use a method of Robertson and Weston **[22]**. Consider an arbitrary f-independent subset U of M and let Φ denote the set of all biunique functions ϕ such that $U \cap A \subset D_\phi \subset U$, $U \cap A \subset R_\phi \subset A$, $\phi(u) = u$ for all $u \in U \cap A$, $\phi(u) \prec u$ for all $u \in D_\phi \sim A$, and the set $(U \sim D_\phi) \cup R_\phi$ is f-independent, where D_ϕ and R_ϕ are respectively the domain and the range of ϕ.

The set Φ is partially ordered by inclusion and is nonempty, for the identity function on $U \cap A$ belongs to Φ. As f-independence is a property of finite character, every linearly ordered subset of Φ admits an upper bound and hence by Zorn's lemma there is a maximal member α of Φ. It remains to show that $D_\alpha = U$.

Suppose that $D_\alpha \neq U$. For each $x \in U \sim D_\alpha$ there is a unique f-circuit C_x such that $x \in C_x \subset x \cup (A \cap P_x)$. As the set $(U \sim D_\alpha) \cup R_\alpha$ is f-independent, it does not contain C_x, and hence there exists $a_x \in C_x \cap (A \sim R_\alpha)$. Let $\phi_x = \alpha \cup (x, a_x)$. As the mapping ϕ_x is a biunique extension of α, it follows from α's maximality that the set

$$(U \sim D_{\phi_x}) \cup R_{\phi_x} = (U \sim (D_\phi \cup x)) \cup (R_\phi \cup a_x)$$

is dependent, whence a_x belongs to a circuit contained in the set. Elimination of a_x from the union of this circuit with C_x leads (by means of condition (γ) of the preceding section) to a circuit that is contained in the set $(U \sim D_\alpha) \cup R_\alpha$. But that is impossible, for the set in question is independent.

To establish the existence of the mapping β as described, it suffices to show that for any g-spanning set V there is a mapping ξ of V *onto* B such that $\xi(v) = v$ for all $v \in V \cap B$ and $\xi(v) \succ v$ for all $v \in V \sim B$. Let Ψ denote the set of all functions ψ such that $V \cap B \subset D_\psi \subset V$, $V \cap B \subset R_\psi \subset B$, $\psi(v) = v$ for all $v \in V \cap B$, $\psi(v) \succ v$ for all $v \in D_\psi \sim B$, and the set $(V \sim D_\psi) \cup R_\psi$ is g-spanning. The set Ψ is partially ordered by inclusion and is nonempty, for the identity function on $V \cap B$ belongs to Ψ. As g-spanning is a property of cofinite character (a set has the property if and only if all of its cofinite supersets have the property), every linearly ordered subset of Ψ admits an upper bound and hence by Zorn's lemma there is a maximal member ξ of Ψ. It remains only to show that $R_\xi = B$.

From (†) there follows

$$(\dagger^*) \qquad m \in B \Rightarrow m \notin g\,(B \sim m) \cup S_m.$$

Now suppose that the set $B \sim R_\xi$ is nonempty, and let b_1 be its first member. As the set $(V \sim D_\xi) \cup R_\xi$ is g-spanning and $R_\xi \subset B \sim b_1$, it follows from (†*) that $V \sim D_\xi \not\subset S_{b_1}$; that is, there exists $v \in V \sim D_\xi$ with $b_1 \succ v$. Let $\eta_1 = \xi \cup (v, b_1)$. From the maximality of ξ it follows that the set

$$(V \sim (D_\xi \cup v)) \cup (R_\xi \cup b_1)$$

is g-nonspanning and hence lies in a g-hyperplane H_1. There is a point $b_2 \in B \sim H_1$, and of course $b_2 \succ b_1 \succ v$, whence the set

$$(V \sim (D_\xi \cup v)) \cup (R_\xi \cup b_2)$$

is also g-nonspanning and lies in a g-hyperplane H_2. As the set $(V \sim D_\xi) \cup R_\xi$ is g-spanning, it is clear that $v \in (H_1 \cup H_2)'$ and hence it follows from an earlier result (dual to part of Theorem 4) that there is a g-hyperplane

$$H_3 \supset (H_1 \cap H_2) \cup v \supset (V \sim D_\xi) \cup R_\xi.$$

The contradiction completes the proof. □

There is a result of Rado **[19]** that is related to Theorem 9 in that it is inspired by Kruskal's construction **[15]** and deals with bases of well-ordered finitary matroids. However, it does not establish the existence of the mappings α and β. Examples of matroids whose bases are not equipollent have been given by Bleicher and Marczewski [3], Hughes **[13]**, and Dlab [7]. Of course they are not finitary or cofinitary.

REFERENCES

1. D. S. Asche, *Minimal dependent sets*, J. Austral. Math. Soc. **6** (1966), 259–262. MR **35** #2749.

2. M. N. Bleicher and G. B. Preston, *Abstract linear dependence relations*, Publ. Math. Debrecen **8** (1961), 55–63. MR **24** #A124.

3. M. N. Bleicher and E. Marczewski, *Remarks on dependence relations and closure operators*, Colloq. Math. **9** (1962), 209–212. MR **26** #58.

4. R. A. Brualdi and E. B. Scrimger, *Exchange systems, matchings, and transversals*, J. Combinatorial Theory **5** (1968), 244–257. MR **38** #55.

5. H. Crapo and G.-C. Rota, *Combinatorial geometries*, mimeographed notes, Massachusetts Institute of Technology, 1969.

6. V. Dlab, *Axiomatic treatment of bases in arbitrary sets*, Czechoslovak Math. J. **15 (90)** (1965), 554–563. MR **32** #4060.

7. ———, *The role of the "finite character property" in the theory of dependence*, Comment. Math. Univ. Carolinae **6 (1)** (1965), 97–104. MR **30** #4706.

8. J. Edmonds, *Minimum partition of a matroid into independent subsets*, J. Res. Nat. Bur. Standards **69B** (1965), 67–72. MR **32** #7441.

9. ———, *Matroids and the greedy algorithm*, Proc. Internat. Sympos. on Mathematical Programming, Princeton, 1967 (to appear).

10. J. Edmonds and D. R. Fulkerson, *Bottleneck extrema*, Memorandum RM-5375-PR, The RAND Corporation, 1968.

11. D. Gale, *Optimal assignments in an ordered set: an application of matroid theory*, J. Combinatorial Theory **4** (1968), 176–180. MR **37** #2624.

12. N. J. S. Hughes, *Steinitz' exchange theorem for infinite bases*, Compositio Math. **15** (1963), 113–118. MR **27** #3646.

13. ———, *Steinitz' exchange theorem for infinite bases*. II, Compositio Math. **17** (1966), 152–155. MR **34** #64.

14. A. Kertesz, *On independent sets of elements in algebra*, Acta Sci. Math. (Szeged) **21** (1960), 260–269. MR **24** #A3114.

15. J. B. Kruskal, Jr., *On the shortest spanning subtree of a graph and the travelling salesman problem*, Proc. Amer. Math. Soc. **7** (1956), 48–49. MR **17**, 1231.

16. E. Marczewski, *Fermeture generalisée et notions d'independence*, C.R. du Sympose Archimedien à Syracuse, 1964.

17. G. J. Minty, *On the axiomatic foundations of the theories of directed linear graphs, electrical networks, and network programming*, J. Math. Mech. **15** (1966), 485–520. MR **32** #5543.

18. R. Rado, *Axiomatic treatment of rank in infinite sets*, Canad. J. Math. **1** (1949), 337–343. MR **11**, 238.

19. ———, *Note on independence functions*, Proc. London Math. Soc. (3) **7** (1956), 300–320. MR **19**, 522.

20. ———, *Abstract linear dependence*, Colloq. Math. **14** (1966), 257–264. MR **32** #2362.

21. ———, *Note on the transfinite case of Hall's theorem on representatives*, J. London Math. Soc. **42** (1967), 321–324. MR **35** #2758.

22. A. P. Robertson and J. D. Weston, *A general basis theorem*, Proc. Edinburgh Math. Soc. **11** (1959), 139–141. MR **22** #17.

23. J. Schmidt, *Mehrstufige Austauschstrukturen*, Z. Math. Logik Grundlagen Math. **2** (1956), 233–249. MR **19**, 377.

24. D. J. Welsh, *Kruskal's theorem for matroids*, Proc. Cambridge Philos. Soc. **64** (1968), 3–4. MR **37** #2623.

25. H. Whitney, *On the axiomatic properties of linear dependence*, Amer. J. Math. **57** (1935), 509–533.

26. D. A. Higgs, *Matroids and duality*, Colloq. Math. **20** (1969), 215–220.

27. ———, *Equicardinality of bases in B-matroids*, Canad. Math. Bull. **12** (1969), 861–862.

UNIVERSITY OF WASHINGTON

COLLECTIONS OF SUBSETS CONTAINING NO TWO SETS AND THEIR UNION

DANIEL KLEITMAN[1]

Suppose we have a collection of subsets of a finite set S, no member of which is the union of two others. Erdös has raised the question: What limitation on the size of this collection can we deduce from this restriction?

It is immediately clear that a collection of subsets which are all the same size automatically satisfies this constraint, and there can be $\binom{n}{[n/2]}$ members of such a collection, if our set S has n elements. Erdös conjectured that the binomial coefficient $\binom{n}{[n/2]}$ is asymptotic to the maximal size of collections having this property, as n increases.

In a previous communication [1], the present author was able to show that no collection having the property indicated above could have more than $2\, 2^{1/2}\binom{n}{[n/2]}$ members. Below we verify Erdös' original conjecture, showing in fact that a bound to the size of a collection containing no two members and their union is given by

$$\binom{n}{[n/2]}(1+2n^{-1/2}(\log n)^{1/2}+O(n^{-1/2})).$$

The present argument makes use of the following ideas. Let F be a collection of subsets satisfying the restriction in question, so that none of its members is the union of two others. Let A be a member of F. Then if $B_1, \ldots, B_p$ are the members of F that are contained in A, we must have $B_i \cup B_j \neq A$ which means that

$$(A-B_i) \cap (A-B_j) \neq \varnothing .$$

Thus the collections $\{A-B_i\}$ consist of nondisjoint subsets of A.

Now Erdös, Ko and Rado [2] have shown that a collection of k-element nondisjoint subsets of an n-element set can have, if $2k \leq n$, at most $\binom{n-1}{k-1}$ members. Thus only a proportion $\binom{n-1}{k-1}/\binom{n}{k}$ or k/n of all k-element subsets can lie in any such collection.

What all this implies here is that if A is in F, then only a very small proportion of the subsets of A which are almost as big as A can lie in F, for such subsets have complements in A whose size will be small compared to that of A, and which

[1] This research was supported in part by NSF GP-7477.

cannot be disjoint. This is the property possessed by members of F which we shall exploit below.

We proceed in three steps. First we remove from F all members which do not have roughly $n/2$ elements in them. Then we define a class of subfamilies of F each having no more than $\binom{n}{[n/2]}$ members. We finally use the results of the last paragraph to show that each member of F with roughly $n/2$ elements in it must lie in most of these subfamilies. From these considerations a size limitation on F naturally emerges.

With F given as above, we define $F(\Delta_1, \Delta_2)$ to consist of those members of F having at least Δ_1 and at most Δ_2 elements in them.

It is a well-known consequence of a theorem of Sperner [3] that the subsets of S may be partitioned into $\binom{n}{[n/2]}$ disjoint chains, where a chain is a collection of subsets that is totally ordered by inclusion. Let P be any such partition. For convenience only we let $\bar{P}$ consist of the chains in P which have a Δ_1-element member along with any collection of chains obtained by adding an appropriate Δ_1-element member to each chain in P which lacks such a member.

Each permutation π of the members of S induces a permutation on the subsets of S and hence on collections of subsets of S. We denote by $\pi(P)$ the partition obtained by acting with permutation π on partition P and we denote the collection obtained when π acts on $\bar{P}$ by $\pi(\bar{P})$. For each A in S we denote the chain in $\pi(\bar{P})$ that contains A already in $\pi(P)$ by $C(\pi, A)$.

Finally, we define $F(\pi, \Delta_1, \Delta_2)$ to consist of the members A of $F(\Delta_1, \Delta_2)$ which are least members of

$$F(\Delta_1, \Delta_2) \cap C(\pi, A).$$

Thus each member of $F(\Delta_1, \Delta_2) - F(\pi, \Delta_1, \Delta_2)$ contains some other member of its chain in $\pi(\bar{P})$ which lies in $F(\Delta_1, \Delta_2)$.

We can immediately conclude since $F(\pi, \Delta_1, \Delta_2)$ can contain at most one member of each chain, that it can have no more than $\binom{n}{[n/2]}$ members. Thus we have

$$\sum_{\pi} |F(\pi, \Delta_1, \Delta_2)| \leq \binom{n}{[n/2]} n!.$$

Let A be a k-element member of $F(\Delta_1, \Delta_2)$. Of the $\binom{k}{\Delta_1}$ Δ_1-element subsets of A at most $\binom{k-1}{\Delta_1}$ can be contained in members of F that are contained in A, by the Erdös-Ko-Rado theorem. (We choose Δ_1 and Δ_2 so that $2\Delta_1 \geq \Delta_2$; their complements in A must be nondisjoint.) Since the permutation group in A is transitive among the Δ_1-element subsets of A, each Δ_1-element subset of A will be in the same chain as A in $\pi(\bar{P})$ for exactly $n!/\binom{k}{\Delta_1}$ permutations π. Thus A can be absent from $F(\pi, \Delta_1, \Delta_2)$ for at most $\binom{k-1}{\Delta_1} n!/\binom{k}{\Delta_1}$ permutations, as A can only be so absent if the Δ_1-element subset in $C(\pi, A)$ is contained in a B in F that is properly contained in A.

Hence each k-element member of F must be in at least $n!(1-\binom{k-1}{\Delta_1}/\binom{k}{\Delta_1})$ collections $F(\pi, \Delta_1, \Delta_2)$. We can conclude, therefore, if the number of k-element members of F is $|F_k|$, that

$$\sum_{k=\Delta_1}^{\Delta_2} n!\left(1-\binom{k-1}{\Delta_1}\Big/\binom{k}{\Delta_1}\right)|F_k| \le n!\binom{n}{[n/2]}$$

which implies since $\binom{k-1}{\Delta_1}/\binom{k}{\Delta_1}=(1-\Delta_1/k)$ that

$$|F(\Delta_1,\Delta_2)| = \sum_{k=\Delta_1}^{\Delta_2} |F_k| \le \binom{n}{[n/2]}\frac{\Delta_2}{\Delta_1}.$$

If $[n/2]<\Delta_1<\Delta_2$ or $\Delta_1<\Delta_2<[n/2]$ then by a similar argument we obtain

$$|F(\Delta_1,\Delta_2)| \le \frac{\Delta_2}{\Delta_1}\cdot\binom{n}{\Delta_1} \quad\text{and}\quad |F(\Delta_1,\Delta_2)| \le \frac{\Delta_2}{\Delta_1}\cdot\binom{n}{\Delta_2}$$

respectively. If we choose Δ_1, Δ_2 to be $[n/2\pm\frac{1}{2}(\log n)^{1/2}n^{1/2}]$ we obtain that

$$|F(\Delta_1,\Delta_2)| \le \binom{n}{[n/2]}(1+2(\log n)^{1/2}n^{-1/2}+O(n^{-1/2})).$$

Similarly for $F-F(\Delta_1,\Delta_2)$ we find that

$$|F-F(\Delta_1,\Delta_2)| \le (2+\varepsilon)\cdot\binom{n}{[n/2-(\log n)^{1/2}n^{1/2}/2]}.$$

This latter number is of the order of $\binom{n}{[n/2]}O(n^{-1/2})$. We conclude that F can have no more than

$$\binom{n}{[n/2]}(1+2(\log n)/n)^{1/2}+O(n^{-1/2}))$$

members. It is apparent that this bound on the $|F|$ can be slightly improved by more detailed argument. Lower bounds for the maximal size of a collection restricted as in F here have so far been of the order of $\binom{n}{[n/2]}(1+O(1/n))$.

BIBLIOGRAPHY

1. D. Kleitman, *On a combinatorial problem of Erdös*, Proc. Amer. Math. Soc. **17** (1966), 139–141. MR **32** #2337.

2. P. Erdös, Chao Ko and R. Rado, *Intersection theorems for systems of finite sets*, Quart. J. Math. Oxford Ser. **72** (48) (1961), 313–320. MR **25** #3839.

3. E. Sperner, *Ein Satz über Untermengen einer endlichen Menge*, Math. Z. **27** (1928), 544–548.

MASSACHUSETTS INSTITUTE OF TECHNOLOGY

A COMBINATORIAL METHOD FOR EMBEDDING A GROUP IN A SEMIGROUP

N. S. MENDELSOHN

1. **Introduction.** A group can always be represented by a set of bijective mappings from a set S to S. A semigroup can always be represented by a set of injective mappings from a set S to S. One such faithful representation can always be taken as the right multiplications in the case of a group—in the case of semigroups it may first be necessary to adjoin a unit element and then use the right multiplications.

A bijective mapping of a finite set S to itself can be represented as a directed graph which consists of a set of disjoint cycles with each vertex of out-valence and in-valence equal 1. An injective mapping of a finite set S to S can be represented as a directed graph with each vertex of out-valence equal 1.

Such a graph is a disjoint union of unilaterally connected directed graphs each of which consists of a cycle (loops are allowed) to which may be attached a number of trees. Each tree has associated with it a *vertex of attachment* to a cycle with the property that from any point on the tree there is a unique directed path along the tree to the vertex of attachment. The following figure illustrates this situation.

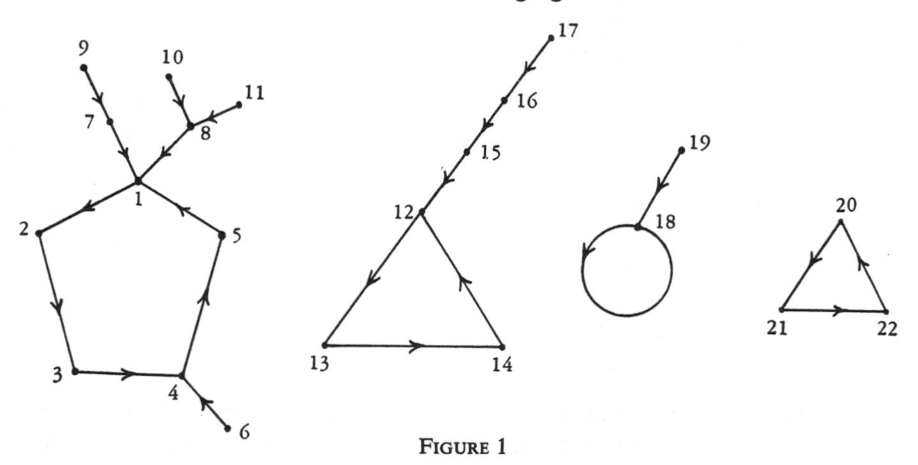

FIGURE 1

It corresponds to the injective mapping $1 \to 2$, $2 \to 3$, $3 \to 4$, $4 \to 5$, $5 \to 1$, $6 \to 4$, $7 \to 1$, $8 \to 1$, $9 \to 7$, $10 \to 8$, $11 \to 8$, $12 \to 13$, $13 \to 14$, $14 \to 12$, $15 \to 12$, $16 \to 12$, $17 \to 12$, $18 \to 18$, $19 \to 18$, $20 \to 21$, $21 \to 22$, $22 \to 20$.

To each vertex of such a graph we assign a *weight*, namely, the length of the unique directed path from the vertex to the point of attachment for points not on a cycle and 0 to points on a cycle. For example the vertices 3, 4, 18 have weight 0, the vertices 6, 7, 8, 19 have weight 1, the vertices 9, 10, 11 have weight 2 and 17 has weight 3. We also note that on the subset of points of weight 0, our mapping is bijective.

In this paper, the above considerations are applied to presentation theory of groups and semigroups. The types of consideration in which we are interested are those which yield readily to computing machine programming. Such a program exists at the University of Manitoba for the IBM 360/65.

2. **Presentations.** Consider a semigroup presentation

$$\text{sgp}\,(a_1, a_2, \ldots, a_r \mid W_1 = W_1^1, W_2 = W_2^1, \ldots, W_n = W_n^1)$$

where $a_1, a_2, \ldots, a_r$ are generators and $W_1, W_2, \ldots, W_n$; $W_1^1, W_2^1, \ldots, W_n^1$ are words in $a_1, a_2, \ldots, a_r$. We may also consider this as a group presentation

$$\text{gp}\,(A_1, A_2, \ldots, A_r \mid \tilde{W}_1 = \tilde{W}_1^1, \tilde{W}_2 = \tilde{W}_2^1, \ldots, \tilde{W}_n = \tilde{W}_n^1)$$

where $\tilde{W}_i$, $\tilde{W}_i^1$ are obtained from W_i, W_i^1 by replacing the small letters by the corresponding capitals. Suppose now that both the semigroup and group defined by the presentation are finite. Then the mapping $a_i \to A_i$ defines an epimorphism, since the elements of a finite group may be written as words in the generators without the use of negative powers. There are ways in which one can go from a presentation of a finite semigroup to a faithful representation of the semigroup by mappings of a finite set into itself. Hence it is natural to ask the question of how to determine which groups are homomorphic images of a semigroup when the generators of the semigroup are given as mappings of a finite set into itself and, in particular, when is the semigroup actually a group. The following theorems yield criteria which are very easy to apply.

THEOREM 1. *Let $x_1, x_2, \ldots, x_r$ be mappings of the set $S=\{1, 2, 3, \ldots, n\}$ into itself. Necessary and sufficient conditions that $x_1, x_2, \ldots, x_n$ generate a group are:*

(1) *The subset of points of weight* 0 *is the same for each of the mappings $x_1, x_2, \ldots, x_r$.*

(2) *The remaining points, if any, have weight* 1, *in each of the mappings $x_1, x_2, \ldots, x_r$.*

(3) *For each point k of weight* 1 *if x_i: $k \to k_i$ and j_i is the unique point of weight* 0 *such that x_i: $j_i \to k_i$, then $j_1 = j_2 = \cdots = j_r$.*

PROOF. Suppose, first, that $x_1, x_2, \ldots, x_r$ generate a group which, of necessity, is finite. Let this group have identity I. If the point k has weight greater than 1 in the mapping x_i, then let x_i: $k \to m$. No other power of x_i maps $k \to m$. Hence $x_i \neq x_i^u$ for $u=2, 3, 4, \ldots$, which contradicts the fact that $x_1, x_2, \ldots, x_r$ generate a group. Hence, all points have weight 1 or 0 in each of the mappings. Let S^* be the set of points of weight 0 in the mapping x_i. The mapping x_i is a bijection on the points of S^*. Let d be the greatest common divisor of the lengths of cycles of the mapping x_i. Then the mappings $x_i, x_i^2, x_i^3, \ldots, x_i^{d-1}$ are all distinct since they are distinct on S^*. Also x_i^d is the identity map on S^* and it is easily verified that

$x_i^{d+1} = x_i$ on the whole set S. Since the x_i generate a group, $I = x_i^d$. Since points of weight 1 are never mapped into themselves by any of the x_i it follows that S^* is the set of all elements in S which are fixed by the identity. Hence, S^* is the same for all generators $x_1, x_2, \ldots, x_n$. Hence $S - S^*$ is the set of all points of weight 1, for each of the generators $x_1, x_2, \ldots, x_r$. Now let $k \in S - S^*$. By the definition of k_i and j_i, $x_i: k \to k_i$ and $j_i \to k_i$. Hence $I = x_i^d$ maps $k \to j_i$. Hence $j_1 = j_2 = \cdots = j_r$. Conversely, let $x_1, x_2, \ldots, x_r$ be mappings which satisfy (1), (2), (3). Denote the set of points of weight 0 by S^*, and for each of the maps $x_1, x_2, \ldots, x_r$ let $(x_i | S^*)$ be the restriction of the map x_i to the set S^*. The mappings $(x_i | S^*)$ are permutations of S^* and thus generate a group. Put $x_i^* = (x_i | S^*)$. The mapping

$$w = w(x_1, x_2, \ldots, x_r) \to w(x_1^*, x_2^*, \ldots, x_r^*) = w^*$$

is an epimorphism of a semigroup to a group. We show that it is also a monomorphism. We need only show that if $w_1(x_1^*, x_2^*, \ldots, x_r^*) = w_2(x_1^*, x_2^*, \ldots, x_r^*)$ then if k is any point in $S - S^*$, $w_1(x_1, x_2, \ldots, x_r)$ and $w_2(x_1, x_2, \ldots, x_r)$ map k into the same point. Let $w_1 = x_i w_2$ and $w_2 = x_m w_3$. Then $x_i^* w_2^* = x_m^* w_3^*$. Now, using the notation kw for the image of k under the mapping w we have $kx_i w_2 = j_i x_i w_2 = j_i x_i^* w_2^* = j_i x_m^* w_3^* = j_m^* x_m^* w_3^* = k x_m w_3$. Hence, our theorem is proved.

EXAMPLE.

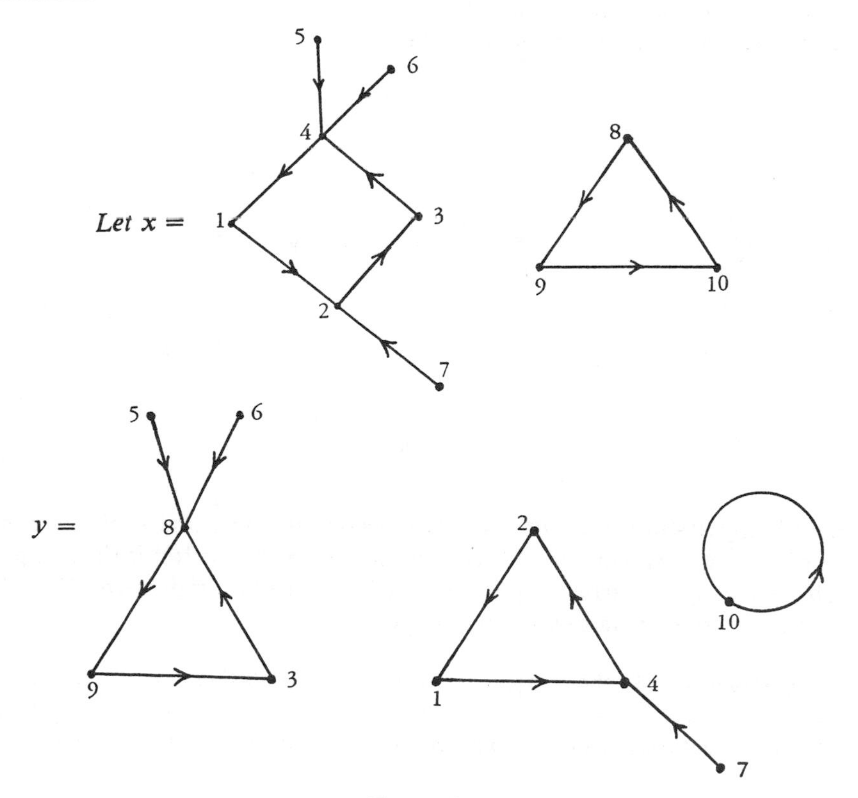

FIGURE 2

The graph makes it instantly clear that conditions (1), (2), (3) are satisfied and that x and y generate a group isomorphic to the permutation group generated by $x^* = (1234)(8910)$, $y^* = (389)(142)(10)$.

THEOREM 2. *If mappings $x_1, x_2, \ldots, x_r$ of the set $S = \{1, 2, 3, \ldots, n\}$ into itself are such that the set S^* of elements of weight 0 is the same for each of $x_1, x_2, \ldots, x_r$, then $x_1, x_2, \ldots, x_r$ generate a semigroup such that $(x_1|S^*), (x_2|S^*), \ldots, (x_r|S^*)$ is a group which is the largest homomorphic image of the semigroup.*

PROOF. If we put $x_i^* = (x_i|S^*)$ it can be shown that any relation in the group of the form $\{w(x_1^*, x_2^*, \ldots, x_n^*)\}^k = I$ comes from a relation $\{w(x_1, x_2, \ldots, x_n)\}^{k+s} = \{w(x_1, x_2, \ldots, x_n)\}^s$ in the semigroup. The minimum value of the integer s is the maximum weight of a point in the mapping of S corresponding to the word $w(x_1, x_2, \ldots, x_n)$. This proves our theorem.

Mappings in which there are points which have 0 weight for one generator and nonzero weight for another generator are extremely complicated. In general, a group which is a homomorphic image of such a semigroup is relatively small, as it is usually possible to obtain a word in the generators for which the corresponding mapping has a small subset of S in its range. When the range can be reduced to a single point, the semigroup has right zeros and hence any group which is a homomorphic image is the identity.

As an illustration consider the mappings

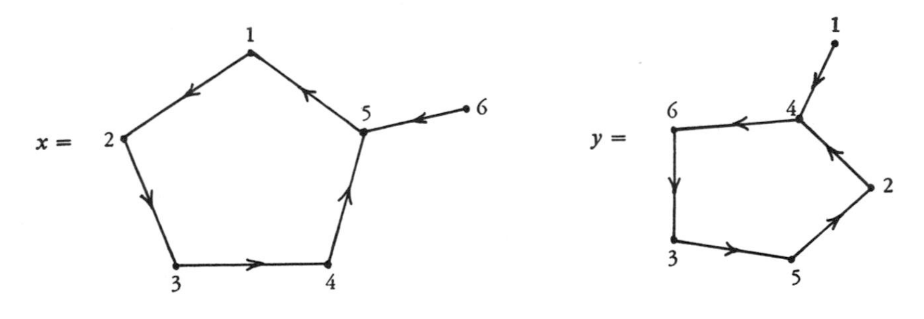

FIGURE 3

Here 1 appears in a cycle of x while its weight is 1 in y and similarly 6 is in a cycle of y but has weight 1 in x. It is easily computed that $(x^5y)^4$ is the mapping in which each integer is carried into 6, and so $(x^5y)^4$ must be a right zero. We will give no further consideration to such mappings.

3. **Construction of the mappings from a presentation.** If a presentation sgp $(a_1, a_2, \ldots, a_r \mid W_1 = W_1^*, \ldots, W_i = W_i^*)$ yields a finite semigroup its right regular representation can be obtained by a modification of the Todd-Coxeter

procedure, see Neumann [2]. We consider only the generalization which corresponds to counting cosets of the identity for groups. In this representation, if the semigroup is of order t, the generators are always given by mappings of a set of $t+1$ points into themselves. These mappings are never onto as 1 does not appear as an image. We then can use Theorem 1 to determine if, in fact, the semigroup is a group. In the case where the semigroup is not a group we can in many cases use Theorem 2 to find a faithful representation of the group given by the presentation.

EXAMPLE 1. In [2] Neumann obtains the representation for

$$\text{sgp}\,(a, b \mid ba = a^2b, b^2 = a^3)$$

as the mappings:

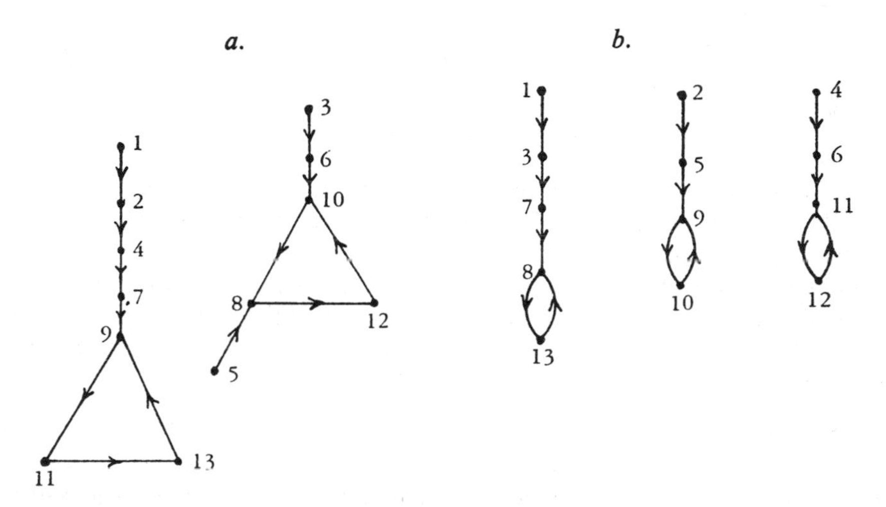

FIGURE 4

The semigroup is of order 12 while the group is of order 6. A regular representation for the group is obtained by restricting the mappings to the set

$$S^* = \{8, 9, 10, 11, 12, 13\}.$$

EXAMPLE 2. In [1], Buck considers the semigroup

$$\text{sgp}\,(a, b \mid aba = b, bab = a^r).$$

A straightforward application of the Neumann modification of the Todd-Coxeter procedure yields that the semigroup is of order $5r+3$. When we represent the mappings graphically we obtain the following. The generator b is represented by $(r+1)$ 4-cycles of which r have a single extra vertex attached to it, while the generator a is represented by two cycles of length $2(r+1)$, one of which has a tail of length r attached to it. The corresponding group is of order $4r+4$. The case $r=3$ is illustrated below.

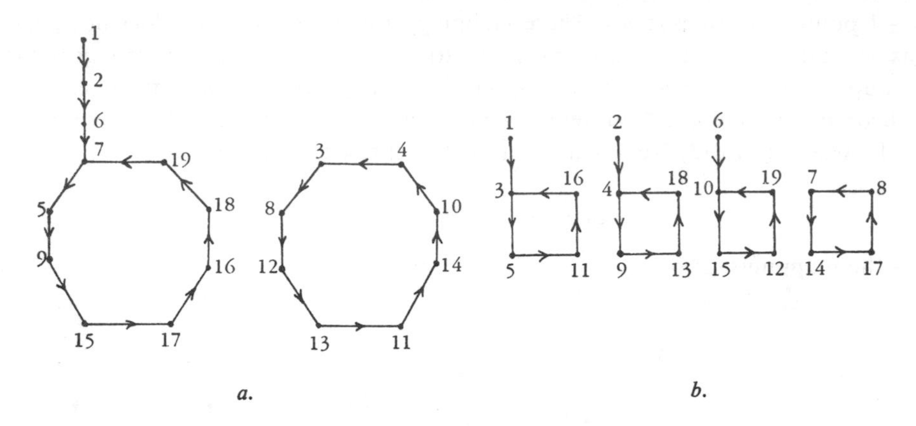

FIGURE 5

The examples of Neumann and Buck illustrate an interesting phenomenon, namely, that the semigroup defined by the presentation has the group defined by the corresponding presentation as a subgroup as well as a homomorphic image. It is interesting to see how one can detect this in the associated graphs. We give here a set of sufficient conditions which are easily verified. In this case it is easy to locate the words which are elements of the subgroup. The following theorem covers the situation.

THEOREM 3. *Let* $A=\text{sgp}\,(x_1, x_2, \ldots, x_r \mid W_1=W_1^1, \ldots, W_n=W_n^1)$ *and let* $B=\text{gp}\,(x_1, x_2, \ldots, x_r \mid W_1=W_1^1, \ldots, W_n=W_n^1)$. *Suppose A is finite and of order $m-1$ and suppose the right regular representation of A is given by mappings of $S=\{1, 2, \ldots, m\}$ into itself (here $2, 3, \ldots, m$ are used instead of the elements of A). Suppose the following conditions hold*:

(a) *The set S^* of points of weight 0 is the same for each of $x_1, x_2, \ldots, x_r$. (We identify the $x_1, x_2, \ldots, x_r$ with their corresponding mappings on S.)*

(b) *Let k be any point in $S-S^*$.*

For each mapping x_i we will associate with k a unique point j_i in S^* as follows. Let k_i be the vertex of attachment of the tree containing k. If the path from k to k_i is of length m, let j_i be the unique vertex in the cycle containing k_i such that the path from j_i to k_i is of length m. Assume now that the mappings $x_1, x_2, \ldots, x_r$ have the property that for each $k \in S-S^*$, $j_1=j_2=\cdots=j_r$.

Then B is a subgroup of A.

PROOF. Let $x_1^*, x_2^*, \ldots, x_r^*$ be the restriction of $x_1, x_2, \ldots, x_r$ to the set S^*. Then $x_1^*, x_2^*, \ldots, x_r^*$ generate a group isomorphic to B. Suppose $x_i^{d_i+m_i}=x_i^{m_i}$ with d_i and m_i minimal. Let $n_i=c_id_i$ be the first multiple of d_i such that $n_i \geqq m_i$. Then the mapping $x_i^{n_i}$ has range S^* and $x_i^{*n_i}$ is the identity map on S^*. Furthermore, from condition (b) it follows that

$$x_1^{n_1} = x_2^{n_2} = \cdots = x_r^{n_r} = \beta \text{ (say)}.$$

β has the property that $\beta^2=\beta$, and furthermore $\beta x_i=\beta x_i^*$ for $i=1, 2, \ldots, r$ and $(\beta w_i)(\beta w_j)=\beta w_i w_j$ for any words w_i, w_j. Also β^* is the identity map on S^*. (Note that the expression βx_i^* makes sense since the range of β is S^*.) Consider the set $\tilde{A}$ of all words in $x_1, x_2, \ldots, x_r$ which are equal to a word of the form $\beta w(x_1, x_2, \ldots, x_r)$. These words, in fact, are all the mappings whose range is S^*. The mapping from $\tilde{A}$ to β given by $\beta w(x_1, x_2, \ldots, x_n) \to w(x_1^*, x_2^*, \ldots, x_n^*)$ is now, trivially, an isomorphism.

COROLLARY. *If the right regular representation of*

$$\operatorname{sgp}(x_1, x_2, \ldots, x_r \mid W_1 = W_1^1, \ldots, W_n = W_n^1)$$

satisfies the conditions of Theorem 3 *and the numbers d_i are defined as in the proof of the theorem then*

$$\operatorname{sgp}(x_1, x_2, \ldots, x_r \mid W_1 = W_1^1, \ldots, W_n = W_n^1, x_1^{d_1+1} = x_1, \ldots, x_r^{d_r+1} = x_r)$$

defines a group isomorphic to $\operatorname{gp}(x_1, x_2, \ldots, x_r \mid W_1 = W_1^1, \ldots, W_n = W_n^1)$.

PROOF. The added relations $x_i^{d_i+1}=x_i$ assure that in the graphs of the mapping no vertex is of weight greater than 1. Theorems 1 and 3 now imply the result.

EXAMPLES. An examination of Figures 4 and 5 reveals that both the Neumann and Buck examples satisfy the conditions of Theorem 3. Note that conditions (a) and (b) are easy to verify. Now let us consider further the Buck example $\operatorname{sgp}(a, b \mid aba=b, bab=a^3)$. Returning to Figure 5, we can associate with each integer i, $i=2, 3, \ldots, 19$, a word in the semigroup which maps $1 \to i$. These words are all the elements of the semigroup. Such a list is $a, b, ab, b^2, a^2, a^3, ba, ab^2, a^2b, b^3, ba^2, ab^3, a^3b, a^2b^2, b^4, a^2b^2a, b^4a, b^4a^2$. If we delete the elements a and a^2, the remaining elements are all the elements of $\operatorname{gp}(a, b \mid aba=b, bab=a^3)$.

Theorem 3 yields a method for embedding a group G of order n in a semigroup H of order $n+m$ such that G is a homomorphic image of H as well as a subgroup. Our construction can be carried out in many ways, and each way is such that G is the largest group which is a homomorphic image of H. Furthermore the construction automatically gives the right regular representation of H.

The construction is as follows. Take a set of generators of G and for each generator draw the graph corresponding to the regular representation. To obtain H and its right regular representation, we add to each graph $m-n+1$ vertices as follows. Start with one of the vertices in the graph of the first generator and adjoin a single chain containing these new vertices to this vertex. These new vertices are then attached to the graphs corresponding to the other generators in such a way that condition (b) of Theorem 3 is satisfied. In general, there are many ways of doing this, but, in any case, there is a unique way if we add the extra condition that the weights of all the new vertices equal 1 for the graphs of the remaining generators. We now have the regular representation for H.

The following figure shows how the quaternion group is embedded in a semigroup of order 12 by means of this construction.

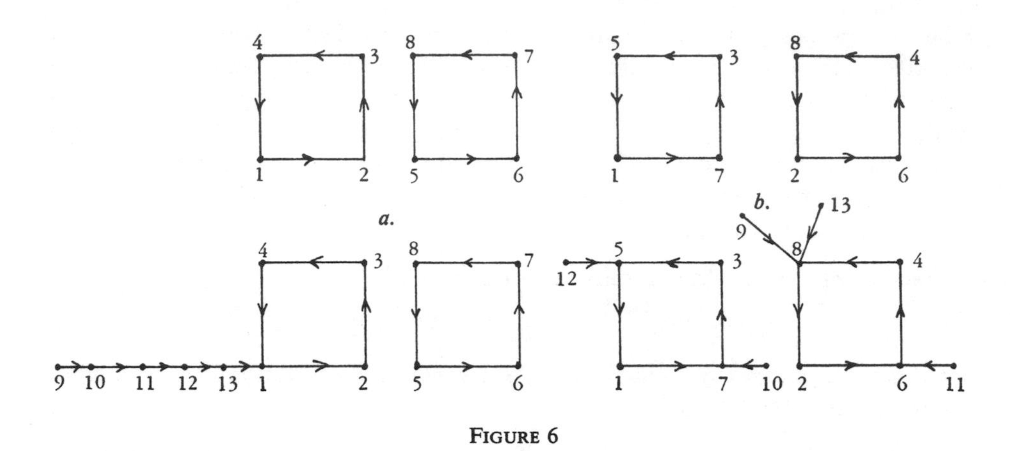

FIGURE 6

REFERENCES

1. R. C. Buck, *On certain decidable semigroups*, Notices Amer. Math. Soc. **15** (1968), 370. Abstract #68T-237.

2. B. H. Neumann, *Some remarks on semigroup presentations*, Canad. J. Math. **19** (1967), 1018–1026. MR **36** #2677.

UNIVERSITY OF MANITOBA, WINNIPEG

ASYMPTOTICS OF TOURNAMENT SCORES

LEO MOSER[1]

Consider a round robin tournament among n players in which each player plays each other player exactly once, scoring one point for a victory and none for a loss. There are no ties. If we total the score of each player and form the nondecreasing sequence of these scores we obtain a sequence of n numbers with the following properties:

1° it is nondecreasing;
2° the sum of the numbers is $\binom{n}{2}$;
3° the sum of the first k numbers is no less than $\binom{k}{2}$.

One can show constructively that any sequence of n integers with these properties is the score sequence for some tournament.

The question that we raise here is—what is the asymptotic behavior of the number of such sequences for large n?

Erdös and myself have found bounds on this number (see [2]), call it S_n: a lower bound of the form $c4^n/n^{3/2}$ and an upper bound of the form $c4^n/n^{9/2}$ (the c's denote different constants). We feel that the asymptotic behavior should be of the form $c4^n/n^{5/2}$ and I hereby offer $25.00 to anyone who will find a proof of this conjecture.

That these bounds are in the right ball park can be seen from the following argument. Consider an n by n chess board, and in it draw a histogram of the score sequence. That is, draw a horizontal line in the kth column a number of squares upward from the base given by the number of victories by the kth weakest player. Insert the connecting vertical lines as well. To each histogram we can correspond a "walk" on the board, starting from the lower left corner and reaching to the upper right. The walk consists of steps either to the right or up. The last step must be up since the best player can win at most $n-1$ games, and there are $2n$ steps in all. There are n steps to the right in all.

The number of such walks clearly cannot exceed $\binom{2n-1}{n}$ which is of the form $c4^n/n^{1/2}$. Now the area under a walk corresponding to a score must be the total

[1] *Editor's note.* L. Moser did not furnish a manuscript. Upon my request from the symposium participants for notes from which the lecture could be reconstructed, Professor D. J. Kleitman kindly sent me the above, writing: "...While I did not take notes on Leo Moser's talk at Los Angeles I happened to work on the problem he described in his talk as a result of hearing it, so that I can provide a partial summary...".

number of won games or $\binom{n}{2}$. If we consider any walk, the area under it in the kth column represents the number of upward steps before the kth sideward step. If we use the analogy of coin tossing and a sideward step is a tail this represents the number of heads already seen at the kth tail.

The area condition says that the average of this number over k must be the average value of $k-1$.

The details of the Erdös-Moser arguments are described in [2] and will not be repeated here.

Note by D. J. Kleitman. The following argument shows that S_n lies between $c4^n/n^2$ and $c4^n/n^3$. (With some effort these limits can be improved to $c4^n/n^{5/2}$, see [1].) The number of sequences $a_1, \ldots, a_n$ of integers ≥ 0 and $\leq n$ fulfilling 1° and 2° above is the coefficient of $x^n y^{n^2}$ in $\prod_{i=1}^{2n} (1+xy^i)$. This coefficient may be extracted asymptotically by standard analytic techniques with a result of the form $c4^n/n^2$.

Given any such sequence $a_1, \ldots, a_n$, we form a new sequence $b_1, \ldots, b_n$ whose sum is zero by subtracting $k-1$ from the kth entry. Now given any such sequence $b_1, \ldots, b_n$, one can obtain by cyclic permutation a new sequence whose partial sums are all nonnegative. (Start after any minimal sized partial sum.) Since each such sequence becomes a nondecreasing sequence upon adding $k-1$ to the kth entry, at least one of each n sequences satisfying 1° and 2° must satisfy 3° as well. Actually of the n cyclic permutations of a sequence $b_1, \ldots, b_n$ only those which make the first entry ≥ 0 and the last entry ≤ 1 correspond to sequences $a_1, \ldots, a_n$.

One can show that this happens on the average for $cn^{1/2}$ of the n cyclical permutations of score sequences, which leads to an estimate of $c(4^n/n^3)n^{1/2}$ for S_n.

References

1. D. Kleitman, *The number of tournament score sequences for a large number of players*, Proc. Calgary 1969 Conf. Combinatorial Structures and their Applications, 1970.

2. J. Moon, *Topics on tournaments*, Holt, New York, 1968.

University of Alberta, Edmonton

SORTING NUMBERS FOR CYLINDERS AND OTHER CLASSIFICATION NUMBERS

THEODORE S. MOTZKIN[1]

0. **Introduction.** Set partitions (corresponding to equivalence relations) are in the following called sortings. Catalan [**3**, pp. 22–23] expressed in 1865 Stirling numbers of the second kind and their sums (now also called Bell-Stirling numbers) as numbers (i.e., cardinals of sets) of sortings. The umbral generating function of these sums, namely e^{e^z-1}, and its coefficients were mentioned in a different context by Boole [**2**, pp. 27, 245] in 1860. (For further references see [**4**] and [**10**].)

When of all sortings obtained from each other by a permutation of the sorted set S only one is counted then the sorting numbers become (arithmetical) partition numbers. As an intermediate step one can use only those permutations that belong to a given subgroup G of the symmetric group on S. We study the two cases where S is a direct product S_1S_2 and G is induced either by the powers of a cyclic permutation of one factor S_1 or by the symmetric group of S_1. In the first case it is natural to call the structure (S, G) a *cylinder*.

Together with sortings we consider similar structures where the sets are replaced by lists (indexed sets), and other classifications. These are structures—i.e., sets (or lists) of sets (or lists) of sets (or lists) etc. (finitely many times) of elements which *pave* a set S, that is, cover *and* pack S. A structure T *covers* S if every element of S occurs *at least* once in T; while if every element of S occurs *at most* once, and if no element outside S occurs, then T is a *packing* (the lists are nonrepetitive and the lists and sets are disjoint).

For numbers of structures so defined we obtain in §§4–7 recurrences, generating functions, values and congruence properties. Many of these results (even for ordinary sorting numbers) are new. Some proofs are omitted or condensed. §§1–3 help systematize the notation.

1. **Mappings.** Let L, M, N be finite, possibly empty sets, with cardinals $|L|=l$, $|M|=m$, $|N|=n$. Let

MN be the direct product of M and N,

L^N the set of mappings from N into L (*N-lists in L*), including as subsets

[1] This work was supported in part by NSF Grant 14026.

$L_<^N$ (in-valence ≤ 1 on L) the set of injections (*N-ads in L*),
$L_>^N$ the set of surjections (*N-lists on L*),
$L_=^N$ the set of bijections,
$M! = M_=^M$ the set of permutations of M.

In analogy to $|MN| = mn$, $|L^N| = l^n$, $|M!| = m!$ we define

$$l_>^n = |L_>^N|, \qquad l_<^n = |L_<^N| = \binom{l}{n} n!,$$

and similarly in the sequel. Since

$$l_>^n = 0 \quad \text{for } l > n \qquad \text{and} \qquad l_<^n = 0 \quad \text{for } n > l,$$

the sums

$$\sum_l l_>^n \qquad \text{and} \qquad \sum_n l_<^n$$

are finite; such sums will be abbreviated to

$$\Sigma_>^n \qquad \text{and} \qquad l_<^\Sigma,$$

and their largest terms to

$$\max{}_>^n \qquad \text{and} \qquad l_<^{\max} = l!.$$

Deemphasizing the individuality of N, the sets L^N, $L_<^N$, $L_>^N$, $L_=^N$ can be regarded as sets of lists of elements of L (*n-lists in L*, *n-ads in L*, etc.) and denoted L^n, $L_<^n$, $L_>^n$, $L_=^n$.

2. **Symmetric merging and subgroupwise merging.** If we merge (identify, collect into equivalence classes) those mappings that are obtained from each other by a permutation of N we replace, in the above notations, N by $!N$ and n by $!n$; similarly for L.

The set $L_<^{!n}$, also denoted $\binom{L}{n}$, is the set of sets of n elements of L (*n-sets in L*), and $L^{!n}$ is the set of *n-tuples in L* (the sum of the "multiplicities" of the elements is n). We have

$$l^{!n} = \binom{l+n-1}{n}, \qquad l_<^{!n} = l_<^n/n! = \binom{l}{n},$$

$$l_>^{!n} = \binom{n-1}{l-1}, \qquad l_=^{!n} = \delta_{l,n},$$

and the largest term of $l_<^{!\Sigma} = 2^l$ is $l_<^{!\max} = \binom{l}{[l/2]}$.

If, for merging of mappings, we use a subgroup G of $N!$ we replace, in the notations, $!N$ by ${}^GN$, $!n$ by ${}^Gn$. If G is the group N_{cy} generated by a cyclic permutation of N we write ${}^{\text{cy}}N$ for ${}^GN$ and ${}^{\text{cy}}n$ for ${}^Gn$.

If only those mappings are admitted that are *invariant* under (every member of) G we write ${}^{\underline{G}}N$ and ${}^{\underline{G}}n$.

3. **Classifications.** Deemphasizing the individuality of L, the set $L^N_>$ becomes $l^N_>$, the set of *listed sortings* of N into l nonempty disjoint classes. While its members are l-ads of sets, those of $!l^N_>$ are l-sets of sets, namely *sortings* of N. We have

$$!l^n_> = l^n_>/l!;$$

$!l^n_>$ is the Stirling number of second kind for n and l.

The members of $!l^{!N}_>$ and $l^{!N}_>$ are respectively sets of cardinals and lists of cardinals, namely *partitions* of n and *listed partitions* (compositions) of n into l nonzero terms, and correspond to the isomorphism classes of sortings and listed sortings. For results on the "cyclic" compositions or partitions ${}^{cy}l^{!N}_>$ and related concepts see [9].

Summing over $\lambda=0,\dots,l$ we have

$$!l^{!n} = \sum_{\lambda \le l} !\lambda^{!n}_>, \qquad !l^n = \sum_{\lambda \le l} !\lambda^n_>.$$

For $\lambda>n$ the terms are 0; the sums become, for $l\ge n$, independent of l and will be denoted by

$$!^{!n} = !\textstyle\sum^{!n}_> \qquad \text{and} \qquad !^n = !\sum^n_>,$$

respectively the partition number and the sorting number (Bell-Stirling number) of n. Their largest terms are

$$!\max{}^{!n}_> \qquad \text{and} \qquad !\max{}^n_>.$$

The set $l^{N+}_>$ of l-ads of nonempty disjoint ads exhausting N and the set $!l^{N+}_>$ of l-sets of such ads cannot be obtained from L^N by subset formation and merging. If we admit empty ads we omit the $>$ sign; again we have

$$!l^{n+} = \sum_{\lambda \le l} !\lambda^{n+}_>, \qquad l^{n+} = \sum_{\lambda \le l} \lambda^{n+}_>,$$

and denote $!l^{n+}$, $l\ge n$, and its largest term by

$$!^{n+} = !\textstyle\sum^{n+}_> \qquad \text{and} \qquad !\max{}^{n+}_>.$$

From $l^{n+}_> = l^{!n}_> n!$ follows for $n\ge 1$

$$\textstyle\sum^{n+}_> = \sum^{!n}_n n! = 2^{n-1}n!,$$

$$\max{}^{n+}_> = \max{}^{!n}_> n! = \binom{n-1}{[(n-1)/2]}n!.$$

If every class of a sorting of N is divided into subclasses, we have an example of a 3-*level classification* of N (levels 0, 1, 2, 3 are the elements, subclasses, classes and the union or set of the classes). The members of $\sum^{N+}_>$, $\sum^N_>$, $!^{N+}$, $!^N$ are various kinds of 2-level classifications of N, N itself and its permutations are 1-level classifications. A classification is *setwise* if it is obtained only by set formation, e.g., a sorting.

A classification of N is *proper* if a class with only one subclass can only have one subsubclass, and if, for $n>1$, not every 1-level class has only one element. The number of proper setwise classifications of N is finite and will be denoted by h_n.

4. **Sortings of a product.** Among numbers connected with mappings from a product MN we mention

$$!\,l^{!m\cdot n}, \qquad !\,l^{\mathrm{cy}\,m\cdot n}, \qquad !\,l^{\underline{!}\,m\cdot n}, \qquad !\,l^{\underline{\mathrm{cy}}\,m\cdot n}. \tag{1}$$

For $l\geq mn$ they are independent of l and can be denoted by

$$!^{!m\cdot n}, \qquad !^{\mathrm{cy}\,m\cdot n}, \qquad !^{\underline{!}\,m\cdot n}, \qquad !^{\underline{\mathrm{cy}}\,m\cdot n}.$$

Some of these latter combinatorial functions arise in the study of the number of identities in semigroups.

For $m=1$ the numbers in (1) become $!\,l^n$. For $m=2$ still $M!=M_{\mathrm{cy}}$, and

$$!\,l^{!2\cdot n} = \tfrac{1}{2}(!\,l^{2n}+!\,l^{\underline{!}2\cdot n}).$$

For $m\geq 3$, each class in a member of $!\,l^{\underline{!}M\cdot N}$ is easily seen to be either MN_1, $N_1\subset N$, or $\{\mu\}N_2$, $\mu\in M$, $N_2\subset N$, and in the latter case the N_2 are the same for all $\mu\in M$. Thus the number of sortings does not depend on m, and

$$!\,l_{>}^{\underline{!}m\cdot n}{}_{m\geq 3} = !\,l_{>}^{\underline{!}n} = \sum_{\lambda\leq l;\nu\leq n}\binom{n}{\nu}\,!\,\lambda_{>}^{\nu}\;!(l-\lambda)_{>}^{n-\nu}.$$

By summation over l we obtain

$$!^{\underline{!}m\cdot n}{}_{m\geq 3} = !^{\underline{!}\cdot n} = \sum_{\nu\leq n}\binom{n}{\nu}\,!^{\nu}\;!^{n-\nu}.$$

For prime m, each class in a member of $!\,l^{\underline{\mathrm{cy}}M\cdot N}$ is easily seen to be either

$$MN_1, \qquad N_1\subset N,$$

or a member of

$$M^{N_2}, \qquad N_2\subset N,$$

and in the latter case its cyclic transforms pave MN_2.

5. **Recurrences.** From the combinatorial definitions one obtains easily

$$n_{<}^{\Sigma} = n(n-1)_{<}^{\Sigma}+1, \qquad n\geq 1, \tag{2}$$

$$2\textstyle\sum_{>}^{n} = (1+\sum_{>}^{|})^{n}, \qquad n\geq 1, \tag{3}$$

$$!^{n+1} = (1\,+\,!^{|})^{n}, \qquad n\geq 0, \tag{4}$$

$$!^{\underline{!}\cdot(n+1)} = 2(1+!^{\underline{!}\cdot|})^{n}, \qquad n\geq 0, \tag{5}$$

$$!^{\underline{\mathrm{cy}}p\cdot(n+1)} = (1+!^{\underline{\mathrm{cy}}p\cdot|})^{n}+(p+!^{\underline{\mathrm{cy}}p\cdot|})^{n}, \qquad n\geq 0,\ p \text{ prime}, \tag{6}$$

$$h_{n+1} = 2\sum_{0}^{n-1}\binom{n}{\nu}h_{n-\nu}h_{\nu+1}-nh_1h_n, \quad n\geq 1, \tag{7}$$

where $^{|}$ is the vicarious exponentiation recipient; e.g., $(1+!^{|})^2=1+2\ !^1+!^2$. (Compare also $b_n=(1+b_|)^n$, $n\geq 2$, for the Bernoulli numbers.)

From the recurrence

$$!^{(n+1)+} = \sum_0^n \binom{n}{\nu}(\nu+1)!\ !^{(n-\nu)+}$$

and the same expansion for $!^{n+}$ we deduce

$$!^{(n+1)+}-n\ !^{n+} = !^{n+}+\sum_1^n n\cdot\cdots\cdot(n-\nu+1)\ !^{(n-\nu)+}$$

and then similarly

$$(!^{(n+1)+}-(n+1)\ !^{n+})-n(!^{n+}-n\ !^{(n-1)+}) = n\ !^{(n-1)+};$$

hence the simpler recurrence

$$!^{(n+1)+} = (2n+1)\ !^{n+}-(n^2-n)\ !^{(n-1)+}, \qquad n\geq 1. \tag{8}$$

6. **Generating functions.** The umbral (exponential) generating functions

$$w = \sum n^{\underset{<}{\Sigma}} z^n/n! = e^z/(1-z), \qquad |z|<1, \qquad (1-z)w' = (2-z)w, \tag{2'}$$

$$w = \sum \textstyle\sum_{>}^{n}\ z^n/n! = 1/(2-e^z), \qquad |z|<\log 2, \qquad w' = 2w^2-w, \tag{3'}$$

$$w = \sum !^n z^n/n! = e^{e^z-1}, \qquad |z|<\infty, \qquad w' = e^z w,\ w''w = w'(w'+w), \tag{4'}$$

$$w = \sum !^{\underline{!}\cdot n} z^n/n! = e^{2(e^z-1)}, \qquad |z|<\infty, \qquad w' = 2e^z w,\ w''w = w'(w'+w), \tag{5'}$$

$$\begin{aligned} w &= \sum !^{\underline{\mathrm{cy}}p\cdot n} z^n/n! = e^{e^z-1+(e^{pz}-1)/p}, \qquad |z|<\infty,\\ w' &= w(e^z+e^{pz}),\quad (w'^2+pw'w-w''w)^p = (p-1)^{p-1}(w''w-w'^2-w'w)w^{2p-2},\\ & \qquad w'''w^2 = 3w''w'w+(p+1)w''w^2-2w'^3-(p+1)w'^2w-pw'w^2, \end{aligned} \tag{6'}$$

$$\begin{aligned} w &= \sum h_n z^n/n!, \qquad z = 2w-1-e^{w-1},\ |z|<\log 4-1,\\ & \qquad\qquad (z+3)w' = 2w'w+1, \end{aligned} \tag{7'}$$

$$w = \sum !^{n+} z^n/n! = e^{z/(1-z)}, \qquad |z|<1, \qquad (1-z)^2 w' = w, \tag{8'}$$

are obtained from the corresponding recurrences, via the first-order differential equations with initial value $w(0) = 1$.

Formula (6′) defines w for every complex p. In particular,

$$w = e^{e^z-1+z} = (e^{e^z-1})' = \sum !^{n+1}z^n/n! \quad \text{for } p=0,$$

$$w \to e^{e^z-1} \quad \text{for } p\to-\infty \text{ and } \operatorname{Re} z\geq 0.$$

The coefficients of w are polynomials in p (see §7) and are, for $n>0$, smaller than $!^{\underline{\mathrm{cy}}p\cdot n}$ when p is composite.

It can be shown that if an umbral series w generates the numbers of classifications of a certain level, then those of the next higher level are generated by e^{w-1} if the new classifications consist of sets of old ones, and by $1/(2-w)$ if they consist of lists (ads).

7. **Values.** The relative size of numbers of lists or sets is apparent from the following table where packing is, but covering is not assumed.

elements	n	0	1	2	3	4	5	For further values or references see
lists	$n^{!\Sigma}_{<}$ (from (2))	1	2	5	16	65	326	**[11]**
	$n^{!\max}_{<}=n!$	1	1	2	6	24	120	**[1**, pp. 272–273]; **[11]**
sets	$n^{!\Sigma}_{<}=2^n$	1	2	4	8	16	32	see below
	$n^{!\max}_{<}=\binom{n}{[n/2]}$	1	1	2	3	6	10	see below

For level 2: numbers of lists (or sets) of lists (or sets or numbers) we reassume covering, and obtain by inspection (i.e., without machines) and use of recurrences

	n	0	1	2	3	4	5	6	7	8	9	For further values or references see
lists of lists	$\Sigma^{n+}=2^{n-1}n!$	1	1	4	24	192	1920	23040	322560	5160960	92897280	
	$\max^{n+}_{>}=\binom{n-1}{[(n-1)/2]}n!$	1	1	2	12	72	720	7200	100800	1411200	24501600	
sets of lists	$!\Sigma^{n+}_{>}=!^{n+}$ (from (8))	1	1	3	13	73	501	4051	37633	394353	4596553	**[11]**
	$!\max^{n+}_{>}$	1	1	2	6	36	240	1800	15120	141120	1693440	
lists of sets	$\Sigma^{n}_{>}$ (from (3))	1	1	3	13	75	541	4683	47293	595835	7088261	**[11]**
	$\max^{n}_{>}$	1	1	2	6	36	240	1800	16800	191520	2328480	
sets of sets	$!\Sigma^{n}_{>}=!^n$ (from (4))	1	1	2	5	15	52	203	877	4140	21147	**; **[11]**
	$!\max^{n}_{>}$	1	1	1	3	7	25	90	350	1701	7770	***; **[1**, p. 835]
lists of numbers	$\Sigma^{!n}_{>}=2^{n-1}$	1	1	2	4	8	16	32	64	128	256	**[1**, pp. 24–44]; **[11]**
	$\max^{!n}_{>}=\binom{n-1}{[(n-1)/2]}$	1	1	1	2	3	6	10	20	35	70	**[1**, pp. 828–830]; **[11**, 1 2 6 20]
sets of numbers	$!\Sigma^{!n}=!!^n$	1	1	2	3	5	7	11	15	22	30	**[1**, pp. 836–839]; **[11]**
	$!\max^{!n}_{>}$	1	1	1	1	2	2	3	4	5	6	*

* continued 9 11 15 18 23 30 37 47 58 71 90.

** The values of $!^n$ for $n \leq 51$ are given in **[8]**. I have the values for $n \leq 200$, obtained (with help from D. Cantor and A. Fraenkel), in 16 sec., by the IBM 360/91 at the UCLA Computing Facility. We give here 4 values to 8 figures:

$$!^{50} = .18572426\ldots\cdot 10^{48},$$
$$!^{100} = .47585391\ldots\cdot 10^{116},$$
$$!^{150} = .68206412\ldots\cdot 10^{193},$$
$$!^{200} = .62474847\ldots\cdot 10^{276}.$$

By **[8]**, $!^n$ is asymptotic to

$$\exp\left(((\log \nu)^2 - \log \nu + 1)\nu - \tfrac{1}{2}\log\log \nu - 1\right) = \nu^n e^{\nu-n-1}/\sqrt{\log n}$$

where $\nu \log \nu = n$. A better approximation seems to be

$$\mu_n = \nu^n e^{\nu-n-1}\left(\frac{n}{\nu}+1\right)^{-1/2}\left(1-\frac{n(2n^2+7n\nu+10\nu^2)}{24\nu(n+\nu)^3}\right),$$

given in [8] without error term or indication of the way in which it might be superior to other asymptotic expressions. Setting $!^n = \mu_n(1-\lambda_n/n)$ we find

$$\begin{aligned} \lambda_{50} &= .0015\ldots, \\ \lambda_{100} &= .0008\ldots, \\ \lambda_{150} &= .0006\ldots, \\ \lambda_{200} &= .0004\ldots. \end{aligned}$$

*** By [7], p. 413 last line (at whose end an exponent 1/2 should be appended) and p. 412, line 17 (both only with hints to proofs), $!\max^n_>$ is asymptotic to

$$\exp\left(((\log \nu)^2 - \log \nu + 1)\nu - \tfrac{1}{2}\log \nu - 1 - \tfrac{1}{2}\log 2\pi\right) = \nu^{n-1/2}e^{\nu-n-1}/\sqrt{2\pi}.$$

Similarly we obtain these numbers of certain sets of subsets of direct products:

n	0	1	2	3	4	5	6	7	8	9
$!^{\underline{!}\cdot n}$	1	2	6	22	94	454	2430	14214	89918	
$!\max_{>}^{!\cdot n}$	1	2	4	12	48	200	1040	5600	33600	
$!^{\underline{!}2\cdot n}$	1	2	7	31	164	999	6841	51790	428131	3929021
$!\max_{>}^{!2\cdot n}$	1	1	3	10	53	265	1700			
$!^{\underline{\mathrm{cy}}3\cdot n}$	1	2	8	42	268	1994	16852			
$!\max_{>}^{\mathrm{cy}3\cdot n}$	1	2	4	24	128	880	7440			

For prime p, $!_{>}^{\mathrm{cy}p\cdot n}$ is a polynomial in p of degree $n-1$ (for $n>0$). The sequence of these polynomials starts

$$\begin{aligned} &1 \\ &2 \\ &5+p \\ &15+6p+p^2 \\ &52+30p+11p^2+p^3 \\ &203+150p+80p^2+20p^3+p^4 \\ &877+780p+525p^2+190p^3+37p^4+p^5 \end{aligned}$$

Finally, for classifications of unbounded level we find

									For further values or references see
h_n	1	1	1	4	26	236	2752	39208	[11]

8. **Congruence properties.** While the linear recurrence (4) for $!^n$ is not of bounded degree, $!^n$ mod a prime p fulfills the linear recurrence

$$!^{n+p} \underset{p}{=} !^n + !^{n+1} \tag{9}$$

of degree p and with constant coefficients [**6**]. A combinatorial proof of (9) is the, by far simplest, case $t=1$ of the proof of (10).

From (9) follows, first letting $n=0$, then using (4) for $n=p-1$,

$$!^p \underset{p}{=} 2, \qquad \sum_{2}^{p-1} (-1)^k \, !^k \underset{p}{=} 2.$$

The recurrence (9) holds also for α^n if $\alpha^p = \alpha + 1$. In the field GF(p^p), the characteristic polynomial $g(x) = x^p - x - 1$ has p roots $\alpha_k = \alpha_p + k$, no proper subset of which has its sum in the prime field GF(p), over which $g(x)$ is therefore irreducible. The α_k^n are the p fundamental solutions of (9); the linear combination that represents $!^n$ can be shown to be

$$!^n \underset{p}{=} \sum f(\alpha_k)\alpha_k^n, \qquad f(x) = x^p - \sum_{0}^{p} !^j x^j.$$

Since $\alpha_k^{p^j} = \alpha_k + j$ we have $\alpha_k^{p'} = 1$ where

$$p' = 1 + p + \cdots + p^{p-1} = (p^p - 1)/(p-1);$$

hence

$$!^{n+p'} \underset{p}{=} !^n.$$

It is unknown where $!^n$ mod p can have a smaller period. The prime decompositions of the first values of p' are, according to J. L. Selfridge,

2	3
3	13
5	11·71
7	29·4733
11	15797·1806113
13	53·264031·1803647
17	10949·1749233·2699538733

For $!^n$ mod p^t, $t \geq 1$, we prove the linear recurrence

$$!^{n+p^t} \underset{p^t}{=} \sum_{0}^{p^t-1} s_{p^{t}-1-j}(p) \binom{p^t-1}{j} \, !^{n+j}, \tag{10}$$

of degree p^t with constant coefficients; $s_\lambda(x)$ denotes the *sorting polynomial* $\sum !(\lambda - k)^\lambda_> x^k$. Indeed consider the set $N + P^t$, $|P| = p$ (+ means disjoint union) and a cyclic permutation τ of P^t, acting indirectly on an arbitrary sorting σ of $N + P^t$. The list

$$\sigma, \sigma\tau, \ldots, \sigma\tau^{p^t-1}$$

has no repetitions unless $\sigma\hat{\tau} = \sigma$ where $\hat{\tau} = \tau^{p^{t-1}}$. If $\sigma\hat{\tau} = \sigma$ then, for each element π of P^t, either

$$\pi, \pi\hat{\tau}, \ldots, \pi\hat{\tau}^{p-1} \tag{11}$$

belong to the same class in σ (case I) or they belong to p different classes (case II); the latter contain no element of N nor any case I-element. Finally it is easily seen that the number of sortings of $N+P^t$ for which exactly j of the p-sets (11) belong to case I ($j=0,\ldots,p^{t-1}$) is

$$s_{p^{t-1}-j}(p)\binom{p^{t-1}}{j}\,!^{n+j}.$$

For $j\neq_p 0$ we have $\binom{p^{t-1}}{j}=_{p^{t-1}}0$, hence $s_{p^{t-1}-j}(p)$ in (10) can then be replaced by its constant coefficient 1. For $j=_p 0$, $\neq_{p^2} 0$ we have $\binom{p^{t-1}}{j}=_{p^{t-2}}0$, but

$$s_{p^{t-1}-j}(p) \underset{p^2}{=} \binom{p^{t-1}-j}{2}p+1 \underset{p^2}{=} 1+0^{p-2}\cdot 2 \quad (\text{unless } p=t=j=2),$$

where $0^0=1$; thus $s_{p^{t-1}-j}(p)$ in (10) can be replaced by $1+0^{p-2}\cdot 2$. There follows in particular

$$!^{n+p^2} \underset{p^2}{=} \sum_0^p \binom{p}{j}\,!^{n+j}+0^{p-2}\cdot 2\;!^n.$$

For $p>2$ this can be written

$$!^{n+p^2} \underset{p^2}{=} (1+\,!^{n+1})^p.$$

A more detailed analysis shows that for every $t\geq 1$ and prime $p>2$ we have

$$!^{n+p^t} \underset{p^t}{=} (1+\,!^{n+1})^{p^{t-1}}. \tag{12}$$

Setting $n=0$ we obtain

$$!^{p^t} \underset{p^t}{=} \,!^{p^{t-1}+1}.$$

The characteristic polynomial of the recurrence (10) is

$$x^{p^t}-\sum s_{p^{t-1}-j}(p)\binom{p^{t-1}}{j}x^j.$$

For $p\neq 2$ we can replace it by

$$x^{p^t}-(x+1)^{p^{t-1}}.$$

For $t=2$, $p=2$ the characteristic polynomial is x^4-x^2-2x-3.

A similar (and simpler) proof than that of (10) shows that

$$!^{\underline{\mathrm{cy}}2\cdot(n+p)} \underset{p}{=} (2-0^{p-2})\;!^{\underline{\mathrm{cy}}2\cdot n}+\,!^{\underline{\mathrm{cy}}2\cdot(n+1)}.$$

For $p>2$ the corresponding polynomial is $x^p-x-2=_p 2g(x/2)$, and p' is again a period.

It follows from the fact that the highest and lowest coefficients of the recurrence (10) are $\neq_p 0$ that $!^n$ is periodic, without preperiod, for p^t and hence for every modulus m, and that the period is $\leq m^\mu$, where μ is the largest prime power dividing m. The periodicity, with possible preperiods, follows also from (4′) and Fujiwara's theorem [5] on the umbral coefficients (if integer) of solutions $w = \sum a_n z^n/n!$ of algebraic differential equations

$$F(z, w, w', \ldots, w^{(d)}) = 0$$

with integer coefficients for which

$$\frac{\partial F}{\partial w^{(d)}}(0, a_0, a_1, \ldots, a_d) = 1.$$

By the same theorem, modular periodicity holds also for the coefficients

$$n^{\Sigma}_{<}, \quad \textstyle\sum^{n}_{>}, \quad !^{\underline{!}\cdot n}, \quad !^{\underline{\mathrm{cy}}\,p\cdot n}, \quad h_n, \quad !^{n+}$$

in (2′), (3′), (5′), (6′), (7′), (8′).

References

1. M. Abramowitz and I. A. Stegun (Editors), *Handbook of mathematical functions, with formulas, graphs and mathematical tables*, Nat. Bur. Standards Appl. Math. Series, 55, Superintendent of Documents, U.S. Government Printing Office, Washington, D.C., 1964; 3rd printing with corrections, 1965. MR **29** #4914; MR **31** #1400.
2. George Boole, *A treatise on the calculus of differences*, Cambridge, 1860.
3. Eugène Catalan, *Mélanges mathématiques*. I, Mém. Soc. Sci. Liège II **12** (1885).
4. H. C. Finlayson, *Numbers generated by* $e^{e^x - 1}$, Masters Thesis, University of Alberta, 1954.
5. M. Fujiwara, *Periodizität der Entwicklungskoeffizienten einer analytischen Funktion nach dem Modul m*, Tôhoku Math. J. **2** (1912), 57–73.
6. Marshall Hall, *Arithmetic properties of a partition function*, Bull. Amer. Math. Soc. **40** (1934) 387. Abstract #200.
7. L. H. Harper, *Stirling behavior is asymptotically normal*, Ann. Math. Statist. **38** (1967), 410–414. MR **35** #2312.
8. Leo Moser and Max Wyman, *An asymptotic formula for the Bell numbers*, Trans. Roy. Soc. Canada Sect. III (3) **49** (1955), 49–54. MR **17**, 1201.
9. T. S. Motzkin, *Ordered and cyclic partitions*, Riveon leMatematika **1** (1947), 61–67 (Hebrew). MR **8**, 566.
10. Gian-Carlo Rota, *The number of partitions of a set*, Amer. Math. Monthly **71** (1964) 498–504. MR **28** #5009.
11. N. J. A. Sloane, *A catalog of sequences* (unpublished).

UNIVERSITY OF CALIFORNIA, LOS ANGELES

PATHOLOGICAL LATIN SQUARES

E. T. PARKER

1. The central open question in finite projective planes is whether one exists of order *not* a prime-power. They have been known to exist for every prime-power order since about the turn of the century. Only one theorem, due to Bruck and Ryser [**1**], shows nonexistence of finite projective planes of infinitely many orders, *not* prime-powers, of course. Considerable partially expository literature is available, such as [**2**], [**3**], [**4**], [**5**].

A latin square of order n is an n by n array of n symbols each present (only once) in each row and each column. A set of latin squares of like order is called *orthogonal* if each pair of distinct members of the set includes all ordered pairs of symbols (only once) among the n^2 positions. A folk theorem of the 1930 decade asserts that existence of a projective plane of order n (n an integer exceeding 1) implies existence of a *complete set* (for no larger set can exist) of $n-1$ orthogonal latin squares of order n, and conversely.

The lowest order for which existence of a projective plane (and equivalently a complete set of orthogonal latin squares) is undecided is ten. The author [**6**] and quite recently John W. Brown have found by using digital computers that few (and quite possibly none) of the numerous distinct latin squares of order ten are extendible to complete sets. Unfortunately for order ten, to say nothing of larger orders, the number of nonisomorphic latin squares (after identifying equivalences by permuting rows, columns, and symbols independently) is astronomical. What seems needed is theorems rejecting wide classes of latin squares as possibilities for inclusion in complete sets. The author knows of only two previously known theorems of this nature: Euler [**7**] proved this and more for *cyclic* latin squares of *even* orders (except for the rather degenerate order two); Mann [**8**] proved two further theorems, in spirit almost a generalization of Euler's. The author in this paper generalizes the fundamental theorem of Euler in another direction.

A latin square not extendible to a complete set will be called *pathological*; thus Euler, Mann, and the author have given sufficient conditions for a latin square to be pathological. While the author is designating these latin squares as pathological, experiments indicate that these constitute the majority for any consequential order. Metaphorically, many more people are ill than sound, and the task is to find the rare healthy individual among the ailing masses—or to show such nonexistent for some orders. Thus it is to be hoped that some genuinely *stringent* and *rapidly* computable pathology properties on latin squares, especially of order ten, may be developed. More theorems are needed. It is the author's hope that those given in this paper may stimulate investigators to develop further results.

The author's first latin square pathology theorem follows. Although it is equally valid for *every even* order $2t$ exceeding four, with a similar triangular array of $2t-2$ rows, columns, and symbols, the author prefers to discuss only order ten. This is the crucial order, which some hope will yield to computer attack. Also, simplicity is gained by using only concrete numbers.

2. THEOREM 1. *A latin square of order* 10 *that includes an array of the form*

$$\begin{array}{cccccccc} 8 & 7 & 6 & 5 & 4 & 3 & 2 & 1 \\ 7 & 6 & 5 & 4 & 3 & 2 & 1 & \\ 6 & 5 & 4 & 3 & 2 & 1 & & \\ 5 & 4 & 3 & 2 & 1 & & & \\ 4 & 3 & 2 & 1 & & & & \\ 3 & 2 & 1 & & & & & \\ 2 & 1 & & & & & & \\ 1 & & & & & & & \end{array}$$

is not extendible to a complete set of orthogonal latin squares (*i.e., it is pathological*).

PROOF. We follow Euler, and define a *transversal* of a latin square of order n as a set of n positions, one in each row, one in each column, and one containing each symbol. It is an observation, hardly worth calling a theorem, that a latin square orthogonal to a latin square of order n is equivalent to a set of n transversals, each pair with none of the n^2 positions in common. Similarly, extension of a latin square of order n to a complete set is equivalent to existence and construction of $n-2$ such sets of n transversals each, such that the above holds on each set of n transversals, and two transversals from different sets of n have at most (actually, exactly) one of the n^2 positions in common. The proof of Theorem 1 depends on obtaining a contradiction when existence of $n(n-2)$ transversals, no two with as many as two common positions, is assumed. This is accomplished without knowledge of the numbers of cells transversals have in common with the array of 36 positions of the one hundred forming the full latin square.

It is verified readily that the triangular array of 36 positions is transformed onto itself by permutation of row, column, symbol coordinate interpretations: if these coordinates are r, c, s the array is given by the symmetric inequality $r+c+s\leq 10$. Thus we merely count the number of pairs of positions in common *rows* among the 36, triple, and arrive at the required contradiction. The number of unordered pairs of distinct positions, among the 36, in common rows is $\binom{8}{2}+\binom{7}{2}+\cdots+\binom{2}{2}+\binom{1}{2}=$ $28+21+15+10+6+3+1+0=84$. Tripling, we have 252 unordered pairs of positions in like row, or like column, or containing like symbol. The displayed triangular array has $\binom{36}{2}=630$ unordered pairs of distinct positions. Thus the

$630-252=378$ pairs of positions remain to be joined on 80 transversals. Each set of ten transversals corresponding to a latin square orthogonal to the original one necessarily includes each of the 36 positions exactly once. The sum of the binomial coefficients $\binom{t}{2}$, t being the number of unordered pairs of the 36 cells joined by the transversal under consideration, is minimum when all numbers t are nearest the average 3.6; i.e., when six transversals include four cells and four transversals include three cells. $6\binom{4}{2}+4\binom{3}{2}=6\cdot 6+4\cdot 3=48$. Multiplying by 8 for extension to a complete set of orthogonal latin squares, we conclude that at least $8\cdot 48=384$ unordered pairs of distinct cells among the 36 must be joined by the transversals generating the complete set of orthogonal latin squares. Since $384>378$, the claimed contradiction has been obtained.[1]

It seems appropriate to demonstrate that the latin squares of order ten shown pathological by the author's argument above are not necessarily instances for which Euler's or Mann's theorems apply. This is requisite to establish that this paper includes something new. An example is now given of an order-ten latin square with the required triangular set of 36 positions:

$$\begin{matrix}
8 & 7 & 6 & 5 & 4 & 3 & 2 & 1 & 9 & 0\\
7 & 6 & 5 & 4 & 3 & 2 & 1 & 0 & 8 & 9\\
6 & 5 & 4 & 3 & 2 & 1 & 9 & 8 & 0 & 7\\
5 & 4 & 3 & 2 & 1 & 8 & 0 & 9 & 7 & 6\\
4 & 3 & 2 & 1 & 0 & 9 & 8 & 7 & 6 & 5\\
3 & 2 & 1 & 8 & 9 & 0 & 7 & 6 & 5 & 4\\
2 & 1 & 9 & 0 & 5 & 7 & 6 & 3 & 4 & 8\\
1 & 0 & 8 & 9 & 7 & 6 & 4 & 5 & 2 & 3\\
9 & 8 & 0 & 7 & 6 & 5 & 3 & 4 & 1 & 2\\
0 & 9 & 7 & 6 & 8 & 4 & 5 & 2 & 3 & 1
\end{matrix}$$

This latin square is *not* isomorphic with Euler's *cyclic* latin square (i.e., table of the cyclic *group*), for the final pair of columns includes a 3-cycle (on symbols 1, 2, 3), not present in a group of order 10.

The formal statement of Mann's [**8**] theorem, mentioned informally early in this paper, is (specialized to order ten): If a latin square of order 10 has a 5 by 5 block having at most two occurrences of symbols *not* in a list of five, then this latin square can be contained in no orthogonal pair of latin squares—and *a fortiori* in no complete set. It will be shown that the displayed latin square does not satisfy the hypothesis of Mann's theorem (the stronger conclusion of Mann's theorem is of no special interest here). Should the hypothesis of Mann's theorem be satisfied for five rows and five columns, it would also be satisfied for the same five columns and the complementary quintuple of rows (shown by counting incidences in

---

[1] (T.S.M.) If any one position is omitted from the triangular array the conclusion still holds: replace one $\binom{4}{2}$ by $\binom{3}{2}$, 384 by 360, 630 by 595, 252 by 238 (each position belongs to 14 pairs), 378 accordingly by 357. Not so for two omitted positions; but if these form a pair then the above numbers become 336, 561, 225, 336, and the equality should lessen the work needed to decide, one way or the other, on extendibility.

columns). Thus there is no loss of generality in assuming that the initial row of the latin square is in the particular set of five. Should Mann's hypothesis hold for the displayed latin square, then there must be five rows, including the first, such that five columns include like digits, with the exception of at most two of these five rows. The cycle structures of the first row with respect to the other nine rows are: (1 0 9 8 7 6 5 4 3 2), (1 8 6 4 2 9 0 7 5 3), (1 9 7 4)(2 0 6 3 8 5 2), (1 7 3 9 6 2 8 4 0 5), (1 6)(2 7)(3 0 4 9 5 8), (1 3 7)(2 6 9 4 5 0 8), (1 5 9 2 4 7 0 3 6 8), (1 4 6 0 2 3 5 7 8 9), and (1 2 5 6 7 9 3 4 8 0). There being no 5-cycles, and no 2-cycle and 3-cycle in the same permutation, it follows that the initial row cannot be among the distinguished three of five rows, having like sets of symbols in the five special columns, needed to satisfy the hypothesis. We continue (repetitive details omitted here) to trace out the permutations for the remaining $\binom{9}{2}$ pairs of rows. Exactly five of these permutations contain 5-cycles, or a 3-cycle and a 2-cycle. None of the quintuples of symbols in any of these 5-cycles, or whose union forms a 3-cycle and a 2-cycle, are the same set. Thus the distinguished set of three rows does not exist; hence the displayed latin square does not satisfy Mann's hypothesis.

3. As noted above, Euler **[7]** showed—and it is an easy exercise in elementary number theory—that a *cyclic group* of *even order* regarded as a latin square has no transversal. Hence *a fortiori* no latin square is orthogonal thereto, and again *a fortiori* such a latin square is not contained in a complete set equivalent to a plane (except for order two, which in the present context is essentially degenerate). Mann's **[8]** theorem apparently was proved with *group* multiplication tables central in his mind, partially generalizing Euler's theorem by relaxing *cyclic* group to any group of order $\equiv 2 \pmod 4$. (This author feels a quarter of a century later that Mann's mention of groups was rather superfluous, since for order 10, and still more for larger relevant orders, the number of groups is small compared with the class of latin squares fulfilling the hypothesis of Mann's theorem.) M. Hall and Paige **[9]**, continuing the emphasis on the limited *group* situation proved (in different terminology) that what Euler showed for cyclic groups of even order, and in turn Mann showed for groups of orders $4t+2$, is valid for groups of even order with cyclic Sylow 2-subgroup. Again, as is the case with Mann's theorem, the hypothesis of the Hall-Paige theorem may be relaxed to include a wider class of latin squares of even order (but almost certainly only a small fraction of all latin squares of any reasonable large even order). The author will now formalize his generalization; the theorems of Euler, Mann (if only *groups* are considered), and Hall-Paige become special cases of this new theorem, both in statements and proofs.

THEOREM 2. *Assume that a latin square L of even order $2^m k$, k an odd integer, has its rows, columns, and symbols all partitioned into 2^m sets of k each such that each k by k minor determined by the partition of rows and columns contains symbols of only one subset of the symbol partition. Assume further that the latin square of order 2^m determined by the k-to-one mappings of rows, columns, and symbols, determined by the partitions, is a table of a cyclic group. Then L has no transversal.*

PROOF. By contradiction. We can, according to the hypothesis, label rows, columns, and symbols of L with (i, j), $1 \leq i \leq 2^m$, $1 \leq j \leq k$, such that: (i) two rows, columns, or symbols with like first indices are in the same subset in the respective partition; (ii) each cell of L has sum of its three first indices $\equiv 0 \pmod{2^m}$. A transversal of L must then have sum of its $2^m k$ first indices on rows, columns, and symbols collectively $\equiv 0 \pmod{2^m}$. However, the sum of integers 0 through $2^m - 1$ inclusive is $\binom{2^m}{2}$, which is $\equiv 2^{m-1} \pmod{2^m}$. Multiplying by three for rows, columns, and symbols collectively, and by odd k, we have the sum of the $3 \cdot 2^n k$ first indices for a transversal of $L \equiv 2^{m-1} \pmod{2^m}$. The contradiction has been obtained.

REFERENCES

1. R. H. Bruck and H. J. Ryser, *The nonexistence of certain finite projective planes*, Canad. J. Math. **1** (1949), 88–93. MR **10**, 319.

2. Marshall Hall, Jr., *Finite projective planes*, Amer. Math. Monthly **62** (1955), no. 7, part II, 18–24. MR **17**, 400.

3. H. J. Ryser, *Geometries and incidence matrices*, Amer. Math. Monthly **62** (1955), no. 7, part II, 25–31. MR **17**, 401.

4. ———, *Combinatorial mathematics*, Carus Math. Monographs, no. 14, Math. Assoc. of America, Wiley, New York, 1963. MR **27** #51.

5. Marshall Hall, Jr., *Combinatorial theory*, Blaisdell, Waltham, Mass., 1967. MR **37** #80.

6. E. T. Parker, *Computer investigation of orthogonal Latin squares of order ten*, Proc. Sympos. Appl. Math., vol. 15, Amer. Math. Soc., Providence, R.I., 1963, pp. 73–81. MR **31** #5140.

7. Leonhard Euler, *Recherches sur une nouvelle espèce des quarrés magiques*, Verh. Zeeuwsch Genoot. Weten. Vliss. **9** (1782), 85–239. (Also in Euler's collected works.)

8. Henry B. Mann, *On orthogonal Latin squares*, Bull. Amer. Math. Soc. **50** (1944), 249–257. (The part of Mann's paper of relevance here is rewritten in Hall, ref. 5, pp. 192–194). MR **6**, 14.

9. Marshall Hall, Jr. and L. J. Paige, *Complete mappings of finite groups*, Pacific J. Math. **5** (1955), 541–549. MR **18**, 109.

UNIVERSITY OF ILLINOIS

SOME PROBLEMS IN THE PARTITION CALCULUS

RICHARD RADO

1. This note deals with work carried out by one or more of the following: P. Erdös, A. Hajnal and R. Rado. For cardinal numbers a, b, c and positive integers r the *partition relation*

$$a \to (b, c)^r, \tag{1}$$

and similarly for more than two entries on the right-hand side of (1), expresses the truth of the following statement. Whenever A is a set of cardinal $|A| = a$ and whenever the set $[A]^r$ of all subsets of A of cardinal r is expressed as $[A]^r = K \cup L$ then there is always a set $X \subseteq A$ such that

$$\text{either} \quad |X| = b \quad \text{and} \quad [X]^r \subseteq K, \quad \text{or} \quad |X| = c \quad \text{and} \quad [X]^r \subseteq L.$$

Ramsey's classical theorem [1] states that $\aleph_0 \to (\aleph_0, \aleph_0)^r$. The logical negation of (1) is written as $a \not\to (b, c)^r$. If the General Continuum Hypothesis (G.C.H.) is assumed, if cases are ignored which involve inaccessible cardinals, if n is finite and every b_ν infinite, then the relation $a \to (b_1, b_2, \ldots, b_n)^r$ may be described as completely analyzed for $r = 2$, and to a considerable extent analyzed for $r \geq 3$ [**2**]. Much progress has been made in obtaining similarly complete results for $r = 2$ without using G.C.H. The following problem seems to block progress in this direction.

PROBLEM 1. *Assume that $a^* = \sum (\nu < \omega) 2^{\aleph_\nu} > \aleph_\omega$ and that $2^{\aleph_\nu}$ is not eventually constant as ν ranges over all finite numbers. Is it then true that*

$$a^* \to (\aleph_\omega, \aleph_\omega)^2 ?$$

Here $\omega = \omega_0$.

2. An obvious extension of the notation (1) is given by a relation of the form

$$a \to (b, (c)_d)^r$$

which has the following meaning. Let $|A| = a$ and

$$[A]^r = K \cup \bigcup (\nu \in D) L_\nu,$$

where $|D| = d$. Then there is always $X \subseteq A$ such that

$$\text{either} \quad |X| = b \quad \text{and} \quad [X]^r \subseteq K,$$
$$\text{or} \quad |X| = c \quad \text{and} \quad [X]^r \subseteq L_\nu \quad \text{for some } \nu \in D.$$

One of the simplest negative relations so far proved only by using G.C.H. is[1]

(2) $$\aleph_{\omega+1} \nrightarrow (\aleph_{\omega+1}, (3)_{\aleph_0})^2.$$

Problem 2. *Can* (2) *be proved without assuming G.C.H.?*

3. For order types α, β, γ and positive integral r the relation

$$\alpha \rightarrow (\beta, \gamma)^r$$

means that whenever the set A is ordered, of type $\operatorname{tp} A = \alpha$, and whenever $[A]^r = K \cup L$ then there is $X \subseteq A$ such that

$$\begin{aligned} &\text{either} \quad \operatorname{tp} X = \beta \quad \text{and} \quad [X]^r \subseteq K, \\ &\text{or} \quad \operatorname{tp} X = \gamma \quad \text{and} \quad [X]^r \subseteq L. \end{aligned}$$

Much less is known about partition relations for order types than about relations for cardinals, even in the special case when $r=2$ and when α, β and γ are in fact ordinal numbers. A relation for initial ordinals is equivalent with the relation for the corresponding cardinals. Thus [3] $\omega_1 \rightarrow (\omega_1, \omega)^2$. In fact, $\omega_1 \rightarrow (\omega_1, \omega+1)^2$ has been proved [4]. In [5] the relations $\omega_1 \rightarrow (\omega 2, \omega k)^2$ for every $k < \omega$ and using G.C.H., $\omega_1 \nrightarrow (\omega_1, \omega+2)^2$ are obtained. All these relations seem to be rather weak in view of the fact that ω_1 is nondenumerable. The simplest problems which suggest themselves are the following.

Problem 3. (i) *To prove, without assuming G.C.H.,* $\omega_1 \nrightarrow (\omega_1, \omega+2)^2$. (ii) *To decide, using G.C.H., which of the following relations hold.*

$$\begin{aligned} &\omega_1 \rightarrow (\omega 2, \omega^2)^2, \\ &\omega_1 \rightarrow (\omega 3, \omega 3)^2, \\ &\omega_1 \rightarrow (\beta, \beta)^2 \qquad (\beta < \omega^2), \\ &\omega_1 \rightarrow (\alpha, \alpha, \alpha)^2 \qquad (\alpha < \omega 2). \end{aligned}$$

4. Let λ denote the order type of the real numbers under magnitude. It is well known that $|\lambda| > \aleph_0$ and $\omega_1, \omega_1^* \not\leq \lambda$. An order type ϕ which satisfies the analogous relations $|\phi| > \aleph_0$; $\omega_1, \omega_1^* \not\leq \phi$ is called a *real* type. It has been shown [4] that for real ϕ,

$$\begin{aligned} &\phi \rightarrow (\alpha, \alpha, \alpha)^2 \qquad (\alpha < \omega 2), \\ &\phi \rightarrow (\alpha, \beta)^2 \qquad (\alpha < \omega 2 \leq \beta < \omega^2), \\ &\phi \rightarrow (\omega, \gamma)^2 \qquad (\gamma < \omega_1), \\ &\phi \rightarrow (4, \alpha)^3 \qquad (\alpha < \omega 2), \end{aligned}$$

and that $\lambda \nrightarrow (\omega, \omega+2)^r$ for $r \geq 3$ and $\lambda \nrightarrow (r+1, \omega+2)^r$ for $r \geq 4$. Also [5], that $\phi \rightarrow (\beta, \gamma)^2$ for $\beta < \omega^2$ and $\gamma < \omega_1$ and every real type ϕ. The problem arises of bridging the gap between the positive and the negative relations. I mention the following specific question.

---

[1] [2], p. 117, Theorem 10. There is a misprint in [2]. The condition $a > a' > \aleph_0$ should read $a > a' \geq \aleph_0$.

PROBLEM 4. *To decide if* $\lambda \to (\gamma, \gamma, \ldots, \gamma)^2$ *for every* $\gamma < \omega_1$ *and for every finite number of entries* γ.

5. It might well be the case that for the truth of positive relations with ordinal numbers the type λ is representative for the whole class of real types ϕ. More specifically, it would be of interest to solve the following problem.

PROBLEM 5. *To decide whether, whenever* $\lambda \to (\sigma, \tau)^2$ *for some ordinals* σ *and* τ, *then always* $\phi \to (\sigma, \tau)^2$ *for every real type* ϕ.

Added in proof. In 1968 F. Galvin proved[2] that every real type ϕ satisfies $\phi \to (\alpha, \alpha)^2$ for every $\alpha < \omega_1$. This constitutes a wide generalization of three relations in §4 and answers Problem 5 in the affirmative.

REFERENCES

1. F. P. Ramsey, *On a problem of formal logic*, Proc. London Math. Soc. (2) **30** (1930), 264–286.
2. P. Erdös, A. Hajnal and R. Rado, *Partition relations for cardinal numbers*, Acta Math. Acad. Sci. Hungar. **16** (1965), 93–196. MR **34** #2475.
3. B. Dushnik and E. W. Miller, *Partially ordered sets*, Amer. J. Math. **63** (1941), 601–610. MR **3**, 73.
4. P. Erdös and R. Rado, *A partition calculus in set theory*, Bull. Amer. Math. Soc **62** (1956), 427–489. MR **18**, 458.
5. A. Hajnal, *Some results and problems on set theory*, Acta Math. Acad. Sci. Hungar. **11** (1960), 277–298. MR **27** #47.

THE UNIVERSITY, READING, ENGLAND

[2] Notices Amer. Math. Soc. **15** (1968), 660; correction, ibid., **16** (1969), 1095.

SOLUTION OF KIRKMAN'S SCHOOLGIRL PROBLEM[1]

D. K. RAY-CHAUDHURI AND RICHARD M. WILSON

0. **Introduction.** Kirkman's schoolgirl problem was introduced in 1850 by Reverend Thomas J. Kirkman as "Query 6" on page 48 of the Ladies and Gentleman's Diary [**14**] (see [**16**] for his statement of the general problem). A teacher would like to take 15 schoolgirls out for a walk, the girls being arranged in 5 rows of three. The teacher would like to ensure equal chances of friendship between any two girls. Hence it is desirable to find different row arrangements for the 7 days of the week such that any pair of girls walk in the same row exactly one day of the week. In the general case one wants to arrange $6n+3$ girls in $2n+1$ rows of three. The problem is to find different row arrangements for $3n+1$ different days such that any pair of girls belong to the same row exactly one day out of the $3n+1$ days. In modern terminology Kirkman's schoolgirl arrangement corresponds to resolvable balanced incomplete block designs with block size 3.

Let b, v, r, k, and λ be positive integers such that $r=\lambda(v-1)/(k-1)$ and $b=\lambda v(v-1)/k(k-1)$. A (v, k, λ)-Balanced Incomplete Block Design (BIBD) consists of a finite set X of v elements called treatments and b subsets (called blocks) $X_1, X_2, \ldots, X_b$ such that

(1) every treatment occurs in exactly r blocks,

(2) every block contains exactly k treatments, and

(3) every pair of treatments occurs together in exactly λ blocks.

A parallel class of blocks consists of a set of disjoint blocks such that every treatment occurs in one block of the class. A (v, k, λ)-BIBD is said to be resolvable if the b blocks can be partitioned into r parallel classes. A $(v, 3, 1)$-resolvable BIBD is said to be a Kirkman design of order v. A necessary condition for the existence of a Kirkman design of order v is that $v \equiv 3 \pmod 6$. A Kirkman arrangement for $6n+3$ schoolgirls is a Kirkman design of order $6n+3$ where the girls correspond to the treatments, the rows correspond to the blocks and the n days correspond to the n parallel classes.

Kirkman's schoolgirl problem generated great interest in the late 19th century and early 20th century. Celebrated mathematicians like Burnside, Cayley, and

[1] This research was supported in part by NSF research grant 7902.

Sylvester contributed to this problem. Oscar Eckenstein [12] gives a bibliography of the problem in Messenger of Mathematics **41** (1912). For historical interest this list of papers ([**13**]–[**60**]) is included at the end of this paper. It was proved that for several infinite families of integers n, Kirkman arrangements for $6n+3$ girls (i.e. Kirkman design of order $6n+3$) exist. However, for an arbitrary nonnegative integer n no solution was known. Marshall P. Hall, Jr., [**5**, p. 242] observes, "Solutions of the schoolgirl problem are known for a number of specific values of $v=6t+3$, but no solution for the general case is known to the writer." In this paper we prove that for any nonnegative integer n, Kirkman arrangements for $6n+3$ girls exist. Stated differently, a $(v, 3, 1)$-resolvable BIBD exists iff $v \equiv 3 \pmod 6$.

1. **Definitions and examples.** We adopt the following conventions, which will be used throughout this paper. For any finite set A, $|A|$ will denote the number of elements of A and for any positive integer n, I_n will denote the set of integers $\{1, 2, \ldots, n\}$.

Let v and λ be positive integers and K be a set of positive integers. Let X be a finite set and $\mathscr{B}$ be a class of subsets of X. For convenience of description, elements of X will be called *treatments* and elements of $\mathscr{B}$ will be called *blocks*. Such a pair $(X, \mathscr{B})$ is called a (v, K, λ)-*pairwise balanced design* (PBD) iff

(i) $|X|=v$,

(ii) $|B| \in K$, for every $B \in \mathscr{B}$,

(iii) for every pair $x, x' \in X$, $x \neq x'$, there are exactly λ blocks which contain both x and x'.

The concept of PBD was introduced independently by Bose and Shrikhande [**3**] and Hanani [**6**]. The quantities v, K, and λ are called the *parameters* of the PBD. It will be convenient to describe a (v, K, λ)-PBD as a λ-PBD on v treatments with block sizes from K. We shall write simply PBD for a 1-PBD.

If the set K consists of a single positive integer k, then a (v, K, λ)-PBD is called a (v, k, λ)-*balanced incomplete block design* (BIBD). The notion of a BIBD is quite old [**2**]. If the number of treatments is equal to the number of blocks, then the BIBD is called a *symmetric* BIBD. It is well known that a $(n^2+n+1, n+1, 1)$-symmetric BIBD exists if and only if a finite projective plane of order n exists.

Let $(X, \mathscr{B})$ be a PBD and let $X_1 \subseteq X$, $\mathscr{B}_1 \subseteq \mathscr{B}$. $(X_1, \mathscr{B}_1)$ is said to be a *sub*-PBD of $(X, \mathscr{B})$ iff every block of $\mathscr{B}_1$ is a subset of X_1 and $(X_1, \mathscr{B}_1)$ is a PBD. If $(X, \mathscr{B})$ is a $(v, k, 1)$-BIBD, a sub-PBD $(X_1, \mathscr{B}_1)$ is said to be a $(v_1, k, 1)$-sub-BIBD of $(X, \mathscr{B})$, where $v_1=|X_1|$.

Again let $(X, \mathscr{B})$ be a PBD. A class of blocks $\mathscr{B}_1 \subseteq \mathscr{B}$ is said to be a *parallel class of blocks* if every treatment occurs in exactly one block of the class $\mathscr{B}_1$. A $(v, k, 1)$-BIBD $(X, \mathscr{B})$ is said to be *resolvable* if $\mathscr{B}$ can be partitioned into r parallel classes $\mathscr{B}_1, \mathscr{B}_2, \ldots, \mathscr{B}_r$ where $r=(v-1)/(k-1)$. Let $(X, \mathscr{B})$ be a $(v, k, 1)$-resolvable BIBD. A sub-BIBD $(X_0, \mathscr{B}_0)$ is said to be a *resolvable sub*-BIBD of $(X, \mathscr{B})$ if it is possible to partition $\mathscr{B}$ into parallel classes $\mathscr{B}_1, \mathscr{B}_2, \ldots, \mathscr{B}_r$ and $\mathscr{B}_0$ into parallel classes $\mathscr{B}'_1, \mathscr{B}'_2, \ldots, \mathscr{B}'_{r_1}$ such that $\mathscr{B}'_i \subseteq \mathscr{B}_i$, $i=1, 2, \ldots, r_1$. Necessarily then, $(X_0, \mathscr{B}_0)$ must itself be a resolvable BIBD. We should note here that every $(v, k, 1)$-resolvable BIBD

$(X, \mathscr{B})$ contains a $(k, k, 1)$-resolvable sub-BIBD, for if $B \in \mathscr{B}$, then $(B, \{B\})$ is a resolvable sub-BIBD of $(X, \mathscr{B})$. A $(v, 3, 1)$-resolvable BIBD is also called a *Kirkman design of order* v [**5**]. It is well known that if a $(v, k, 1)$-BIBD exists, then $v-1 \equiv 0 \pmod{(k-1)}$ and $v(v-1) \equiv 0 \pmod{k(k-1)}$. If a $(v, k, 1)$-BIBD is resolvable, then we must also have $v \equiv 0 \pmod{k}$. These conditions reduce to $v \equiv k \pmod{k(k-1)}$, so that a necessary condition for the existence of a Kirkman design of order v is $v \equiv 3 \pmod 6$. In this paper we prove that a Kirkman design of order v exists iff $v \equiv 3 \pmod 6$.

$B[K, \lambda]$ will denote the set of integers v for which a (v, K, λ)-PBD exists and $B[k, \lambda]$ will denote the set of integers v for which a (v, k, λ)-BIBD exists. This notation is due to Hanani [**6**]. $B^*[k, 1]$ will denote the set of integers v for which a $(v, k, 1)$-resolvable BIBD exists. Similarly we define

$$R_k = \{t \mid ((k-1)t+1, k, 1)\text{-BIBD exists}\},$$
$$R_k^* = \{t \mid ((k-1)t+1, k, 1)\text{-resolvable BIBD exists}\}.$$

Let $(X, \mathscr{B})$ be a PBD. A set of blocks $\mathscr{B}'$ is said to be a *clear set of blocks* if no treatment occurs more than once among the blocks $\mathscr{B}'$. The concept of a clear set of blocks is due to Bose and Shrikhande [**3**]. We define $B[K, K', 1]$ to be the set of integers v for which there exists a PBD $(X, \mathscr{B} \cup \mathscr{B}')$ such that $|X| = v$, blocks of $\mathscr{B}$ have sizes from K, and $\mathscr{B}'$ is a clear set of blocks with sizes from K'. A PBD $(X, \mathscr{B})$ is said to be a (k, t)-*completed resolvable design* iff (i) $|X| = kt+1$ and (ii) $\mathscr{B}$ consists of blocks of size $(k+1)$ and a single block of size t. The block of size t will be called the *block at infinity*. The relation of (k, t)-completed resolvable designs to $((k-1)t+1, k, 1)$-resolvable BIBD's is the same as that of finite projective planes to finite affine planes (see Lemma 2.1).

An (m, n, d, λ)-*orthogonal array is* a quadruple (S_1, S_2, S_3, A) where S_1, S_2, and S_3 are finite sets with cardinalities m, λn^d, and n respectively, and A is a mapping from $S_1 \times S_2$ to S_3 such that for distinct elements $a_1, a_2, \ldots, a_d$ of S_1 and arbitrary elements $b_1, b_2, \ldots, b_d$ of S_3 the following system of equations has exactly λ solutions for x:

$$\begin{aligned} A(a_1, x) &= b_1, \\ \vdots \quad & \quad \vdots \\ A(a_d, x) &= b_d, \text{ and} \\ x &\in S_2. \end{aligned}$$

Usually S_1 is regarded as the set of rows of the array, S_2 as the set of columns and S_3 as the set of symbols, and in the ith row and jth column one writes down the symbol $A(i, j)$. The use of orthogonal arrays in the construction of Kirkman designs is found in a rudimentary form in Harrison [**1**]. Formally, the concept was first introduced by Rao [**10**]. It is well known that a set of $m-2$ mutually orthogonal Latin squares of order n exists iff an $(m, n, 2, 1)$-orthogonal array exists. If we let $N(n)$ denote the maximum number of mutually orthogonal Latin squares of order

n, then all the results we shall need on the existence of orthogonal arrays are furnished by the following well-known result due to MacNeish [3]:

LEMMA 1.1. *If* $n = p_1^{e_1} p_2^{e_2} \cdots p_r^{e_r}$ *is the factorization of the integer* n *into powers of distinct primes, then* $N(n) \geqq \min (p_i^{e_i} - 1)$, $i = 1, 2, \ldots, r$.

Let (S_1, S_2, S_3, A) be an (m, n, d, λ)-orthogonal array. A subset S_2^1 of columns is said to be a *parallel class of columns* if for all $a \in S_1$ and $b \in S_3$ the following equation has exactly one solution: $A(a, x) = b$, $x \in S_2^1$, in other words iff in the set of columns S_2^1 in every row every symbol occurs exactly once. The orthogonal array is said to be a *resolvable orthogonal array* if S_2 can be partitioned into parallel classes of columns. It is well known that an $(m, n, 2, 1)$-orthogonal array exists iff an $(m-1, n, 2, 1)$-resolvable orthogonal array exists. We now proceed to give a few examples to illustrate some of the definitions. The following is a resolvable orthogonal array with $m=3$, $n=3$, $d=2$, $\lambda=1$, and $S_3 = \{a, b, c\}$:

$$\begin{array}{ccc} S_2^1 & S_2^2 & S_2^3 \\ a\ b\ c & a\ b\ c & a\ b\ c \\ a\ b\ c & b\ c\ a & c\ a\ b \\ a\ b\ c & c\ a\ b & b\ c\ a \end{array} \tag{1}$$

The following are the blocks of resolvable BIBD with $v=9$, $k=3$, $\lambda=1$, $r=4$, and $X = \{1, 2, 3, 4, 5, 6, 7, 8, 9\}$

$$\begin{array}{cccc} \mathscr{B}_1 & \mathscr{B}_2 & \mathscr{B}_3 & \mathscr{B}_4 \\ (1, 2, 3) & (1, 4, 7) & (1, 6, 8) & (1, 5, 9) \\ (4, 5, 6) & (2, 5, 8) & (2, 4, 9) & (2, 6, 7) \\ (7, 8, 9) & (3, 6, 9) & (3, 5, 7) & (3, 4, 8) \end{array} \tag{2}$$

The following are the blocks of a (3,4)-completed resolvable design, or equivalently, a (13, 4, 1)-BIBD, with $X = \{1, 2, 3, 4, 5, 6, 7, 8, 9, \theta_1, \theta_2, \theta_3, \theta_4\}$.

$$\begin{array}{cccc} (1, 2, 3, \theta_1) & (1, 4, 7, \theta_2) & (1, 6, 8, \theta_3) & (1, 5, 9, \theta_4) \\ (4, 5, 6, \theta_1) & (2, 5, 8, \theta_2) & (2, 4, 9, \theta_3) & (2, 6, 7, \theta_4) \\ (7, 8, 9, \theta_1) & (7, 8, 9, \theta_2) & (3, 5, 7, \theta_3) & (3, 4, 8, \theta_4) \\ & \multicolumn{2}{c}{(\theta_1, \theta_2, \theta_3, \theta_4)} & \end{array} \tag{3}$$

The following is a resolvable BIBD with $v=27$, $k=3$, $\lambda=1$, and a resolvable sub-BIBD with $v_1=9$, $k=3$, $\lambda=1$ where $X = \{(i, a), (i, b), (i, c), i = 1, 2, 3, 4, 5, 6, 7, 8, 9\}$ and $X_1 = \{(i, a), i = 1, 2, 3, \ldots, 9\}$: Eight parallel classes of the BIBD are obtained as

$$\begin{aligned} \mathscr{B}_i^* = \{((x_1, \theta_1), (x_2, \theta_2), (x_3, \theta_3)) \mid {} & (x_1, x_2, x_3) \text{ is a block of } \mathscr{B}_i \\ & \text{and } (\theta_1, \theta_2, \theta_3) \text{ is a column of } S_2^j\}, \quad i = 1, 2, 3, 4 \text{ and } j = 2, 3. \end{aligned} \tag{4}$$

$\mathscr{B}_i$ and S_2^j are as defined in (1) and (2). Four more parallel classes are given by $((x_1, \theta), (x_2, \theta), (x_3, \theta))$ where

$$(x_1, x_2, x_3) \in \mathscr{B}_i, \qquad i = 1, 2, 3, 4 \quad \text{and} \quad \theta = a, b, c.$$

These parallel classes contain the 4 parallel classes of the subdesign. Finally the 13th parallel class is

$$\{((x, a), (x, b), (x, c)) \mid x = 1, 2, \ldots, 9\}.$$

2. **Composition theorems for balanced incomplete block designs.** In this section we prove several theorems of the following nature. One starts with a number of designs of small orders and perhaps some orthogonal arrays and produces as a result, a design of larger order.

LEMMA 2.1. *$t \in R_k^*$ iff a (k, t)-completed resolvable design exists.*

PROOF. Assume that $t \in R_k^*$ and let $(X, \mathscr{B})$ be a $((k-1)t+1, k, 1)$-resolvable BIBD. To construct a (k, t)-completed resolvable design, for the treatment set we take $X^* = X \cup \{\theta_1, \theta_2, \ldots, \theta_t\}$. Let $\mathscr{B}_i$, $i=1, 2, \ldots, t$, be the t parallel classes of the resolvable BIBD. For each $B \in \mathscr{B}_i$, define $B^* = B \cup \{\theta_i\}$. Let $B_\infty = \{\theta_1, \theta_2, \ldots, \theta_t\}$ and define

$$\mathscr{B}^* = \bigcup_{B \in \mathscr{B}} \{B^*\} \cup \{B_\infty\}.$$

Then it is easily checked that $(X^*, \mathscr{B}^*)$ is a (k, t)-completed resolvable design.

Conversely, assume that $(X^*, \mathscr{B}^*)$ is a (k, t)-completed resolvable design. From the definition of a completed resolvable design, $\mathscr{B}^*$ contains a block $B_\infty = \{\theta_1, \theta_2, \ldots, \theta_t\}$ of size t. Let x be a treatment other than θ_i, $i=1, 2, \ldots, t$. Then there is a unique block $B_i \in \mathscr{B}^*$ of size $(k+1)$ containing both x and θ_i, $\theta = 1, 2, \ldots, t$. For different choices of i, the sets $B_i - \{x\}$ are disjoint and each consists of k elements. Thus $|\bigcup_{i=1}^t B_i| = kt+1 = |X^*|$, so there can be no other blocks containing x. This argument shows that every block B^* of $\mathscr{B}^*$, other than B_∞, intersects B_∞ in precisely one treatment. To construct a $((k-1)t+1, k, 1)$-resolvable BIBD, for the treatment set take $X = X^* - B_\infty$. For each block $B^* \in \mathscr{B}^*$, $B^* \neq B_\infty$, define $B = B(B^*) = B^* - (B^* \cap B_\infty)$ and put $\mathscr{B} = \{B \mid B^* \in \mathscr{B}^*, B^* \neq B_\infty\}$. Clearly $(X, \mathscr{B})$ is a $((k-1)t+1, k, 1)$-BIBD. Let $\mathscr{B}_i^* = \{B^* \mid \theta_i \in B^*, B^* \neq B_\infty\}$ and let $\mathscr{B}_i = \{B \mid B^* \in \mathscr{B}_i^*\}$ for each $i=1, 2, \ldots, t$. Then we see $\mathscr{B}_1, \mathscr{B}_2, \ldots, \mathscr{B}_t$ are t parallel classes and this completes the proof of the lemma.

THEOREM 1. $B[R_k^*, 1] = R_k^*$.

PROOF. In general, we always have $B[K, 1] \supseteq K$ since given $n \in K$, the PBD $(I_n, \{I_n\})$ shows $n \in B[K, 1]$. Now let $v \in B[R_k^*, 1]$ be given. That is, there exists a PBD $(X, \mathscr{S})$ on v treatments such that $|S| \in R_k^*$ for all $S \in \mathscr{S}$. To prove the theorem, because of Lemma 2.1 it will be sufficient to construct a (k, v)-completed resolvable design. For the treatment set we take $X^* = X \times I_k \cup \{\theta\}$. Since $\forall S \in \mathscr{S}$, $|S| \in R_k^*$, there exists by Lemma 2.1, a $(k, |S|)$-completed resolvable design. Construct such a design on the treatment set $S \times I_k \cup \{\theta\}$. Write $S = \{a_1, a_2, \ldots, a_t\}$. Without loss of generality, we can take the block at infinity to be $\{(a_1, 1), (a_2, 1), \ldots, (a_t, 1)\}$, and the blocks through θ to be $\{\theta, (a_i, 1), (a_i, 2), \ldots, (a_i, k)\}$, $i=1, 2, \ldots, t$. In other words we assume the following configuration:

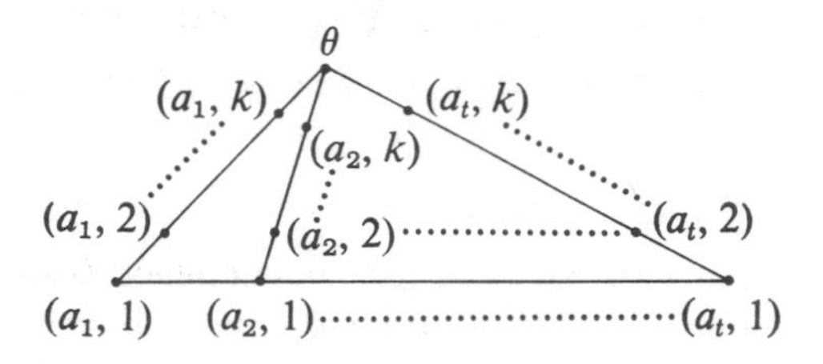

By $\mathscr{B}_S$, we denote the remaining blocks of the $(k, |S|)$-completed resolvable design on $S \times I_k \cup \{\theta\}$. This is done for each block $S \in \mathscr{S}$.

Let $B_\infty = \{(a, 1) \mid a \in X\}$ and let $\mathscr{B}_0$ be the class of blocks $\{\theta, (a, 1), (a, 2), \ldots, (a, k)\}$, $a \in X$. Put $\mathscr{B}^* = \bigcup_{S \in \mathscr{S}} \mathscr{B}_S \cup \{B_\infty\} \cup \mathscr{B}_0$. We now check that $(X^*, \mathscr{B}^*)$ is the desired (k, v)-completed resolvable design. The pairs of treatments of the type $\{(a, i), (a', j)\}$, $a \neq a'$, $i \neq j$, occur in one block of $\mathscr{B}_S$ where S is the unique block containing a and a'. The pairs of the type $\{(a, i), (a, j)\}$ occur in a block of $\mathscr{B}_0$. Pairs of the type $\{(a, 1), (a', 1)\}$ occur only in B_∞, and pairs of the type $\{(a, i), (a', i)\}$, $i \neq 1$, occur in one block of $\mathscr{B}_S$, where S is the block of $\mathscr{S}$ containing a and a'. Finally, all pairs of the type $\{\theta, (a, i)\}$ occur only in one block of $\mathscr{B}_0$. This completes the proof of the theorem.

The (k, v)-completed resolvable design constructed in the proof of Theorem 1 leads to a $((k-1)v+1, k, 1)$-resolvable BIBD which contains resolvable sub-BIBD's of order $(k-1)|S|+1$ for all $S \in \mathscr{S}$.

We pause now to indicate how Theorem 1 is useful in the construction of Kirkman designs. In example (2), we have a Kirkman design of order 9. Hence $4 \in R_3^*$, so $B[4, 1] \subseteq R_3^*$. Hanani has proved [6] that $v \in B[4, 1]$ iff $v \equiv 1$ or 4 (mod 12). That is, all positive integers congruent to 1 or 4 modulo 12 are elements of R_3^* and consequently there exist Kirkman designs of order v for all positive integers $v \equiv 3$ or 9 (modulo 24). Our treatment of the problem, however, will be independent of Hanani's result.

We now consider the problem of finding PBD's with block sizes in R_k^*. To this end, Theorem 2 will be helpful. Theorems 1, 2, and Lemma 2.2 will be applied in §4.

THEOREM 2. *If* $v \in B[R_k, K', 1]$, *then* $(k-1)v+1 \in B[K'', 1]$ *where* $K'' = \{k\} \cup \{(k-1)y+1 \mid y \in K'\}$.

PROOF. From the hypothesis there exists a pairwise balanced design $(X, \mathscr{E} \cup \mathscr{S})$ such that $|X| = v$ and blocks of $\mathscr{E}$ have sizes from R_k and $\mathscr{S}$ consists of a set of clear blocks with sizes from K'. To construct the required design on $(k-1)v+1$ treatment, for the treatment set we take $X^* = X \times I_{k-1} \cup \{\theta\}$. Let $B \in \mathscr{E}$. Since $|B| \in R_k$, $a((k-1)|B|+1, k, 1)$-BIBD exists. Let $B \times I_{k-1} \cup \{\theta\}$ be the treatment set of such a design. Without loss of generality, we can take the blocks through θ to be $\{\theta, (a, 1), \ldots, (a, k-1)\}$, $a \in B$. In other words we assume the existence of the following configuration:

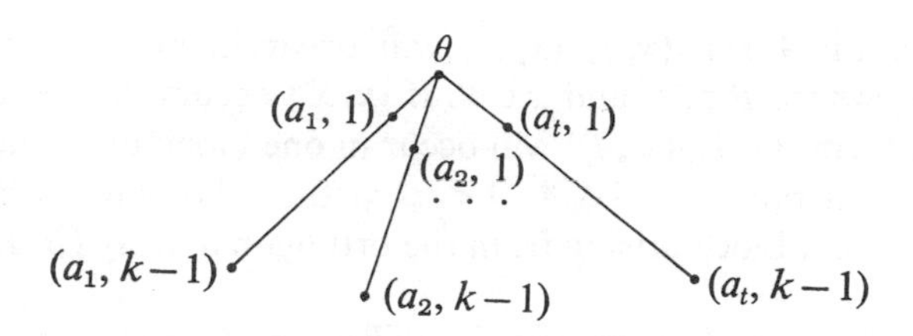

where $B=\{a_1, a_2, \ldots, a_t\}$ and $t \in R_k$. Let $\mathscr{B}_B$ be the remaining blocks of the design on $B \times I_{k-1} \cup \{\theta\}$. Let $\mathscr{B}_0=\{(\theta, (a, 1), \ldots, (a, k-1)) \mid a \in X-\bigcup_{S\in\mathscr{S}} S\}$, $\mathscr{B}_{\mathscr{S}}=\{(S \times I_{k-1} \cup \{\theta\}) \mid S \in \mathscr{S}\}$ and $\mathscr{B}^*=\bigcup_{B\in\mathscr{E}} \mathscr{B}_B \cup \mathscr{B}_0 \cup \mathscr{B}_{\mathscr{S}}$. It is now easily checked that $(X^*, \mathscr{B}^*)$ is the required design on $(k-1)v+1$ treatments.

Consider an $(n, m, 2, 1)$-orthogonal array (S_1, S_2, S_3, A). Take $S_1 \times S_3$ as a set of treatments. The orthogonal array gives a class of m^2 blocks $\{(x, A(x, y)) \mid x \in S_1\}$, $y \in S_2$. We shall denote this class of m^2 blocks by $\mathscr{P}_{n,m}$. Suppose S_2 can be partitioned as $S_2=S_2^1 \cup S_2^2 \cup \cdots \cup S_2^q \cup S_2'$ where S_2^i, $i=1, 2, \ldots, q$, are parallel classes of columns. Then $\mathscr{P}_{n,m}$ can be partitioned as $\mathscr{P}_{n,m}=\mathscr{P}_{n,m}^1 \cup \mathscr{P}_{n,m}^2 \cdots \cup \mathscr{P}_{n,m}^q \cup \mathscr{P}_{n,m}'$ where $\mathscr{P}_{n,m}^i$ is a parallel class of blocks on $S_1 \times S_3$, $i=1, 2, \ldots, q$.

LEMMA 2.2. *If an $(n, m, 2, 1)$-orthogonal array with q parallel classes exists, then $nm+q \in B[\{n, n+1\}, \{m, q\}, 1]$.*

PROOF. For our treatment set, we take $X=I_n \times I_m \cup \{\theta_1, \theta_2, \ldots, \theta_q\}$. From the hypothesis it follows that there exists an orthogonal array (I_n, I_{m^2}, I_m, A). Let $\mathscr{P}_{n,m}$ be the corresponding class of blocks. For $B \in \mathscr{P}_{n,m}^i$ define $B^*=B \cup \{\theta_i\}$, $i=1, 2, \ldots, q$, and for $B \in \mathscr{P}_{n,m}'$ set $B^*=B$. Let $\mathscr{P}^*=\bigcup_{B\in\mathscr{P}_{n,m}} \{B^*\}$. Finally define $B_\infty=\{\theta_1, \theta_2, \ldots, \theta_q\}$, $B_i=\{(i, 1), (i, 2), \ldots, (i, m)\}$, $i=1, 2, \ldots, n$, and $\mathscr{P}=\{B_\infty\} \cup \{B_i \mid i=1, 2, \ldots, n\}$. Now it is easily checked that $(X, \mathscr{P}^* \cup \mathscr{P})$ is the required PBD with $\mathscr{P}$ as a clear set of blocks of size m and q.

The next theorem for the special case $k=3$ was proved by G. D. Harrison **[1]** as early as 1916.

THEOREM 3 (HARRISON'S THEOREM). *If $kl_i \in B^*[k, 1]$, $i=1, 2$, and a $(k, l_2, 2, 1)$-resolvable orthogonal array exists, then $kl_1l_2 \in B^*[k, 1]$.*

PROOF. Let $(X, \mathscr{B})$ be the $(kl_1, k, 1)$-resolvable BIBD and $\mathscr{B}_i$, $i=1, 2, \ldots, r_1$, be the r_1 parallel classes.

To construct the $(kl_1l_2, k, 1)$-resolvable BIBD, we take $X^*=X \times I_{l_2}$ as the treatment set. For $B \in \mathscr{B}_1$, construct a $(kl_2, k, 1)$-resolvable BIBD on the treatment set $B \times I_{l_2}$. Let $\mathscr{S}_{B,j}$, $j=1, 2, \ldots, r_2$, be the r_2 parallel classes of the design. Define $\mathscr{S}_j^*=\bigcup_{B\in\mathscr{B}_1} \mathscr{S}_{B,j}, j=1, 2, \ldots, r_2$. Clearly $\mathscr{S}_j^*$ is a parallel class on $X^*, j=1, 2, \ldots, r_2$. For $B \in \mathscr{B}_i$, $i=2, 3, \ldots, r_1$, construct an orthogonal array $(B, I_{l_2}^2, I_{l_2}, M)$ with B as the set of rows, I_{l_2} as the set of symbols, $I_{l_2}^2$ as the set of columns and M as the mapping. Let $\mathscr{P}_{B,j}$, $j=1, 2, .., l_2$, be the l_2 parallel classes of blocks on $B \times I_{l_2}$ generated from the resolvable orthogonal array. Define $\mathscr{S}_{i,j}^*=\bigcup_{B\in\mathscr{B}_i} \mathscr{P}_{B,j}$, $i=2, 3, \ldots, r_1$, $j=1, 2, \ldots, l_2$. Clearly $\mathscr{S}_{i,j}^*$ is a parallel class of blocks on X^*. Finally set $\mathscr{S}^*=\bigcup_{j=1}^{r_2} \mathscr{S}_j^* \cup \bigcup_{i=2}^{r_1} \bigcup_{j=1}^{l_2} \mathscr{S}_{i,j}^*$ and check that $(X^*, \mathscr{S}^*)$ is the required design. Pairs of the form $\{(x, i), (x, j)\}$ will occur in exactly one block of the BIBD on $B \times I_{l_2}$ where $B \in \mathscr{B}_1$ and $x \in B$. If $\{x, x'\}$ occurs in a block $B \in \mathscr{B}_1$, then the pairs of the form $\{(x, i), (x', i)\}$ will occur in one block of the design on $B \times I_{l_2}$. If $\{x, x'\}$ occur in a block B, $B \in \mathscr{B}_i$, $1 < i \leqq r_1$, then the pairs of the form $\{(x, i), (x', j)\}$ will occur in a block arising from the orthogonal array $(B, I_{l_2}^2, I_{l_2}, M)$.

THEOREM 4. *If a $(v_1, k, 1)$-resolvable BIBD, a $((k-1)m+v_2, k, 1)$-resolvable BIBD with a $(v_2, k, 1)$-resolvable sub-BIBD (or $v_2=1$), and a $(k, m, 2, 1)$-resolvable orthogonal array exist, then an $(m(v_1-1)+v_2, k, 1)$-resolvable BIBD exists.*

PROOF. Let $(X, \mathscr{B})$ be the $(v_1, k, 1)$-resolvable BIBD and $\mathscr{B}_i$, $i=1, 2, \ldots, r_1$, denote the r_1 parallel classes. Let θ be a fixed treatment and B_i, $i=1, 2, \ldots, r_1$, denote the blocks which contain θ, where $B_i \in \mathscr{B}_i$.

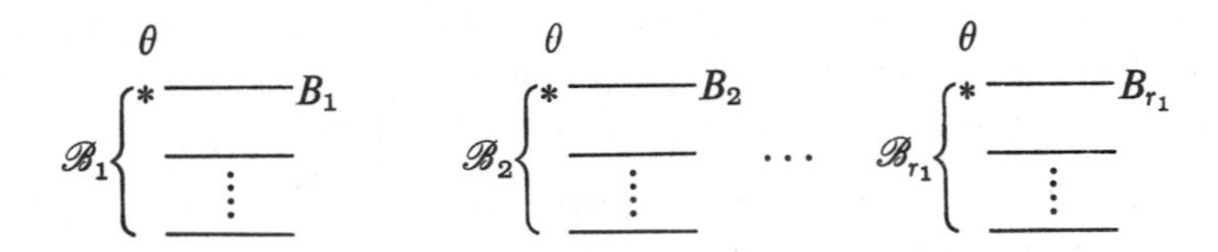

Let $X' = X - \{\theta\}$. To construct our resolvable design, we shall take the treatment set to be $X^* = X' \times I_m \cup Y$ where Y is a set of v_2 treatments. Let $B_i' = B_i - \{\theta\}$, $i=1, 2, \ldots, r_1$, and construct a $((k-1)m+v_2, k, 1)$-resolvable BIBD on the treatment set $B_i' \times I_m \cup Y$ containing a $(v_2, k, 1)$-resolvable sub-BIBD (or $v_2=1$) on the treatment set Y. Let the parallel classes of the BIBD be $\mathscr{S}_j^i$, $j=1, 2, \ldots, (r_2+m)$ and if $v_2 > 1$, let the parallel classes of the sub-BIBD be $\mathscr{S}_j'$, $j=1, 2, \ldots, r_2$ where $\mathscr{S}_j' \subset \mathscr{S}_j^i$, $j=1, 2, \ldots, r_2$. Let $\mathscr{S}_j^{*i} = \mathscr{S}_j^i - \mathscr{S}_j'$, $j=1, 2, \ldots, r_2$.

This is to be done for each $i=1, 2, \ldots, r_1$. The following diagram is suggestive:

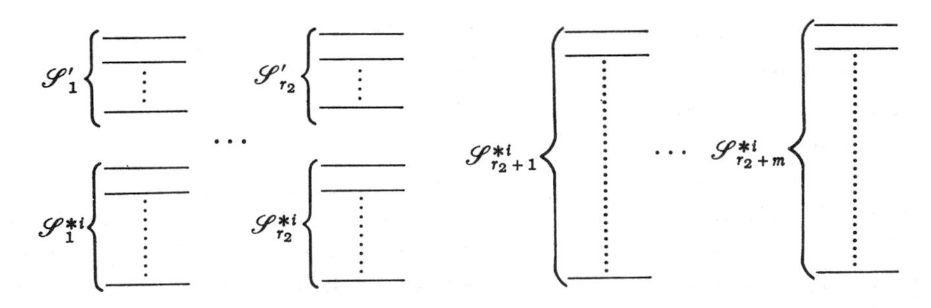

$\mathscr{S}_j^{*i}$ is a parallel class on $B_i' \times I_m$. Since $\bigcup_{i=1}^{r_1} B_i'$ is a parallel class on X', $\bigcup_{i=1}^{r_1} \mathscr{S}_j^{*i}$ is a parallel class on $X' \times I_m$. It follows that $\mathscr{E}_j = \mathscr{S}_j' \cup \bigcup_{i=1}^{r_1} \mathscr{S}_j^{*i}$, $j=1, 2, \ldots, r_2$, is a parallel class on $X^* = X' \cup Y$. For $B \in \mathscr{B}_i$, $B \neq B_i$, $i=1, 2, \ldots, r_1$, construct a $(k, m, 2, 1)$-resolvable orthogonal array with B as the set of rows and I_m as the set of symbols. Let $\mathscr{P}_{B,j}^i$, $j=r_2+1, \ldots, m$, denote the m parallel classes of blocks on

$B \times I_m$, $B \in \mathscr{B}_i$ derived from the columns of the orthogonal array. $\mathscr{S}^i_j, j=r_2+1, \ldots, r_2+m$, is a parallel class of blocks on $B'_i \times I_m \cup Y$.

Note that $\bigcup_{B \in \mathscr{B}_i; B \neq B_i} \{B\}$ is a parallel class on $X' - B'_i$. Hence

$$\mathscr{E}^i_j = \bigcup_{B \in \mathscr{B}_i; B \neq B_i} \mathscr{P}_{B,j} \cup \mathscr{S}^i_j \qquad j = (r_2+1), \ldots, (r_2+m), \quad i = 1, 2, \ldots, r_1$$

is a parallel class on $X' \times I_m \cup Y$. Finally let $\mathscr{E}^*$ denote the union of the parallel classes $\mathscr{E}_j, j=1, 2, \ldots, r_2$, and $\mathscr{E}^i_j, i=1, 2, \ldots, r_1, j=(r_2+1), \ldots, (r_2+m)$. $(X^*, \mathscr{E}^*)$ is the required resolvable BIBD. Pairs of the form $\{y, a\}$, $y \in Y$, $a \in X' \times I_m$, occur exactly once in a block belonging to $\mathscr{E}_j$ for some j. Pairs of the form $\{(x, l), (x, l')\}$, $l \neq l'$, $x \in X'$, occur in a block coming from the BIBD on $B'_i \times I_m \cup Y$ where B'_i contains x. Such a block of course belongs to the class $\mathscr{E}_j$ for some j. Pairs of the form $\{(x, l), (x', l')\}$, $x \neq x' \in X'$, will occur in a block coming from the BIBD on $B'_i \times I_m \cup Y$ if B'_i contains $\{x, x'\}$. If the block containing (x, x') is B, $B \neq B_i$, then the pair considered above will occur in a block arising from the orthogonal array with B as the set of rows and I_m as the set of symbols. This completes the proof of the theorem.

It is worth noticing that the $(m(v_1-1)+v_2, k, 1)$-resolvable BIBD constructed in Theorem 4 contains resolvable sub-BIBD's on v_2 treatments and $(k-1)m+v_2$ treatments.

3. **Direct construction of Kirkman designs.** The theorems of the previous section are of a recursive nature in the sense that we hypothesize the existence of certain resolvable BIBD's and other designs in order to construct resolvable BIBD's of larger order. Clearly, it is necessary to know the existence of a number of Kirkman designs of small order if we are to recursively construct, and so prove the existence of, Kirkman designs of arbitrary order $v \equiv 3 \pmod 6$. In this section we give methods of constructing Kirkman designs from certain algebraic structures, and we illustrate the applications of these constructions and Theorems 3 and 4 of §2 by establishing the existence of a class of Kirkman designs which will be useful in §4.

These methods of direct construction are called "difference methods" or "base block methods". They were first introduced by Bose [2] and are discussed by Hall [5]. Let G be an additive abelian group. We consider as possible treatment sets for designs, the two sets $X_1 = G \times I_n$ and $X_2 = G \times I_n \cup \{\infty\}$. For convenience of notation, we write the ordered pair $(a, i) \in G \times I_n$ as $(a)_i$ or simply as a_i if there is no danger of confusion.

Let $B = \{(a_1)_{i_1}, (a_2)_{i_2}, \ldots, (a_k)_{i_k}\} \subseteq G \times I_n$. For $g \in G$, we let $B+g$ denote the block $\{(a_1+g)_{i_1}, (a_2+g)_{i_2}, \ldots, (a_k+g)_{i_k}\}$. Similarly, for $B = \{(a_1)_{i_1}, \ldots, (a_k)_{i_k}, \infty\} \subseteq G \times I_n \cup \{\infty\}$, we let $B+g = \{(a_1+g)_{i_1}, \ldots, (a_k+g)_{i_k}, \infty\}$. For $B \subseteq X_1$ or X_2, we write ΔB for the class of blocks $\{B+g \mid g \in G\}$. A list on G is a mapping l for G to nonnegative integers. It is convenient to regard a list as a collection of elements of G in which elements can occur more than once. The union of two lists l and l' is the mapping $m = l + l'$ where $m(g) = l(g) + l'(g)$, $g \in G$. Let $B \cap (G \times \{i\}) = \{g^i_1, g^i_2, \ldots, g^i_{c_i}\}$. These are the elements in B which have the subscript i, $i=1, 2, \ldots, n$. By the "list of $[i, i]$ differences of B" we mean the list of the elements

$$\{g_r^i - g_{r'}^i \mid r \neq r'; r, r' = 1, 2, \ldots, c_i\}.$$

For $i \neq j$, by the "list of $[i, j]$-differences of B" we mean the list of elements

$$\{g_r^i - g_{r'}^j \mid r = 1, 2, \ldots, c_i; r' = 1, 2, \ldots, c_j\}.$$

By the list of $[i, j]$-differences of $\mathscr{B}$ we mean the union of the lists of $[i, j]$-differences of B for all $B \in \mathscr{B}$.

We use the phrase "all nonzero pure differences occur in B λ times" to mean that for every nonzero $g \in G$ and for every $i \in I_n$, g occurs λ times in the list of $[i, i]$ differences of $\mathscr{B}$. The phrase "all mixed differences occur in $\mathscr{B}$ λ times" will mean that for every $g \in G$ and for every pair $i, j \in I_n$, $i \neq j$, the element g occurs λ times in the list of $[i, j]$-differences of $\mathscr{B}$. The following two lemmas are due to Bose **[2]**. A proof can be found also in Hall **[5]**.

LEMMA 3.1. *Let $\mathscr{B}$ be a class of subsets of $X_1 = G \times I_n$. Then $(X_1, \bigcup_{B \in \mathscr{B}} \Delta B)$ is a PBD iff*
(i) *all nonzero pure differences occur once in $\mathscr{B}$ and*
(ii) *all mixed differences occur once in $\mathscr{B}$.*

When $\mathscr{B}$ is a class of subsets of $X_2 = G \times I_n \cup \{\infty\}$, we must also consider the pairs $\{(a)_i, \infty\}$. But noting that $(a)_i$ and ∞ occur together in λ blocks of ΔB iff $\infty \in B$ and the number of elements $g \in G$ such that $(g)_i \in B$ is precisely λ, we have

LEMMA 3.2. *Let $\mathscr{B}$ be a class of subsets of $X_2 = G \times I_n \cup \{\infty\}$. Then $(X_2, \bigcup_{B \in \mathscr{B}} \Delta B)$ is a PBD iff*
(i) *all nonzero pure differences occur once in $\mathscr{B}$,*
(ii) *all mixed differences occur once in $\mathscr{B}$,*
(iii) *for each $i \in I_n$, there is a unique $g \in G$ such that ∞ and $(g)_i$ occur together in some block of $\mathscr{B}$.*

We now give an example of the direct construction of a Kirkman design of order 69. (For examples of direct constructions of Kirkman designs with orders $v \equiv 3 \pmod 6$, $9 \leqq v \leqq 99$, by essentially the same techniques, the reader is referred to **[1]**.) We let $G = Z_{23}$, the additive group of integers modulo 23, and we take as our treatment set $Z_{23} \times I_3$. As base blocks we take

$$B_0: (0_1, 0_2, 0_3)$$

$B_1: (18_1, 19_1, 15_1)$ $\quad B_4: (2_1, 4_1, 17_1)$ $\quad B_7: (1_1, 8_1, 13_1)$
$B_2: (18_2, 19_2, 22_2)$ $\quad B_5: (2_2, 4_2, 17_2)$ $\quad B_8: (1_2, 8_2, 13_2)$
$B_3: (18_3, 19_3, 22_3)$ $\quad B_6: (2_3, 15_3, 17_3)$ $\quad B_9: (1_3, 8_3, 13_3)$

$B_{10}: (3_1, 9_1, 14_2)$ $\quad B_{13}: (7_1, 21_1, 6_2)$
$B_{11}: (3_2, 9_2, 14_3)$ $\quad B_{14}: (7_2, 21_2, 6_3)$
$B_{12}: (3_3, 9_3, 14_1)$ $\quad B_{15}: (7_3, 21_3, 6_1)$

$B_{16}: (5_1, 20_2, 11_3)$ $B_{19}: (10_1, 16_2, 12_3)$
$B_{17}: (5_2, 20_3, 11_1)$ $B_{20}: (10_2, 16_3, 12_1)$ $B_{22}: (22_1, 15_2, 4_3)$
$B_{18}: (5_3, 20_1, 11_2)$ $B_{21}: (10_3, 16_1, 12_2)$

$B_{23}: (0_1, 18_2, 11_3)$ $B_{24}: (0_2, 18_3, 11_1)$

$B_{25}: (0_1, 13_2, 16_3)$ $B_{28}: (0_1, 10_2, 14_3)$ $B_{31}: (0_1, 2_2, 22_3)$
$B_{26}: (0_2, 13_3, 16_1)$ $B_{29}: (0_2, 10_3, 14_1)$ $B_{32}: (0_2, 2_3, 22_1)$
$B_{27}: (0_3, 13_1, 16_2)$ $B_{30}: (0_3, 10_1, 14_2)$ $B_{33}: (0_3, 2_1, 22_2)$.

It may be verified that all nonzero pure differences and all mixed differences occur precisely once in the blocks $B_0, B_1, \ldots, B_{33}$. Thus by our previous discussion, $(Z_{23} \times I_3, \bigcup_{i=1}^{33} \Delta B_i)$ will be a (69, 3, 1) BIBD. To show this design is in fact a Kirkman design, it remains to exhibit a partition of $\bigcup_{i=0}^{33} \Delta B_i$ into parallel classes. We note that the blocks $B_0, B_1, \ldots, B_{22}$ form a parallel class, so that for each $g \in Z_{23}$, B_0+g, $B_1+g, \ldots, B_{22}+g$ is a parallel class. Hence $\bigcup_{i=0}^{22} \Delta B_i$ may be partitioned into 23 parallel classes. Finally we note that for each i, $23 \leqq i \leqq 33$, ΔB_i is itself a parallel class, so indeed $\bigcup_{i=0}^{33} \Delta B_i$ may be partitioned into 34 parallel classes.

Let q be a prime power, $q \equiv 1 \pmod 6$, and let $GF(q)$ denote the finite field with q elements. If we take G to be the additive group of $GF(q)$, we may take advantage of the multiplicative structure to find base blocks for Kirkman designs.

THEOREM 5. *Let $q=6t+1$ be a prime power. Then a Kirkman design of order $3q$ exists.*

PROOF. We shall construct a Kirkman design on the set of treatments $GF(q) \times I_3$. Let g be a generator of the multiplicative group of $GF(q)$ and take as base blocks

$$\begin{aligned}
A_0 &= (0_1, 0_2, 0_3) \\
B_1^i &= (g_1^i, g_1^{i+2t}, g_1^{i+4t}) && i = 0, 1, \ldots, t-1 \\
B_2^i &= (g_2^i, g_2^{i+2t}, g_2^{i+4t}) && i = 0, 1, \ldots, t-1 \\
B_3^i &= (g_3^i, g_3^{i+2t}, g_3^{i+4t}) && i = 0, 1, \ldots, t-1 \\
A^i &= (g_1^i, g_2^{i+2t}, g_3^{i+4t}) && i = 0, 1, \ldots, 6t-1.
\end{aligned}$$

Let $\mathscr{B} = \{A_0\} \cup \{B_j^i \mid 0 \leqq i < t, 1 \leqq j \leqq 3\} \cup \{A^i \mid 0 \leqq i < 6t\}$. We need to show first that $(GF(q) \times I_3, \bigcup_{C \in \mathscr{B}} \Delta C)$ is a $(3q, 3, 1)$-BIBD, and second, we need to exhibit the parallel classes.

Consider first the $[j, j]$ pure differences from $\mathscr{B}$. All $[j, j]$ pure differences are found only in the blocks B_j^i, $i=0, 1, \ldots, t-1$, and the differences that occur are $\pm g^i(g^{2t}-1)$, $\pm g^i(g^{4t}-1)$, $\pm g^i(g^{4t}-g^{2t})$, $i=0, 1, \ldots, t-1$. Writing $a=g^{2t}-1$ and noting that $-1=g^{3t}$ and $g^{4t}-1=g^t a$, we see that the $[j, j]$ differences that occur in $\mathscr{B}$ are $g^i a$, $g^{i+3t}a$, $g^{i+t}a$, $g^{i+4t}a$, $g^{i+2t}a$, $g^{i+5t}a$ for $i=0, 1, \ldots, t-1$. Thus all nonzero pure differences occur once in $\mathscr{B}$.

We now consider the $[j_1, j_2]$, $j_1 \neq j_2$, mixed differences. The 0 mixed differences all occur in A_0. All other mixed differences occur in the blocks A^i, $i=0, 1, \ldots, 6t-1$, and it is clear that all nonzero mixed differences occur once in these blocks, e.g. the [2, 1] differences that occur are $g^i(g^{2t}-1)$, $i=0, 1, \ldots, 6t-1$. Thus by Lemma 3.1, $(GF(q)\times I_3, \bigcup_{C\in\mathscr{B}} \Delta C)$ is a $(3q, 3, 1)$-BIBD.

Let us write $\mathscr{B}=\mathscr{B}_1 \cup \mathscr{B}_2$ where $\mathscr{B}_1=\{A_0\} \cup \{B_j^i \mid 0\leqq i<t,\ 1\leqq j\leqq 3\} \cup \{A^i \mid t\leqq i<2t,\ 3t\leqq i<4t,\ 5t\leqq i<6t\}$ and $\mathscr{B}_2=\{A^i \mid 0\leqq i<t,\ 2t\leqq i<3t,\ 4t\leqq i<5t\}$. Noting that $\mathscr{B}_1$ is a parallel class, it follows that for each $a \in GF(q)$, $\{C+a \mid C\in\mathscr{B}_1\}$ is a parallel class, so we may partition $\bigcup_{C\in\mathscr{B}_1} \Delta C$ into q parallel classes. For each block $A^i \in \mathscr{B}_2$, ΔA^i is itself a parallel class. Thus we may partition $\bigcup_{C\in\mathscr{B}} \Delta C$ into $q+3t=9t+1$ parallel classes, and this completes the proof.

THEOREM 6. *Let $q=6t+1$ be a prime power. Then a Kirkman design of order $2q+1$ exists.*

PROOF. We shall construct a Kirkman design on the set of treatments $GF(q) \times I_2 \cup \{\infty\}$. Let g be a generator of the multiplicative group of $GF(q)$ and select m such that $2g^m=g^t+1$. We take as base blocks

$$\begin{aligned}
A_0 &= (0_1, 0_2, \infty) & & \\
B_1^i &= (g_1^i, g_1^{i+t}, g_2^{i+m}) & i &= 0, 1, \ldots, t-1\\
B_2^i &= (g_1^{i+2t}, g_1^{i+3t}, g_2^{i+m+2t}) & i &= 0, 1, \ldots, t-1\\
B_3^i &= (g_1^{i+4t}, g_1^{i+5t}, g_2^{i+m+4t}) & i &= 0, 1, \ldots, t-1\\
A^i &= (g_2^{i+m+t}, g_2^{i+m+3t}, g_2^{i+m+5t}) & i &= 0, 1, \ldots, t-1
\end{aligned}$$

and let $\mathscr{B}=\{A_0\} \cup \{B_j^i \mid 1\leqq j\leqq 3, 0\leqq i<t\} \cup \{A^i \mid 0\leqq i<t\}$. The [1, 1] differences that occur in $\mathscr{B}$ are $\pm g^i(g^t-1)$, $\pm g^{i+2t}(g^t-1)$, $\pm g^{i+4t}(g^t-1)$, $i=0, 1, \ldots, t-1$, and recalling $-1=g^{3t}$, we see that all nonzero [1, 1] differences occur once in $\mathscr{B}$. The [2, 2] differences all occur in the blocks A^i and are $\pm g^{i+m+t}(g^{2t}-1)$, $\pm g^{i+m+t}(g^{4t}-g^{2t})$, $\pm g^{i+m+t}(g^{4t}-1)$, which gives us every nonzero [2, 2] difference precisely once.

It suffices now to consider only the [2, 1] mixed differences, since the [1, 2] mixed differences are just the negatives of the [2, 1] difference. The [2, 1] difference 0 occurs only in A_0 and the other [2, 1] differences are $g^i(g^m-1)$, $g^i(g^m-g^t)$, $g^{i+2t}(g^m-1)$, $g^{i+2t}(g^m-g^t)$, $g^{i+4t}(g^m-1)$, $g^{i+4t}(g^m-g^t)$, $i=0, 1, \ldots, t-1$. Now since $2g^m=g^t+1$, we have $g^m-g^t=-(g^m-1)=g^{3t}(g^m-1)$, and it then follows that all [2, 1] differences occur once in $\mathscr{B}$. Finally, we note that ∞ occurs with an element of each subscript precisely once. Thus by Lemma 3.2, $(GF(q)\times I_2 \cup \{\infty\}, \bigcup_{C\in\mathscr{B}} \Delta C)$ is a $(2q+1, 3, 1)$-BIBD.

We now observe that $\mathscr{B}$ is itself a parallel class, so we have a natural partition of $\bigcup_{C\in\mathscr{B}} \Delta C$ into q parallel classes. This proves the theorem.

We now let U be the set of integers u for which a Kirkman design of order $6u+3$ exists. Our objective then is to prove that U is in fact the set of all nonnegative integers.

u	$6u+3$	Proof
0	3	Trivial
1	9	Example 2
2	15	Theorem 6
3	21	Theorem 5
4	27	Theorem 3: $l_1=3$, $l_2=3$
		Theorem 4: $v_1=9$, $m=3$, $v_2=3$
		Theorem 6
5	33	Theorem 4: $v_1=9$, $m=4$, $v_2=1$
6	39	Theorem 5
		Theorem 6
7	45	Theorem 3: $l_1=3$, $l_2=5$
		Theorem 4: $v_1=15$, $m=3$, $v_2=3$
8	51	Theorem 6
9	57	Theorem 4: $v_1=15$, $m=4$, $v_2=1$
		Theorem 5
10	63	Theorem 3: $l_1=3$, $l_2=7$
		Theorem 4: $v_1=21$, $m=3$, $v_2=3$
		Theorem 6
11	69	Example given on p. 196.
12	75	Theorem 3: $l_1=5$, $l_2=5$
		Theorem 4: $v_1=9$, $m=9$, $v_3=3$
		Theorem 5
		Theorem 6
13	81	Theorem 3: $l_1=3$, $l_2=9$
		Theorem 4: $v_1=21$, $m=4$, $v_2=1$
		Theorem 4: $v_1=9$, $m=9$, $v_2=9$
14	87	Theorem 6
15	93	Theorem 5
16	99	Theorem 3: $l_1=3$, $l_2=11$
		Theorem 4: $v_1=15$, $m=7$, $v_2=1$
		Theorem 4: $v_1=9$, $m=12$, $v_2=3$
		Theorem 6
17	105	Theorem 3: $l_1=5$, $l_2=7$
		Theorem 4: $v_1=9$, $m=13$, $v_2=1$
		Theorem 4: $v_1=9$, $m=12$, $v_2=9$
18	111	Theorem 5
19	117	Theorem 3: $l_1=3$, $l_2=13$
26	159	Theorem 6
27	165	Theorem 3: $l_1=5$, $l_2=11$

LEMMA 3.3. *If* $0 \leqq u \leqq 19$, *or* $u=26$ *or* 27, *then* $u \in U$.

PROOF. We apply our previous results as indicated. To illustrate the applications of Theorems 3 and 4, many cases are proved in various ways. One should note that the existence of the orthogonal arrays required in the hypotheses of these theorems is guaranteed in all cases by Lemma 1.1.

4. **Main Theorem.** In this section we shall establish the main theorem of this paper.

THEOREM 7. *If* $t, q \in U$, *and there exists a* (4, t, 2, 1)-*orthogonal array with* q *parallel classes, then* $4t+q \in U$.

PROOF. By Lemma 2.2 there exists a PBD on $4t+q$ treatments with blocks of size 4 and 5 and a clear set of blocks of size q and t. Since 4 and 5 belong to R_4, by Theorem 2 there exists a PBD on $3(4t+q)+1$ treatments with blocks of size 4, $3t+1$ and $3q+1$. Since 4, $3t+1$ and $3q+1 \in R_3^*$, by Theorem 1, $3(4t+q)+1 \in R_3^*$, i.e. there exists a Kirkman design of order $2(3(4t+q)+1)+1$ which in turn implies $4t+q \in U$.

LEMMA 4.1. *For* $l=0, 1, 2$ *and* $0 \leqq m \leqq 23$, $24l+m \in U$.

PROOF. By Lemma 3.3, $u \in U$ if $0 \leqq u \leqq 19$ or $u=26$ or $u=27$. We complete the proof of the lemma by applying Theorem 7, with choices of t and q as indicated in the following table:

u	t	q
$20 \leqq u \leqq 25$	5	$u-4t$
$28 \leqq u \leqq 35$	7	$u-4t$
$36 \leqq u \leqq 45$	9	$u-4t$
$46 \leqq u \leqq 55$	11	$u-4t$
$56 \leqq u \leqq 65$	13	$u-4t$
$66 \leqq u \leqq 80$	16	$u-4t$
$81 \leqq u \leqq 95$	19	$u-4t$

THEOREM 8. *A Kirkman design of order* v *exists iff* $v \equiv 3 \bmod 6$.

PROOF. It is easily seen that if a Kirkman design of order v exists, then $v \equiv 3 \pmod 6$. To prove the converse, we need to show that U is the set of all nonnegative integers. By Lemma 4.1, $u \in U$, if $u=24l+m$, $0 \leqq l \leqq 2$, $0 \leqq m \leqq 23$. We shall complete the proof by induction on l. Assume: $u \in U$ if $u=24l'+m$, $0 \leqq l' \leqq l-1$, $0 \leqq m \leqq 23$, $l \geqq 3$. By applying Theorem 8 with choices of t and q given by the following table we show that $u \in U$ if $u=24l+m$, $0<m<23$:

$u=4t+q$	t	q
$24l$	$6l-1$	4
$24l+1$	$6l-1$	5
2	$6l-1$	6
3	$6l-1$	7
4	$6l+1$	0
5	$6l+1$	1
6	$6l+1$	2
7	$6l+1$	3
8	$6l+1$	4
9	$6l+1$	5
10	$6l+1$	6
11	$6l+1$	7
12	$6l+1$	8
13	$6l+1$	9
14	$6l+1$	10
15	$6l+1$	11
16	$6l+1$	12
17	$6l+1$	13
18	$6l+1$	14
19	$6l+1$	15
20	$6l+5$	0
21	$6l+5$	1
22	$6l+5$	2
23	$6l+5$	3

Note that if $t \geqq 5$ and, $t \equiv 1$ or 5 (mod 6), then the least prime factor of t is not smaller than 5; hence a (5, t, 2, 1)-orthogonal array exists by MacNeish's Theorem (Lemma 1.1), and thus a (4, t, 2, 1)-resolvable orthogonal array with t parallel classes exists.

REFERENCES

1. W. W. Rouse Ball (revised by H. S. M. Coxeter), *Mathematical recreations and essays*, Macmillan, New York, 1947, pp. 267–298. MR **8**, 440.

2. R. C. Bose, *On the construction of balanced incomplete block design*, Ann. Eugenics **9** (1939), 353–399. MR **1**, 199.

3. R. C. Bose and S. S. Shrikhande, *On the construction of sets of mutually orthogonal latin squares and the falsity of a conjecture of Euler*, Trans. Amer. Math. Soc. **95** (1960), 191–209. MR **22** #2557.

4. ———, *On the composition of balanced incomplete block designs*, Canad. J. Math. **12** (1960), 177–188. MR **22** #1046.

5. Marshall Hall, Jr., *Combinatorial theory*, Blaisdell, Waltham, Mass., 1967. MR **37** #80.

6. H. Hanani, *The existence and construction of balanced incomplete block designs*, Ann. Math. Statist. **32** (1961), 361–386. MR **29** #4161.

7. H. B. Mann, *Analysis and design of experiments*, Dover, New York, 1949. MR **11**, 262.

8. E. Parker, *On collineations of symmetric designs*, Proc. Amer. Math. Soc. **8** (1957), 350–351. MR **19**, 163.

9. C. R. Rao, *A study of* BIB *designs with replications* 11 *to* 15, Sankhyā **23** (1961), 117–127. MR **25** #732.

10. ———, *Factorial experiments derivable from combinatorial arrangements of arrays*, Suppl. J. Roy. Statist. Soc. **9** (1947), 128–139. MR **9**, 264.

11. H. J. Ryser, *Combinatorial mathematics*, Carus Mathematical Monograph No. 14, 1963. MR **27** #51.

12. Oscar Eckenstein, *Bibliography of Kirkman's schoolgirl problem*, Messenger of Mathematics **41–42** (1912), 33–36.

13. T. P. Kirkman, *On a problem in combinations*, Cambridge and Dublin Mathematical J. **2** (1847), 191–204.

14. ———, *Query*, Lady's and Gentleman's Diary (1850), p. 48.

15. A. Cayley, *On the triadic arrangements of seven and fifteen things*, London, Edinburgh, and Dublin Philosophical Magazine and Journal of Science **37** (1850), 50–53.

16. T. P. Kirkman, *Note on an unanswered prize question*, Cambridge and Dublin Mathematical J. **5** (1850), 255–262.

17. ———, *On the triads made with fifteen things*, London, Edinburgh, and Dublin Philosophical Magazine and Journal of Science **37** (1850), 169–171.

18. *Solutions to Query*, VI, Lady's and Gentleman's Diary (1851), p. 48.

19. R. R. Anstice, *On a problem in combinations*, Cambridge and Dublin Mathematical J. **7** (1852), 279–292.

20. ———, *On a problem in combinations* (continued from **7**, p. 292), Cambridge and Dublin Mathematical J. **8** (1853), 149–154.

21. J. Steiner, *Combinatorische Aufgabe*, Crelle's Journal für die reine und angewandte Mathematik **45** (1853), 181–182.

22. T. P. Kirkman, *Theorems on combinations*, Cambridge and Dublin Mathematical J. **8** (1853), 38–45.

23. ———, *On the perfect r partitions of* r^2-r+1, Transactions of the Historic Society of Lancashire and Cheshire **9** (1856–1857), 127–142.

24. B. Peirce, *Cyclic solutions of the school-girl puzzle*, Astronomical Journal (U.S.A.) **6** (1860), 169–174.

25. J. J. Sylvester, *Note on the historical origin of the unsymmetrical six-valued function of six letters*, London, Edinburgh, and Dublin Philosophical Magazine and Journal of Science **21** (1861), ser. 4, 369–377.

26. W. S. B. Woolhouse, *On the Rev. T. P. Kirkman's problem respecting certain triadic arrangements of fifteen symbols*, London, Edinburgh, and Dublin Philosophical Magazine and Journal of Science **21** (1861), 510–515.

27. J. J. Sylvester, *On a problem in tactics which serves to disclose the existence of a four-valued function of three sets of three letters each*, London, Edinburgh, and Dublin Philosophical Magazine and Journal of Science **21** (1861), ser. 4, 515–520.

28. W. S. B. Woolhouse, *On triadic combinations of* 15 *symbols*, Lady's and Gentleman's Diary (1862), 84–88.

29. Paper [16] is reprinted in the Assurance Magazine **10** (1862), pt. v., No. 49, pp. 275–281.

30. T. P. Kirkman, *On the puzzle of the fifteen young ladies*, London, Edinburgh, and Dublin Philosophical Magazine and Journal of Science **23** (1862), ser. 4, 198–204.

31. A. Cayley, *On a tactical theorem relating to the triads of fifteen things*, London, Edinburgh, and Dublin Philosophical Magazine and Journal of Science, **25** (1863), ser. 4, 59–61.

32. W. S. B. Woolhouse, *On triadic combinations*, Lady's and Gentleman's Diary (1863), pp. 79–90.

33. J. Power, *On the problem of the fifteen schoolgirls*, Quarterly Journal of Pure and Applied Mathematics **8** (1867), 236–251.

34. S. Bills, *Solution of problem proposed by W. Lea*, Educational Times Reprints **8** (1867), 32–33.

35. W. Lea, *Solution of problem proposed by himself*, Educational Times Reprints **9** (1868), 35–36.

36. T. P. Kirkman, *Solution of three problems proposed by W. Lea*, Educational Times Reprints **11** (1869), 97–99.

37. A. Frost, *General solution and extension of the problem of the fifteen schoolgirls*, Quarterly Journal of Pure and Applied Mathematics **11** (1871), 26–37.

38. W. Lea, *Solution of problem proposed by himself*, Educational Times Reprints **22** (1874), 74–76.

39. J. J. Sylvester, *Proposed problem*, Educational Times, November 1, 1875, p. 193.

40. Appendix note, Proceedings of the London Mathematical Society **7** (1875), 235–237.

41. E. Carpmael, *Some solutions of Kirkman's 15-school-girl problem*, Proceedings of the London Mathematical Society **12** (1881), 148–156.

42. *A fifteen puzzle*, Knowledge **1** (1881), 80.

43. A. Bray, *The fifteen schoolgirls*, Knowledge **2** (1882), 80–81.

44. E. Marsden, *The school-girls' problem*, Knowledge **3** (1883), 183.

45. A. Bray, *Twenty-one school-girl puzzle*, Knowledge **3** (1883), 268.

46. J. J. Sylvester, *Note on a nine schoolgirls problem*, Messenger of Mathematics **22** (1893), 159–160.

47. *Correction to the note on the nine schoolgirls problem*, Messenger of Mathematics **22** (1893), 192.

48. A. C. Dixon, *Note on Kirkman's problem*, Messenger of Mathematics **23** (1893), 88–89.

49. W. Burnside, *On an application of the theory of groups to Kirkman's problem*, Messenger of Mathematics **23** (1894), 137–143.

50. E. H. Moore, *Tactical memoranda*, American J. Math. **18** (1896), 264–303.

51. E. W. Davis, *A geometric picture of the fifteen school-girl problem*, Ann. of Math. **2** (1896–1897), 156–157.

52. A. F. H. Mertelsmann, *Das Problem der 15 Pensionatsdamen*, Z. Math. Phys. **43** (1898), 329–334.

53. W. Ahrens, Review of *Schubert's Mathematische Mussestunden*, Zeitschrift für mathematischen und naturwissenschaftlichen Unterricht **31** (1900), 386–388.

54. H. E. Dudeney, *Solution*, Educational Times Reprints **14** (1908), 97–99.

55. ———, *Solution* (continued), Educational Times Reprints **15** (1909), 17–19.

56. O. Eckenstein, *Note*, Educational Times Reprints **16** (1909), 76–77.

57. H. E. Dudeney, *Solution* (continued), Educational Times Reprints **17** (1910), 35–38.

58. O. Eckenstein, *Solutions*, Educational Times Reprints **17** (1910) 38–39.

59. ———, *Note*, Educational Times Reprints **17** (1910), 49–53.

60. W. W. Rouse Ball, *Proposed problem*, Educational Times (February 1, 1911), p. 82.

Ohio State University

A GENERALIZATION OF RAMSEY'S THEOREM

BRUCE ROTHSCHILD[1]

1. **Introduction.** F. P. Ramsey proved [3] the following combinatorial theorem:

THEOREM 1 (RAMSEY). *For each three positive integers $k, r, t, k \geq r$, there is a number $N(k, r, t)$, depending only on k, r, t, such that if S is any set with at least $N(k, r, t)$ elements, and if all the r-subsets of S (i.e., the subsets with r elements) are divided into t classes $\mathscr{A}_1, \mathscr{A}_2, \ldots, \mathscr{A}_t$ in any manner, then there is some k-subset T and some j, $1 \leq j \leq t$, such that all the r-subsets of T are in $\mathscr{A}_j$.*

In the same paper, he proved independently an infinite version of the same theorem:

THEOREM 2 (RAMSEY). *Let S be an infinite set, and let r and t be any two positive integers. Let all the r-subsets of S be divided into t classes $\mathscr{A}_1, \mathscr{A}_2, \ldots, \mathscr{A}_t$ in any manner. Then there is some infinite subset $T \subseteq S$ and some j such that all the r-subsets of T are in $\mathscr{A}_j$.*

By a well-known "compactness" argument, Theorem 2 actually implies Theorem 1. This argument has the disadvantage, however, that it provides no estimate for the numbers $N(k, r, t)$ guaranteed by Theorem 1. There has been considerable interest in estimating and evaluating these numbers. However, in this paper the main concern will be generalizing Ramsey's theorems and not estimating the numbers. For this purpose it is very convenient to use the compactness argument to obtain finite versions of theorems like Ramsey's from infinite versions (Theorem 5 below). This approach provides a natural way to state some finite theorems in terms of the corresponding infinite versions.

The point of departure from Ramsey's theorem is a theorem (actually two theorems) due recently to T. S. Motzkin [1]. We say that a *chain* (resp. *ascending chain*) is a sequence of sets $S_0, S_1, \ldots, S_n$ such that $|(S_i \cup S_{i+1}) - (S_i \cap S_{i+1})| = 1$ (resp. $S_i \subseteq S_{i+1}$ and $|S_{i+1} - S_i| = 1$) for all i. A set S is *obstructed* (resp. *blocked*) by a set C of subsets of S if every chain (resp. ascending chain) with $S_0 = \varnothing$ and $S_n = S$ contains a member of C.

THEOREM 3 (MOTZKIN). *For each three positive integers $k, r, t, k \geq r$, there is a number $M^*(k, r, t)$ (resp. $M(k, r, t)$), depending only on k, r, t, such that for any set*

[1] This work partially supported by a National Science Foundation Graduate Fellowship.

S with $|S| \geq M^*(k, r, t)$ (resp. $M(k, r, t)$), any set C of subsets $F \subseteq S$ with $|F| \leq r$ such that C obstructs (resp. blocks) S, and any division of C into t classes $\mathscr{A}_1, \mathscr{A}_2, \ldots, \mathscr{A}_t$, there is some k-subset $T \subseteq S$, and some j, $1 \leq j \leq t$, such that T is obstructed (resp. blocked) by a subset of $\mathscr{A}_j$.

Since the subset diagram for a finite set S (Example 1 below) has exactly the subsets of S for vertices, there is an obvious translation of Theorem 3 to a corresponding theorem about subset diagrams. The (ascending) chains are just (directed) paths in the subset diagram for S. A set C of subsets of S which (blocks) obstructs S is just a cut-set for the set of all (directed) paths from $\varnothing$ to S in the subset diagram. Then this translated version of Theorem 3 can be generalized by permitting a greater variety of sets of paths, that is, sets of paths other than the set of all directed paths or the set of all paths from $\varnothing$ to S. It is generalizations such as this with which this paper is concerned.

We first make a general statement (statement T below) about a class of infinite graphs, called here "graded" graphs. Some special cases of this statement are true, including Theorem 2 above and some infinite versions of Motzkin's theorem (Corollaries 4.1 and 4.2 below), and some special cases are false, including a conjecture of Erdös. Using the compactness argument (Theorem 5) we can obtain from some instances of the general statement T which are true, some theorems about finite graded graphs, including, as special cases, Theorems 1 and 3 above.

DEFINITION 1. G is a *graded graph* if it satisfies the following conditions:

(i) The set of vertices of G is the union of disjoint, nonempty subsets $l_i(G)$, $i=0, 1, 2, \ldots$.

(ii) $|l_0(G)|=1$.

(iii) For each $v \in l_{r+1}(G)$, there are some $u \in l_r(G)$ with edges in G connecting the u's and v, $r=0, 1, 2, \ldots$.

(iv) The edges of (iii) are the only edges of G.

Let P be a *path* in a graded graph G, that is, a sequence of vertices $v_0, v_1, v_2, \ldots$ such that $v_k v_{k+1}$ is an edge of G. Then P is an *infinite directed path* if for each $i=0, 1, 2, \ldots, v_i \in l_i(G)$. A *directed path* is any path contained in an infinite directed path. If S is any subgraph of G or subset of vertices of G, then $l_r(S)$ is the set of vertices $l_r(G) \cap S$. $H(S)$ denotes the subgraph of all edges and vertices lying on directed paths with one endpoint in S and the other $l_0(G)$. We write $p_r(S)=l_r(H(S))$. Two graded graphs G and H are *isomorphic*, and we write $G \cong H$, if there exists a one-to-one mapping ϕ of the vertices $u \in G$ onto the vertices $v \in H$ such that $\phi(l_0(G))=l_0(H)$, and u, u' are connected by an edge of G if and only if $\phi(u), \phi(u')$ are connected by an edge of H. Finally, we say that a graded graph G which is *locally finite* is one in which $|l_r(G)|<\infty$ for all r.

DEFINITION 2. A *finite graded graph* is a graph satisfying (ii), (iii), (iv) of Definition 1, together with the condition:

(i′) The vertices of G are the union of disjoint sets $\bigcup_{i=0}^{n} l_i(G)$ for some n, and $|l_i(G)|<\infty$ for all i.

EXAMPLE 1. The *subset diagram* for a set S is a graph G in which:

(1) $l_r(G)=\{F \mid F\subset S, |F|=r\}, r=0, 1, 2, \ldots.$

(2) Two vertices E and F in G are joined by an edge if and only if $|E \cup F-E \cap F| =1$.

That is, G is the diagram of the subset lattice. Then if S is a finite set, G is a finite graded graph. Since $l_1(G)=\{\{x\} \mid x \in S\}$, we sometimes abuse terminology for convenience by saying that a subset $T\subseteq S$ is a subset of $l_1(G)$, or that a subset $L\subseteq l_1(G)$ is a subset of S. For instance, if P is an infinite directed path in G, we say that $l_1(H(P))$ is an infinite subset of S.

EXAMPLE 2. The *subspace diagram* of a vector space V over a field F is the graph G in which:

(1) $l_r(G)=\{U \mid U\subseteq V$ is a subspace, and $\dim U=r\}, r=0, 1, 2, \ldots.$

(2) Two vertices W, U in G are joined by an edge of G if and only if either $U\subseteq W$ or $W\subseteq U$, and $|\dim U-\dim W|=1$.

That is, G is the diagram of the lattice of subspaces of V. If V is finite dimensional, and F is finite, then G is a finite graded graph.

2. **Generalizations of the infinite theorem.** Let G be the infinite subset diagram, described in Example 1. For each infinite subset $T\subseteq l_1(G)$, there is some infinite directed path P in G such that $l_1(H(P))\subseteq T$ (in fact there are uncountably many such paths). And $l_1(H(P))$ is, in fact, an infinite subset of $l_1(G)$. Thus Theorem 2 above is equivalent to the statement that for G the infinite subset diagram and for each pair of positive integers r, t the following holds:

> $S_\infty(r, t)$: Let all the vertices of $l_r(G)$ be divided into t subsets $\mathscr{A}_1, \mathscr{A}_2, \ldots, \mathscr{A}_t$ in any manner. Then there is some infinite directed path P in G and some j, $1\le j\le t$, such that $p_r(P)\subseteq\mathscr{A}_j$.

We can find some simple examples of other graded graphs, besides the subset diagram, which satisfy the statement $S_\infty(r, t)$ for all r and t, but examples which are neither trivial nor directly derived from the subset diagram are not known. In particular, whether or not the infinite-dimensional subspace lattice (Example 2) satisfies $S_\infty(r, t)$ for finite fields F is not known.

EXAMPLE 3. Let G be any tree containing an infinite path. Let v be any vertex of G, and for each $r\ge 0$ let $l_r(G)$ be the set of vertices which are distance r from v. Then G is a graded graph which trivially satisfies $S_\infty(r, t)$ for all t, since $p_r(P)$ consists of exactly one vertex for any infinite path P.

EXAMPLE 4. We define G inductively. Let $l_0(G)=\{v_0\}$, $l_1(G)=\{v_1, v_2, \ldots\}$. Let $k_1, k_2, \ldots$ be positive integers. For each $r>1$, let the vertices of $l_r(G)$ be all the k_{r-1}-subsets of the vertices of $l_{r-1}(G)$. If $v \in l_r(G)$, $u \in l_{r-1}(G)$, then let there be an edge between u and v if and only if $u \in v$, or $r=1$. Then G satisfies $S_\infty(r, t)$ for all r, t. (This follows directly from Theorem 2 above.)

$S_\infty(r, t)$ is a special case of a more general statement, which we now consider. Let G be a graded graph, $\mathscr{P}$ a set of paths in G, C a subset of the vertices of G, and for each path $P \in \mathscr{P}$, let $C(P)$ be a set of subsets of $C \cap H(P)$. Then the statement is:

> T: Let the vertices of C be divided into t subsets $\mathscr{A}_1, \mathscr{A}_2, \ldots, \mathscr{A}_t$ in any manner. Then for some path $P \in \mathscr{P}$, some j, $1 \leq j \leq t$, and some $C' \in C(P)$, we have $C' \subseteq \mathscr{A}_j$.

If $C = l_r(G)$, $\mathscr{P}$ the set of infinite directed paths in G, and $C(P) = \{l_r(H(P))\}$, then this becomes precisely the statement $S_\infty(r, t)$. We consider a few special cases of statement T below.

DEFINITION 3. If $\mathscr{P}$ is a set of paths in a graph G, then a *cut-set* for $\mathscr{P}$ is a set S of vertices of G such that each $P \in \mathscr{P}$ has some vertex in S.

DEFINITION 4. Let G be a graded graph, and let P be an infinite path in G with initial vertex $l_0(G)$. Then P is an *infinite almost directed path* if for each $r \geq 0$ there are only a finite number of vertices of P in $l_r(G)$.

DEFINITION 5. Let C be a set of vertices in a graded graph G. Then C is *bounded* if for some $m \geq 0$ we have $C \leq \bigcup_{i=0}^{m} l_r(G)$. In this case we say that C is bounded by m.

We now prove a theorem for graded graphs G which satisfy both $S_\infty(r, t)$ for all r, t, and $H(P) \cong G$ for all infinite directed paths P in G. The subset diagram for an infinite, countable set S (Example 1) satisfies both of these conditions. The subspace diagram for a countably infinite-dimensional vector space V (Example 2) also satisfies $H(P) \cong G$.

THEOREM 4. *Let G be a graded graph satisfying $S_\infty(r, t)$ for all r, t, and $H(P) \cong G$ for all infinite directed paths in G. Let $\mathscr{P}$ be a set of infinite almost directed paths such that $\mathscr{P}$ contains all the infinite directed paths. For each $P \in \mathscr{P}$ let $\mathscr{P}(P)$ be the subset of $\mathscr{P}$ consisting of those paths lying entirely in $H(P)$. Let C be a bounded cut-set for $\mathscr{P}$, and for each $P \in \mathscr{P}$, let $C(P)$ be those subsets of $C \cap H(P)$ which are cut-sets for $\mathscr{P}(P)$. Then statement T is true for all t.*

PROOF. The proof is like that of Ramsey for Theorem 2. Let $r(C)$ denote the maximum r such that $l_r(G) \cap C \neq \varnothing$. We prove the theorem by induction on $r(C)$. If $r(C) = 1$, then the theorem is equivalent to $S_\infty(1, t)$, which is assumed to be true. Assume, then, that the theorem is true for $r(C) < r$ and let $r(C) = r$. Let the vertices of C be divided into t sets, $\mathscr{A}_1, \mathscr{A}_2, \ldots, \mathscr{A}_t$. We must show that for some $P \in \mathscr{P}$, some j, $1 \leq j \leq t$, and some $C' \in C(P)$, we have $C' \subset \mathscr{A}_j$.

Define $(t+1)$ sets $\mathscr{B}_1, \mathscr{B}_2, \ldots, \mathscr{B}_{t+1}$ for the vertices of $l_r(G)$ as follows:

$$\mathscr{B}_i = \mathscr{A}_i \cap l_r(G), \qquad i = 1, 2, \ldots, t,$$

$$\mathscr{B}_{t+1} = l_r(G) - (\mathscr{B}_1 \cup \mathscr{B}_2 \cup \cdots \cup \mathscr{B}_t).$$

Then by $S_\infty(r, t+1)$, there is an infinite directed path, $P (P \in \mathscr{P})$, such that $l_r(H(P)) \subseteq \mathscr{B}_j$, $1 \leq j \leq t+1$. If $1 \leq j \leq t$, then we are done, since $l_r(H(P)) \in C(P)$, and $l_r(H(P)) \subseteq \mathscr{B}_j \subseteq \mathscr{A}_j$. So we may assume that $l_r(H(P)) \subseteq \mathscr{B}_{t+1}$. Now consider statement T applied to $H(P)$, $\mathscr{P}(P)$, $C \cap H(P)$ rather than to G, $\mathscr{P}$, C. We have $H(P) \cong G$; $\mathscr{P}(P)$ contains all infinite directed paths in $H(P)$; $C \cap H(P)$ is a bounded cut-set for $H(P)$, and by definition of $\mathscr{B}_{t+1}$, $r(C \cap H(P)) \leq r-1$. Thus by induction, statement T is true, and there is some path P' in $\mathscr{P}(P) \subseteq \mathscr{P}$, some cut-set $C' \in C(P')$, and some j, $1 \leq j \leq t$, such that $C' \subseteq \mathscr{A}_j$. Then P' and C' are the desired path and cut-set, and the theorem is proved.

COROLLARY 4.1. *Let G be the subset diagram for a countable, infinite set, $\mathscr{P} = \mathscr{P}_0$, the set of infinite directed paths in G, C and $C(P)$ as in Theorem* 4. *Then T is true for all t.*

COROLLARY 4.2. *Let G be the subset diagram for a countable, infinite set, and $\mathscr{P} = \mathscr{P}_1$, the set of all almost directed paths in G. Then T is true for all t, where C, $C(P)$ are as in Theorem* 4.

Corollary 4.1 is the infinite version of Motzkin's theorem (Theorem 3) for "blocking", and Corollary 4.2 is the infinite version for "obstructing", in the sense that the finite versions are obtained from these corollaries by Theorem 5 below.

Erdös conjectured that another version of statement T is true. It is described below, and a counterexample is given.

DEFINITION 6. Let $\{a_1, a_2, \ldots\} = A$ be a set of elements. A *permutation* of A is a mapping $\pi: A \to A$ which is one-to-one and onto. We may denote such a permutation by a sequence $b_1, b_2, \ldots$, where $\pi(a_i) = b_i$, $i = 1, 2, \ldots$. We say that such *a permutation is blocked* by an unordered set $S \subseteq A$ if for some k, $S = \{b_1, \ldots, b_k\}$. The set A is *blocked* by a set C of subsets of A if every permutation of A is blocked by some member of C.

Erdös' conjecture is as follows: Let C be a set of subsets of the set S of positive integers, $S = \{1, 2, 3, \ldots\}$, such that C blocks S. Let the elements of C be divided into t subsets in any manner, $\mathscr{A}_1, \mathscr{A}_2, \ldots, \mathscr{A}_t$. Then there is an infinite subset $U \subseteq S$, and some j, $1 \leq j \leq t$, such that U is blocked by $\mathscr{A}_j \cap 2^U$.

To see how this is a special case of statement T let G be the subset diagram for the set $S = \{1, 2, 3, \ldots\}$ and let $\mathscr{P}$ be the set of infinite directed paths in G. Let C be any set of vertices of G which blocks S. For each $P \in \mathscr{P}$ let $C(P)$ be the subsets of $C \cap H(P)$ which block $l_1(H(P))$. Since every infinite subset $U \subseteq S$ is $l_1(H(P))$ for some $P \in \mathscr{P}$, the conjecture is just the statement T.

We get a counterexample (the particular one exhibited here was found in part by M. Perles and E. G. Straus) by finding sets C, $\mathscr{A}_1$, $\mathscr{A}_2$, of subsets of S with $C = \mathscr{A}_1 \cup \mathscr{A}_2$ blocking S, such that every infinite subset U is blocked neither by $\mathscr{A}_1 \cap 2^U$ nor by $\mathscr{A}_2 \cap 2^U$. That is, for each infinite subset U we find permutations P_1 and P_2 of U blocked by no subsets in $\mathscr{A}_1$ or $\mathscr{A}_2$, respectively.

Let $C = \{F \subseteq S \mid |F| < \infty, |F| \in F\}$. For each $F = \{a_1, a_2, \ldots, a_m\} \in C$ let $m(F)$ be the number of $a_j \leq m$. Then for $i = 1, 2$ define

$$\mathscr{A}_i = \{F \in C \mid m(F) \equiv i \ (\mathrm{mod}\ 2)\}.$$

Let U, then, be any infinite subset of S, say $U = \{i_1, i_2, i_3, \ldots\}$ with $i_1 < i_2 < i_3 < \cdots$.

Let P_1 be the permutation $a_1, a_2, a_3, \ldots$ of U defined as follows, where $A_k = \{a_1, a_2, \ldots, a_k\}$:

$$a_1 = i_2,$$
$$a_k = \begin{cases} i_{2k} & \text{if } k \notin A_{k-1}, \\ \min(U - A_{k-1}) & \text{if } k \in A_{k-1}, \end{cases} \qquad k > 1.$$

The first few terms of this sequence are: $i_2, i_4, \ldots, i_{2(i_2-1)}, i_1, i_{2(i_2+1)}, \ldots, i_{2(i_4-1)}, i_3, \ldots$. Since obviously $k \in A_{k-1}$ an infinite number of times, we have $\min(U - A_{k-1}) < \min(U - A_k)$ an infinite number of times. That is, $\bigcup_{k=1}^{\infty} A_k = U$, and P_1 is indeed a permutation.

Now $m(A_k) = m(A_{k-1})$ or $m(A_k) = m(A_{k-1}) + 2$. For $m(A_{k-1})$ counts the number of pairs (j, a_j) with $j \leq k-1$ and $a_j \leq k-1$, and $m(A_k)$ counts the number of pairs (j, a_j) with $j \leq k$ and $a_j \leq k$. Thus these two are equal unless $a_j = k$ for some $j < k$. But then, by definition, $a_k = \min(U - A_{k-1})$, which is clearly less than k. In this case, then, $m(A_k)$ counts two more than $m(A_{k-1})$, namely a_j and a_k, and $m(A_k) = m(A_{k-1}) + 2$. Since $m(A_1) = 0$, this means $m(A_k) \equiv 0 \pmod 2$ for all k, and $A_k \notin \mathscr{A}_1$ for all k. P_1 has the desired property.

Let P_2 be the permutation $b_1, b_2, b_3, \ldots$ of U defined as follows, where $B_k = \{b_1, b_2, \ldots, b_k\}$:

The first few terms of P_2 are $i_2, i_4, \ldots, i_{2(i_2-1)}, i_{2i_2}$.

For $k > i_2$ let

$$b_k = \begin{cases} i_{2k} & \text{if } k \notin B_{k-1}, \\ \min(U - B_{k-1}) & \text{if } k \in B_{k-1}. \end{cases}$$

Now B_k is neither in $\mathscr{A}_1$ nor in $\mathscr{A}_2$ for $k < i_2$, since $|B_k| \notin B_k$. But $B_{i_2} \in \mathscr{A}_1$. Then arguing exactly as with P_1 above, we see that $m(B_k) \equiv m(B_{k-1}) \pmod 2$ for $k > i_2$. Hence $B_k \notin \mathscr{A}_2$ for all k. P_2 is a permutation of U (for the same reasons as P_1 is). Hence P_2 is the desired sequence, and the counterexample is established.

Suppose we weaken the conjecture a bit by requiring that the set C blocking S be bounded. Then C must be a cut-set for the set $\mathscr{P}$ of infinite directed paths in the subset diagram G for the set S. For suppose on the contrary that $C \subseteq \bigcup_{i=0}^{r} l_i(G)$, and that P is a path in $\mathscr{P}$ with no vertex in C. If the first $r+1$ vertices of P are $\varnothing$, $\{i_1\}, \{i_1, i_2\}, \ldots, \{i_1, i_2, \ldots, i_r\}$, then none of these sets is in C. The permutation $i_1, i_2, \ldots, i_r, 1, 2, 3, \ldots, i_1 - 1, i_1 + 1, \ldots, i_2 - 1, i_2 + 1, \ldots, i_r - 1, i_r + 1, i_r + 2, i_r + 3, \ldots$ is blocked by none of the sets in C, contradicting the definition of C. So C is indeed a cut-set for $\mathscr{P}$. Similarly, any set $C' \subseteq C \cap H(P)$ which blocks $l_1(H(P))$ must be a cut-set for all infinite directed paths in $H(P)$. Erdös' conjecture with the additional requirement that C be bounded is exactly the statement for Corollary 4.1, and is therefore true. On the other hand, if we strengthen the statement of Corollary 4.1 to permit C to be unbounded, then the counterexample for the conjecture of Erdös is also a counterexample for this statement, since for any infinite directed path P in G, there are the infinite directed paths P_1 and P_2 in $H(P)$ such that no vertex of P_1 is in $\mathscr{A}_1$, and no vertex of P_2 is in $\mathscr{A}_2$, as we saw above.

3. Generalizations of the finite theorem. We consider here a class of generalizations of Ramsey's theorem which includes Motzkin's theorem. Let G be a graded graph, and let $l_0(G)=G_0\subset G_1\subset G_2\subset\cdots$ be a nested sequence of finite graded graphs such that $H(G_i)=G_i$, and $\bigcup_{i=0}^{\infty} G_i=G$. Let $\mathscr{P}$ be a set of paths in G and $\mathscr{C}$ a set of cut-sets for $\mathscr{P}$. For each $v\in G$ let $\mathscr{P}(v)$ be those paths in $H(v)$ which contain both $l_0(G)$ and v, and which are contained in $P\cap H(v)$ for some $P\in\mathscr{P}$. Let $\mathscr{C}(v)$ be the class of those cut-sets for $\mathscr{P}(v)$ which are subsets of $C\cap H(v)$, for suitable $C\in\mathscr{C}$. Finally, let t and k be integers. Then the statement corresponding to Ramsey's theorem is:

R: There is an integer N, depending only on k, $\mathscr{C}$, t such that if C is any member of $\mathscr{C}$, and $C\cap G_N$ is divided into t classes $\mathscr{A}_1, \mathscr{A}_2, \ldots, \mathscr{A}_t$ in any manner, then there is some $v\in l_k(G_N)$ such that for some j, $1\le j\le t$, $\mathscr{A}_j\cap H(v)\in\mathscr{C}(v)$.

In particular, if G is the subset diagram for $S=\{0, 1, 2, \ldots\}$, G_i the subset diagram for $\{0, 1, 2, \ldots, i\}$, $i=0, 1, 2, \ldots$, $\mathscr{P}=\mathscr{P}_0$, all infinite directed paths, and $\mathscr{C}=\{l_r(G)\}$, then R is just Theorem 1 for k, r, t. If $\mathscr{C}$ is the set of all cut-sets for $\mathscr{P}_0$ which are bounded by r, then R is just Motzkin's theorem for "blocking". If $\mathscr{P}=\mathscr{P}_1$, the set of all infinite almost directed paths, and $\mathscr{C}$ is the set of all cut-sets for $\mathscr{P}_1$ bounded by r, then R is Motzkin's theorem for "obstructing". These theorems are all proved simultaneously in Theorem 5 below.

DEFINITION 7. Let $\mathscr{C}$ be a set of subsets of the vertices of G. Let $F(\mathscr{C})$ be the set $\{C\cap G_n \mid C\in\mathscr{C}, n=0, 1, \ldots\}$. Then $\mathscr{C}$ is *closed* if for every infinite sequence $F_1\subset F_2\subset F_3\subset\cdots$ of members of $F(\mathscr{C})$ we have $\bigcup_{i=1}^{\infty} F_i\in\mathscr{C}$.

DEFINITION 8. A set of paths $\mathscr{P}$ is complete if whenever $P, Q\in\mathscr{P}$ have a vertex v in common, and P_1 is the part of P preceding v, Q_2 the part of Q following v, then P_1+v+Q_2 is in $\mathscr{P}$. (The set $\mathscr{P}_0$ of all infinite directed paths, and the set $\mathscr{P}_1$ of all infinite almost directed paths are both complete.)

Now let $\mathscr{P}$ be a set of infinite almost directed paths in G which is complete and which contains $\mathscr{P}_0$. Let $\mathscr{C}$ be a closed set of cut-sets for $\mathscr{P}$ bounded by r. For each $P\in\mathscr{P}$, let $\mathscr{P}(P)$ be the set of those paths in $\mathscr{P}$ lying in $H(P)$. For each $C\in\mathscr{C}$ and $P\in\mathscr{P}$ let $C(P)$ be the cut-sets for $\mathscr{P}(P)$ in $C\cap H(P)$. Then we have:

THEOREM 5. *With the assumptions above, if statement T is true for all $C\in\mathscr{C}$, then statement R is true for all $k\ge r$ and all t.*

PROOF. Assume T is true for all $C\in\mathscr{C}$, and suppose R is false for some k and t. This means that for each n there are counterexamples. That is, there is some $D\in\mathscr{C}$ and some partition of $D\cap G_n$ into t classes $\mathscr{A}_j(D), j=1, 2, \ldots, t$, such that $\mathscr{A}_j(D)\cap H(v)\notin\mathscr{C}(v)$ for all $v\in l_k(G_n)$. Each such counterexample can be denoted by a $(t+2)$-tuple $[G_n; G_n\cap D; \mathscr{A}_1(D), \ldots, \mathscr{A}_t(D)]$. Further, if $[G_n; G_n\cap D; \mathscr{A}_1(D), \ldots, \mathscr{A}_t(D)]$ is a counterexample, so is $[G_{n-1}; G_{n-1}\cap D; \mathscr{A}_1(D)\cap G_{n-1}, \ldots, \mathscr{A}_t(D)\cap G_{n-1}]$, for if $\mathscr{A}_j(D)\cap G_{n-1}\cap H(v)\in\mathscr{C}(v)$, so is $\mathscr{A}_j(D)\cap H(v)$.

Now we form a graded graph K by letting $l_0(K)=l_0(G)$, and $l_n(K)$ be the set of counterexamples for G_n. Let there be an edge from $[G_{n-1}; G_{n-1}\cap D; \mathscr{A}_1(D), \ldots,$

$\mathscr{A}_t(D)]$ to $[G_n; G_n \cap D'; \mathscr{A}_1(D'), \ldots, \mathscr{A}_t(D')]$ if and only if $\mathscr{A}_j(D) \subseteq \mathscr{A}_j(D')$ for $j=1, 2, \ldots, t$. Then by the observation above every vertex in $l_n(K)$ is connected to some vertex in $l_{n-1}(K)$. Since G_n is finite, K is locally finite. Thus, by König's theorem [2, p. 29], there is an infinite directed path Q in K.

Let the vertices of Q be $[G_n; D_n \cap G_n; \mathscr{A}_1(D_n), \ldots, \mathscr{A}_t(D_n)]$, $n=0, 1, 2, \ldots$. Let $E=\bigcup_{i=0}^{\infty} (D_n \cap G_n)$, and for each j let $\mathscr{A}_j=\bigcup_{n=1}^{\infty} \mathscr{A}_j(D_n)$. The $\mathscr{A}_j$ partition E into t classes. E is in $\mathscr{C}$ since $\mathscr{C}$ is closed. Then by statement T, there is some infinite path $P \in \mathscr{P}$, and some subset $C' \in E(P)$ such that $C' \subseteq \mathscr{A}_j$ for some j.

Let $v \in l_k(G) \cap P$. For sufficiently large n, $H(v) \subseteq H(G_n)=G_n$, and $C' \cap H(v) \subseteq \mathscr{A}_j \cap (H(v) \cap G_n)=(\mathscr{A}_j \cap G_n) \cap H(v)=\mathscr{A}_j(D_n) \cap H(v)$. If $C' \cap H(v) \in \mathscr{C}(v)$, then $\mathscr{A}_j(D_n) \cap H(v) \in \mathscr{C}(v)$, which would contradict the choice of $\mathscr{A}_j(D_n)$ and prove the theorem. But $C' \in E(P)$ is a cut-set for all paths in $\mathscr{P}(P)$. Also C' is bounded by r. Since (a) $v \in l_k(G) \cap P$, (b) $\mathscr{P}$ is complete, and (c) $\mathscr{P}_0 \subseteq \mathscr{P}$, $C' \cap H(v)$ is in fact a cut-set for $\mathscr{P}(v)$. This completes the proof.

As we saw above, if the G and G_i are the subset diagrams, $\mathscr{P}=\mathscr{P}_0$, and $\mathscr{C}=\{l_r(G)\}$ (which is closed), then Theorem 5 simply says that Theorem 2 (statement T here) implies Theorem 1 (statement R here). Similarly if $\mathscr{P}=\mathscr{P}_0$ and $\mathscr{C}$ is the set of all cut-sets for $\mathscr{P}$ bounded by r, which is closed (by the lemma below), then Theorem 2 says that Corollary 4.1 implies Motzkin's theorem for "blocking". If $\mathscr{P}=\mathscr{P}_1$ and $\mathscr{C}$ is all the cut-sets for $\mathscr{P}$ bounded by r (closed by the lemma), then Theorem 5 says that Corollary 4.2 implies Motzkin's theorem for "obstructing".

However, from Theorem 4 and Theorem 5 we can produce a variety of generalizations of Ramsey's theorem along these lines by choosing different complete sets of paths and closed sets of cut-sets for them. Finally, if $S^{\infty}(r, t)$ can be established for the subspace diagram of an infinite-dimensional vector space over a finite field F, then we have the finite theorems like Motzkin's also.

LEMMA. *Let $\mathscr{P}$ be any set of infinite almost directed paths in $G=\bigcup_{i=0}^{\infty} G_i$. Let $\mathscr{C}$ be the set of all cut-sets for $\mathscr{P}$ which are bounded by r. Then $\mathscr{C}$ is closed.*

PROOF. Let $F_1, F_2, F_3, \ldots$ be an infinite sequence of members of $\mathscr{C}$ for which $F_1 \cap G_1 \subseteq F_2 \cap G_2 \subseteq F_3 \cap G_3 \subseteq \cdots$. We must show that $E=\bigcup_{m=1}^{\infty} (F_m \cap G_m)$ is in $\mathscr{C}$. E is certainly bounded by r, since each F_m is bounded by r. So it remains to show that E is a cut-set for $\mathscr{P}$.

Let P be any path in $\mathscr{P}$. Since P is infinite almost directed, P has only a finite number of vertices in $\bigcup_{m=0}^{r} l_m(G)$. Thus for a sufficiently large n all of these vertices of P are in G_n. That is, $P \cap \bigcup_{m=0}^{r} l_m(G) \subseteq P \cap G_n$. Since F_n is a cut-set for $\mathscr{P}$ bounded by r, we have $\varnothing \neq F_n \cap P=(F_n \cap \bigcup_{m=0}^{r} l_m(G)) \cap P$. Therefore $\varnothing \neq F_n \cap (P \cap \bigcup_{m=0}^{r} l_m(G)) \subseteq F_n \cap (P \cap G_n)=P \cap (G_n \cap F_n) \subseteq P \cap E$. So $P \cap E$ is not empty, E is a cut-set for $\mathscr{P}$, and $\mathscr{C}$ is closed.

REFERENCES

1. T. S. Motzkin, *Combinatorial functions and theorems like Ramsey's*, Notices Amer. Math. Soc. **13** (1966), 630; and *Shadows of finite sets*, J. Combinatorial Theory **4** (1968), 40–48. MR **36** #1337.

2. O. Ore, *Theory of graphs*, Amer. Math. Soc. Colloq. Publ., vol. 38, Amer. Math. Soc., Providence, R.I., 1962. MR **27** #740.

3. F. P. Ramsey, *On a problem of formal logic*, Proc. London Math. Soc. (2) **30** (1930), 264–286.

4. B. L. Rothschild, *A generalization of Ramsey's theorem and a conjecture of Rota*, Doctoral thesis, Yale University, New Haven, Conn., 1967.

MASSACHUSETTS INSTITUTE OF TECHNOLOGY

NONAVERAGING SETS[1]

E. G. STRAUS

1. **Introduction.** We wish to consider sets of integers $A=\{a_1, \ldots, a_n\}$ so that $0 \le a_1 < a_2 < \cdots < a_n \le x$ and no a_i is the average of any subset of A consisting of two or more elements. Thus our restriction is more severe than the one that no three elements are in arithmetic progression, which is sometimes referred to as "nonaveraging."

Let $f(x) = \max |A|$ for all nonaveraging sets in the interval $[0, x]$. Our main result is to establish a lower bound for $f(x)$.

1.1. THEOREM.

$$\log_2 f(x) > (2 \log_2 x)^{1/2} + \tfrac{1}{2} + O(1/\sqrt{\log x}).$$

Thus $f(x)$ grows more rapidly than any power of $\log x$, though it presumably grows less rapidly than any positive power of x.

We may even permit weighted averages with nonnegative integral weights $\le M$ by defining $f_M(x)$ as

$$f_M(x) = \max |A_M|,$$

where $A_M = \{a_1, \ldots, a_n\}$ so that $0 \le a_1 < a_2 < \cdots < a_n \le x$ and

$$a_j \sum_{i=1}^{n} m_i = \sum_{i=1}^{n} m_i a_i, \qquad 0 \le m_i \le M, \qquad \sum m_i > 0$$

implies $m_i = \delta_{ij}$.

Then

1.2. THEOREM.

$$\log_2 f_M(x) > (2 \log_2 x + (\log_2 M - \tfrac{1}{2})^2)^{1/2} - (\log_2 M - \tfrac{1}{2}) + O(1/\sqrt{\log x}).$$

This formula remains valid even if M is allowed to grow with x.

The problem arose from a question raised by P. Erdös on the maximum number, $g(x)$, of integers in the interval $[0, x]$ so that none divides the sum of any others. We shall call such sets *nondividing sets*.

---

[1] The preparation of this paper was sponsored in part by NSF Grant GP-5497.

Since every nondividing set is obviously nonaveraging we have $g(x) \leq f(x)$. On the other hand, given a nonaveraging set, $0 \leq a_1 < a_2 < \cdots < a_n \leq x/n$ (x an integer), then the set $\{x-a_n, x-a_{n-1}, \ldots, x-a_1\}$ is nondividing. To see this assume $x-a_i$ divides

$$S = (x-a_{j_1})+\cdots+(x-a_{j_k}) = kx-(a_{j_1}+\cdots+a_{j_k}).$$

Now

$$S-(k-1)(x-a_i) = x-(a_{j_1}+\cdots+a_{j_k})+(k-1)a_i > 0,$$

while

$$S-(k+1)(x-a_i) = -x-(a_{j_1}+\cdots+a_{j_k})+(k+1)a_i < 0.$$

Thus we must have $S=k(x-a_i)$ which means that a_i is the average of $\{a_{j_1}, \ldots, a_{j_k}\}$ contrary to hypothesis.

We have therefore proved

1.3. THEOREM.

$$\max\{f(x/f(x)), f(\sqrt{x})\} \leq g(x) \leq f(x).$$

In view of Theorem 1.1 this gives a lower bound for $g(x)$.

1.4. COROLLARY.

$$\log_2 g(x) > (2\log_2 x)^{1/2}+O(1/\sqrt{\log x}).$$

In an analogous manner we could define $g_M(x)$ as the maximum number of integers in the interval $[0, x]$ so that none divides linear combinations of others with integral coefficients in the interval $[1, M]$. We omit details.

R. Rado has raised the question of finding the maximum number, $h(x)$, of integers in the interval $[0, x]$ so that different sets have different averages. In this generality we get $h(x)$ much smaller than $f(x)$. In fact

1.4. THEOREM.[2]

$$(1+O(1))\log x/\log\log x < h(x) < \log_2 x+O(\log\log x).$$

However we can generalize our method of proof for Theorems 1.1 and 1.2 to give a large lower bound for the maximum number $h^*(x)$ of integers in the interval $[0, x]$ such that any two subsets with a *relatively prime number of elements* have different averages.

1.5. THEOREM.

$$\log_2 h^*(x) \geq (\log_2 x)^{1/2}-1+O(1/\sqrt{\log x}).$$

This is in sharp contrast to the small upper bound in Theorem 1.4.

[2] The lower bound is improved to $-1+\log_4 x$ in **[1]**.

2. **Proof of Theorems 1.1, 1.2 and 1.5.** Since Theorem 1.1 is a special case of Theorem 1.2 it suffices to prove the latter.

2.1. LEMMA.

$$f_M((M/2)xf_M(x)+2x+1) \geq 2f_M(x).$$

PROOF. Assume we are given a set A_M of $f_M(x)$ integers in the interval $[0, x]$ with the nonaveraging property discussed in Theorem 1.2. Since the mapping $a \to [x]-a$ maps A_M into such a nonaveraging set we may assume

$$\frac{1}{f_M(x)} \sum_{a \in A_M} a \geq \frac{x}{2}.$$

We now choose y so large that the weighted average of y with any subset of A_M exceeds x. Since no element of A_M exceeds x, this condition will certainly be satisfied if it is satisfied when we attach maximal weight to each element of A_M and weight 1 to y, that is

$$\frac{M \sum_{a \in A_M} a+y}{Mf_M(x)+1} \geq \frac{Mf_M(x)x/2+y}{Mf_M(x)+1} > x,$$

or

$$y > Mf_M(x)x/2+x.$$

Thus we may choose

$$y = [Mf_M(x)x/2+x+1].$$

Now consider the set

$$A'_M = A_M \cup (y+A_M)$$

where $y+A_M=\{y+a \mid a \in A_M\}$. Then A'_M is a set of integers contained in the interval $[0, y+x]$ and hence in the interval $[0, (M/2)xf_M(x)+2x+1]$. By hypothesis no element of A_M is the weighted average with positive integral weights no greater than M of other elements of A_M and by construction every such average which includes elements of $y+A_M$ exceeds the elements of A_M. By symmetry it is clear that no element $y+A_M$ is a weighted average of the proscribed type of other elements of A'_M. Now $|A'_M|=2|A_M|=2f_M(x)$ which completes the proof.

PROOF OF THEOREM 1.2. We obviously have $f_M(1)=2$. Using Lemma 2.1 repeatedly we get

$$\begin{aligned} &f_M(M+3) = f_M((M+2)+1) \geq 4 \\ &f_M((2M+3)(M+3)) \geq f_M((2M+2)(M+3)+1) \geq 8 \\ &\vdots \\ &f_M\left(\prod_{k=0}^{n-2} (2^k M+3)\right) \geq 2^n. \end{aligned} \tag{2.2}$$

Now

$$\prod_{k=0}^{n-2}(2^kM+3) < 2^{\binom{n-1}{2}}M^{n-1}\prod_{k=0}^{\infty}\left(1+\frac{3}{2^kM}\right) \tag{2.3}$$
$$= c2^{\binom{n-1}{2}}M^{n-1}.$$

Thus

$$f_M(c2^{\binom{n-1}{2}}M^{n-1}) \geq 2^n.$$

Now, given x, let n be the integer so that

$$cM^{n-1}2^{\binom{n-1}{2}} < x \leq cM^n2^{\binom{n}{2}}.$$

In other words

$$\binom{n}{2}+n\log_2 M+\log_2 c \geq \log_2 x$$

so that

$$\log_2 f_M(x) \geq n \geq (2\log_2 x+(\log_2 M-\tfrac{1}{2})^2-2\log_2 c)^{1/2}-(\log_2 M-\tfrac{1}{2})$$
$$= (2\log_2 x+(\log_2 M-\tfrac{1}{2})^2)^{1/2}-(\log_2 M-\tfrac{1}{2})+O(1/\sqrt{\log x}).$$

There are various other ways of obtaining constructions analogous to the one used in Lemma 2.1. For example one could start with an admissible set A_M in the interval $[0, x]$ and then construct an admissible set

$$A'_M = 2Mf_M(x)A_M \cup (2Mf_M(x)A_M-1)$$

where A'_M is in the interval $[0, 2Mxf_M(x)]$ and $|A'_M|=2f_M(x)$.

In order to prove Theorem 1.5 we make a construction analogous to that in Lemma 2.1.

2.4. LEMMA.

$$h^*(2x(h^*(x))^2+x) \geq 2h^*(x).$$

PROOF. We consider a set S of $h^*(x)$ integers in the interval $[0, x]$ so that no two subsets with *relatively prime* numbers of elements have equal averages. We may assume that

$$\frac{1}{h^*(x)}\sum_{s\in S} s \leq \frac{x}{2}.$$

We now wish to find an integer y sufficiently large so that the set $S'=S\cup(y+S)$ has the same nonaveraging property. In other words, if $(a+b, c+d)=1$ and $\max\{a+b, c+d\}>1$, then

$$(c+d)[(y+s_{i_1})+\cdots+(y+s_{i_a})+s_{j_1}+\cdots+s_{j_b}] \tag{2.5}$$
$$\neq (a+b)[(y+s_{k_1})+\cdots+(y+s_{k_c})+s_{l_1}+\cdots+s_{l_d}],$$

where the s with various subscripts are elements of S. If $a=c=0$ or if $b=d=0$ then inequality (2.5) is true by hypothesis. Otherwise we have $a(c+d)-c(a+b)\neq 0$. Assuming, without loss of generality, that $a(c+d)-c(a+b)\geq 1$ we get that (2.5) is satisfied whenever

$$y > (a+b)(s_{k_1}+\cdots+s_{k_c}+s_{l_1}+\cdots+s_{l_d}) \\ -(c+d)(s_{i_1}+\cdots+s_{i_a}+s_{j_1}+\cdots+s_{j_b}) \tag{2.6}$$

which certainly holds whenever

$$y \geq 2(h^*(x))^2x \geq 2h^*(x)\cdot 2\sum_{s\in S} s.$$

Thus we get an admissible set S' with $|S'|=2|S|=2h^*(x)$ in the interval $[0, 2x(h^*(x))^2+x]$.

PROOF OF THEOREM 1.5. We have $h^*(1)=2$. Applying Lemma 2.4 repeatedly we get

$$h^*(2\cdot 2^2+1) \geq 4$$

$$h^*((2\cdot 2^2+1)(2\cdot 4^2+1)) \geq 8$$

$$h^*\left(\prod_{k=1}^{n-1}(2\cdot 2^{2k}+1)\right) \geq 2^n.$$

Now

$$\prod_{k=1}^{n-1}(2^{2k+1}+1) = 2^{n^2}\prod_{k=1}^{n-1}\left(1+\frac{1}{2^{2k+1}}\right) < c\cdot 2^{n^2}$$

so that

$$h^*(c\cdot 2^{n^2}) \geq 2^n.$$

For a given x we choose n so that

$$c\cdot 2^{n^2} \leq x < c2^{(n+1)^2}$$

or

$$n^2+\log_2 c \leq \log_2 x < (n+1)^2+\log_2 c$$

which yields

$$\log_2 h^*(x) \geq n > (\log_2 x-\log_2 c)^{1/2}-1 \\ = \sqrt{\log_2 x}-1+O(1/\sqrt{\log x}).$$

3. **Proof of Theorem 1.4.** We could obtain a lower bound by the methods of §2. We try a slightly different approach by choosing the set of powers of a sufficiently large integer A, that is $S=\{1, A, \ldots, A^{n-1}\}$ where

$$n-1 = [\log x/\log A]. \tag{3.1}$$

Assume

$$(1/k)(A^{i_1}+\cdots+A^{i_k}) = (1/l)(A^{j_1}+\cdots+A^{j_l}) \tag{3.2}$$

with $0 \le i_1 < \cdots < i_k < n$; $0 \le j_1 < \cdots < j_l < n$ and $\{i_1, \ldots, i_k\} \ne \{j_1, \ldots, j_l\}$. Then clearly $k \ne l$. Let m be the maximal exponent in (3.2) and assume without loss of generality that $j_l = m$. Thus if we solve (3.2) for A^m we get that kA^m (if $i_k < m$) or $|k-l|A^m$ (if $i_k = m$) as a linear combination of lower powers of A. Thus

$$A^m \le \max\{l, k\}(1 + A + \cdots + A^{m-1}) \le n(A^m - 1)/(A-1)$$

which is impossible if

$$n = A - 1 = [\log x/\log A] + 1. \tag{3.3}$$

Equation (3.3) leads to

$$A \sim \log x/\log \log x$$

and

$$h(x) \ge n = A - 1 \sim \log x/\log \log x. \tag{3.4}$$

The upper bound for $h(x)$ follows from the fact that a set with n elements has

$$\binom{n}{[n/2]}$$

different subsets of cardinality $[n/2]$. These sets have different averages if and only if they have different sums. If all elements are in $[0, x]$ all such sums are in the interval $[0, nx/2]$ so that

$$c\frac{2^n}{\sqrt{n}} \le \binom{n}{[n/2]} < \frac{nx}{2},$$

or

$$n - (3/2)\log_2 n \le \log_2 x + c_1$$

leading to

$$h(x) = n \le \log_2 x + O(\log \log x). \tag{3.5}$$

Theorem 1.4 follows from (3.4) and (3.5).

P. Erdös [1] has proved that we get an upper bound $O(\log^2 x)$ even when only sets with a different number of elements are assumed to have different averages.

4. **Equivalent and related questions.** The question of obtaining good upper bounds for nonaveraging sets seems to be more difficult. We conjecture that our lower bound for $\log f(x)$ is at least of the correct order of magnitude, but the only available upper bounds[3] for $f(x)$ are the ones obtained by the deep methods of K. F. Roth for sets with no three elements in arithmetic progression which appear far from the truth. We therefore state some related problems which are of interest in their own right.

4.1. THEOREM. *A set of integers $\{a_1, \ldots, a_n\}$ is nonaveraging if and only if the antisymmetric matrix*

[3] An upper bound, $f(x) = O(x^{2/3})$, is obtained in [1] using the ideas in this section.

$$A = (a_i - a_j) = \begin{pmatrix} 0 & a_1-a_2 & a_1-a_3 & \cdots & a_1-a_n \\ a_2-a_1 & 0 & a_2-a_3 & \cdots & a_2-a_n \\ \vdots & & & & \vdots \\ a_n-a_1 & a_n-a_2 & \cdot & \cdot \quad \cdot & 0 \end{pmatrix}$$

has the property that for every vector $\varepsilon = (\varepsilon_1, \ldots, \varepsilon_n)$ *with components* 0 *or* 1 *we have that the vector* $\eta = (\eta_1, \ldots, \eta_n)$,

$$(\eta_1, \ldots, \eta_n) = A \begin{pmatrix} \varepsilon_1 \\ \vdots \\ \varepsilon_n \end{pmatrix}$$

has no 0 *component unless* ε *is the zero vector or a unit vector.*

PROOF. The condition

$$0 = \eta_i = \varepsilon(a_i - a_1) + \cdots + \varepsilon_n(a_i - a_n)$$

leads to

$$a_i = (\varepsilon_1 a_1 + \cdots + \varepsilon_n a_n)/(\varepsilon_1 + \cdots + \varepsilon_n).$$

If we wish to consider weighted averages we would get the analogous condition on the matrix A applied to vectors $(m_1, \ldots, m_n)$ with integral components m_i in the interval $[0, M]$.

Assume, for the moment, that the number n in Theorem 4.1 is odd and that the a_i are in ascending order. Then the nonvanishing of $\eta_{n/2}$ leads to the following.

4.2. *Problem.* What is the maximum number $n = F(x)$ such that there exist two sets of integers $A = \{a_1, \ldots, a_n\}$, $B = \{b_1, \ldots, b_n\}$ in the interval $[0, x]$ so that the sums of nonempty subsets of A are different from the sums of nonempty subsets of B?

It is again rather easy to establish lower bounds.

4.3. *Conjecture and theorem.* We conjecture that the maximum number $F(x)$ is attained when we choose the two sets at opposite ends of the interval $[0, x]$. That is $A = \{0, 1, \ldots, n-1\}$ and $B = \{[x], [x]-1, \ldots, [x]-n+1\}$. This construction leads to

$$F(x) \geq [\sqrt{(2x)}] - 1.$$

As a corollary to Theorem 4.1 we have

$$f(x) \leq 2F(x) + 1. \tag{4.4}$$

The number on the right is a bound for odd sets of integers in $[0, x]$ whose middle element is not the average of other elements.

One could generalize Problem 4.2 to that of finding the maximum number $n = F_k(x)$ so that there are k sets of integers $A_i = \{a_{1i}, a_{2i}, \ldots, a_{ni}\}$; $i = 1, \ldots, k$ in the interval $[0, x]$ so that the sums of nonempty subsets of A_i are different from the sums of nonempty subsets of A_j for $i \neq j$.

4.5. *Conjecture and theorem.* We conjecture that the maximal value $F_k(x)$ is attained if the sets A_i are sets of consecutive integers so that the sum of all elements of A_i is less than the minimal element of A_{i+1}. This construction leads to

$$A_i = \{a_i, a_i+1, \ldots, a_i+n-1\}; \qquad a_i = \frac{n^{i-1}-1}{n-1}\left(\binom{n}{2}+1\right), \qquad i = 1, \ldots, k.$$

Thus we can choose n as the largest integer satisfying

$$\frac{n^{k-1}-1}{n-1}\left(\binom{n}{2}+1\right)+n-1 \leq x$$

or

$$F_k(x) \geq n \geq [(2x)^{1/k}]-1. \tag{4.6}$$

Reference

1. Paul Erdös and E. G. Straus, *Nonaveraging sets* II, Proceedings of the Colloquium on Combinatorial Mathematics, Balatonfüred, 1969 (to appear).

University of California, Los Angeles

ON (k, l)-COVERINGS AND DISJOINT SYSTEMS

J. D. SWIFT[1]

1. **Introduction.** We shall study a generalization of tactical systems. A special case of covering systems was first studied by Fort and Hedlund [**1**]. More recently, general definitions and extremal bounds have been given by Schönheim [**4**]. The latter paper contains a bibliography and more detailed historical remarks.

The principal result of the present paper will be to show that a bound given by Schönheim is not attainable for certain cases of a generalized Steiner problem.

Other, more special, results are presented in the final section.

Certain results of Schönheim have been quoted for completeness. These are indicated either by name or denoted by an asterisk (*) before the statement.

2. **Definitions. Persistence theorems.**

DEFINITION 1. A (k, l)-disjoint system on n objects is a set of k-tuples of the n objects (without repeats) such that no l-tuple appears in more than one k-tuple of the set. Such a system will be denoted by $D(k, l, n)$. When it is desired to call attention to s, the number of k-tuples in the system, the notation will be $D(k, l, n, s)$.

A (k, l)-covering on n objects is a set of k-tuples of the n objects such that every l-tuple appears in at least one k-tuple of the set. Such a system will be denoted by $C(k, l, n)$ or $C(k, l, n, s)$ following the procedure used for disjoint systems.

DEFINITION 2. A system which is both (k, l)-disjoint and (k, l)-covering is called a tactical system. It will be denoted by $S(k, l, n)$.

DEFINITION 3. We denote by $m(k, l, n)$ the maximum number of n-tuples in a $D(k, l, n)$. I.e., $D(k, l, n, m(k, l, n))$ exists and for $D(k, l, n, s)$, $s \leq m(k, l, n)$. Similarly $M(k, l, n)$ is the minimum for $C(k, l, n)$.

It is obvious that $D(k, l, n, s)$ exist for all $s \leq m(k, l, n)$ and $C(k, l, n, s)$ exist for all $s \geq M(k, l, n)$. Similarly $M(k, l, n) \geq m(k, l, n)$ and equality is equivalent to the existence of an $S(k, l, n)$. We assume without loss of generality that $n \geq k > l$.

A well-known necessary condition for the existence of an $S(k, l, n)$ is that $\binom{n-h}{l-h}/\binom{k-h}{l-h}$ be integral for $h=0, \ldots, l-1$. These conditions are also sufficient for $k \leq 4$ [**2**]. When an $S(k, l, n)$ exists it contains $\binom{n}{l}/\binom{k}{l}$ k-tuples.

[1] The preparation of this paper was supported in part by National Science Foundation Grant #GP-5497.

Two basic, but rather simple, theorems give limiting conditions for disjoint and covering systems. In their statements and the sequel, $D(k, l, n, s)$ and $C(k, l, n, s)$ will be used to assert the existence of such systems as well as to denote the systems themselves.

THEOREM 1 (PERSISTENCE OF FAILURE).

(i) $D(k+1, l+1, n+1, s) \Rightarrow D(k, l, n, [s(k+1)/(n+1)])$ *where* $[x] = -\lfloor -x \rfloor$ *is the smallest integer not less than* x.

(ii) $C(k+1, l+1, n+1, s) \Rightarrow C(k, l, n, [s(k+1)/(n+1)])$.

PROOF. (i) Consider the total number of objects in the $(k+1)$-tuples of $D(k+1, l+1, n+1, s)$. There are $s(k+1)$ such appearances. The average number of appearances of an individual object is $s(k+1)/(n+1)$. Therefore one object appears in at least $\lfloor s(k+1)/(n+1) \rfloor$ $(k+1)$-tuples. Select these $(k+1)$-tuples and delete the specified object. The result is a $D(k, l, n)$.

(ii) Similarly, in the $C(k+1, l+1, n+1, s)$ there is an object which appears in no more than $[s(k+1)/(n+1)]$ $(k+1)$-tuples. If these $(k+1)$-tuples are reduced to k-tuples by deletion of the specified object, the result is a $C(k, l, n)$ since every l-tuple on the remaining objects must appear.

COROLLARY 1.

(i) $$m(k+1, l+1, n+1) \leq \left[\frac{n+1}{k+1} m(k, l, n)\right],$$

(ii) $$M(k+1, l+1, n+1) \geq \left\lfloor\frac{n+1}{k+1} M(k, l, n)\right\rfloor.$$

The statements of the corollary are cases of the contrapositive of the statements in the theorem. If

$$s = m(k+1, l+1, n+1) > \left[\frac{n+1}{k+1} m(k, l, n)\right],$$

$$s > \frac{n+1}{k+1} m(k, l, n) \quad \text{or} \quad s\frac{k+1}{n+1} > m(k, l, n)$$

so that $\lfloor s(k+1)/(n+1) \rfloor > m(k, l, n)$ and a contradiction results.

The proof for part (ii) is entirely similar.

*COROLLARY 2.

(i) $$m(k, l, n) \leq \left[\frac{n}{k}\left[\frac{n-1}{k-1}\cdots\left[\frac{n-l+1}{k-l+1}\right]\cdots\right]\right] = \phi(k, l, n).$$

(ii) $$M(k, l, n) \geq \left\lfloor\frac{n}{k}\left\lfloor\frac{n-1}{k-1}\cdots\left\lfloor\frac{n-l+1}{k-l+1}\right\rfloor\cdots\right\rfloor\right\rfloor = \psi(k, l, n).$$

PROOF. An immediate induction from Corollary 1.

The bounds φ and ψ will be referred to as the Schönheim bounds in the sequel.

An obvious question is: "How good are they?" When a tactical system exists they are equal and perfect.

The result of Fort and Hedlund [1] is that $M(3, 2, n)=\psi(3, 2, n)$. The result of Schönheim [4] is that $m(3, 2, n)=\varphi(3, 2, n)$ if $n\not\equiv 5 \pmod 6$ and $m(3, 2, n)=\varphi(3, 2, n)-1$ for $n\equiv 5 \pmod 6$. Somewhat more general statements will follow from Theorem 2.

THEOREM 2 (PERSISTENCE OF SUCCESS).
(i) $D(k, l, n, s) \Rightarrow D(k, l, n-1, s-[sk/n])$,
(ii) $C(k, l, n, s) \Rightarrow C(k, l, n+1, s+M(k-1, l-1, n))$.

PROOF. (i) In $D(k, l, n, s)$ the average number of appearances of an object in sk/n. Hence at least one object appears in no more than $[sk/n]$ k-tuples. Delete these k-tuples and a $D(k, l, n-1)$ results.

(ii) To $C(k, l, n, s)$ adjoin an $(n+1)$st object and the k-tuples formed by adding that object to all the $(k-1)$-tuples of a $C(k-1, l-1, n, M(k-1, l-1, n))$. The resulting system is a $C(k, l, n+1)$ since, given any l-tuple, it is already in $C(k, l, n, s)$ if it does not contain the new object.

*COROLLARY. (i) $S(k, l, n) \Rightarrow D(k, l, n-1, \varphi(k, l, n-1))$;
(ii) $S(k, l, n) \Rightarrow C(k, l, n+1, \psi(k, l, n+1))$.

The proof of (i) is a simple calculation from part (i) of the theorem. Part (ii) requires an additional induction to establish the size of $M(k-1, l-1, n)$. The details may be found in [3].

The designation of Theorem 2 as "Persistence of Success" is illustrated by the corollary. A tactical system $S(k, l, n)$ guarantees another incidence of the φ-bound and of the ψ-bound. The use of "Persistence of Failure" for Theorem 1 refers to the contrapositive of the statement and will be fully discussed in the next section.

3. **Successive failure of the φ-bound.** It has already been noted that $m(3, 2, n) < \varphi(3, 2, n)$ for $n\equiv 5 \pmod 6$. Corollary 1(i) of Theorem 1 then provides that $m(4, 3, n)<\varphi(4, 3, n)$ for $n\equiv 0 \pmod 6$, $m(5, 4, n)<\varphi(5, 4, n)$ for $n\equiv 1 \pmod 6$, etc. In this section we shall examine these and similar failures of the Schönheim φ-bound.

If $n\equiv 5 \pmod 6$,

$$\varphi(3, 2, n) = \left[\frac{n}{3}\left[\frac{n-1}{2}\right]\right] = \frac{n(n-1)-2}{6} = \frac{1}{3}\left(\frac{n(n-1)}{2}-1\right).$$

Thus if $m(3, 2, n)$ were $\varphi(3, 2, n)$, all pairs but one would be covered by triples. A parity argument can dispose of this possibility. If we consider unordered pairs in an $n\times n$ incidence matrix with diagonal elements omitted, we see that each triple covers six positions, 2 in each of 3 rows and columns. Thus, since each row (and column) had an even number of positions to begin with, an even number of blanks remain after the adjunction of any number of triples. But a pair corresponds to only one incidence in a row or column and we obtain a contradiction.

More generally, consider a $D(p, p-1, n)$ for p an odd prime and determine n as follows: $n \equiv 1(2)$; $n \not\equiv p-1 \pmod{p_i}$, $3 \le p_i < p$, $n \equiv p-1 \pmod{p}$. These conditions guarantee that

$$\varphi(p, p-1, n) = \frac{1}{p}\left(\binom{n}{p-1} - 1\right).$$

Again, if $m(p, p-1, n) = \varphi(p, p-1, n)$ precisely one $(p-1)$-tuple will remain uncovered. In the generalized $(p-1)$-dimension incidence matrix an ordered $(p-1)$-tuple occupies $(p-1)!$ positions, $(p-2)!$ in each applicable file. However, a p-tuple occupies $p!$ positions, $(p-1)!$ in each applicable file. Now, if $2^r \| (p-2)!$, where $\|$ is used for "divides exactly," $2^{r+1} | (p-1)!$ and a file contains $(n-1)(n-2)\cdots(n-p+2)$ slots which is also divisible by 2^{n+1}. Thus, by the extended parity argument we again have that it is impossible to cover all but one $(p-1)$-tuple.

We thus get new cases of impossibility to meet the φ-bound for each odd prime. These cases are indeed different from exclusions inherited from the prime p_i since the latter are congruent to $p-1$ modulo p_i and are thus not included in values for p. The fraction of newly excluded residues for p_i is

$$\left(1-\frac{1}{2}\right)\left(1-\frac{1}{3}\right)\cdots\left(1-\frac{1}{p_{i-1}}\right)\frac{1}{p_i}.$$

Thus the total fraction excluded at p_i is

$$\sum_{j=2}^{i} \prod_{k=1}^{j-1} \left(1-\frac{1}{p_k}\right)\frac{1}{p_j} = \frac{1}{2} - \prod_{j=1}^{i}\left(1-\frac{1}{p_j}\right).$$

Finally, integers n which meet the necessary condition for tactical systems for $S(p, p-1, n)$ satisfy the following congruences: $n \equiv 1 \pmod 2$, $n \not\equiv p-1 \pmod{p_i}$ for odd $p_i \le p$. The proportion of such possible numbers at p_i is clearly $\prod_{j=1}^{i}(1-1/p_j)$. Thus, any odd residue modulo p is in precisely one of the following classes:

(1) "Inherited" rejects from preceding primes.
(2) New rejects for p.
(3) Possible values for a tactical system.

We sum up the results in the following theorem.

THEOREM 3. *If $k \ge p_i$ (the ith prime) the Schönheim φ-bound for $D(k, k-1, n)$ cannot be attained for at least*

$$\left(\frac{1}{2} - \prod_{j=1}^{i}\left(1-\frac{1}{p_i}\right)\right)$$

of the residues modulo $\prod p_j$. If $k = p_i$, every odd residue is either such that the φ-bound is not attained or is a possible candidate for a tactical system.

There is no general theory as yet for even residues for odd values of n although it can be shown in special cases that the φ-bound cannot be attained. Also, aside from the "Steiner" case in which $l = k-1$, no general results are known.

4. Miscellaneous results and conjectures. There exists no comparable argument for that given in the last section for the case of coverings. The first case of failure of the ψ-bound is that $M(4, 3, 7)=12$ while $\psi(4, 3, 7)=11$. The proof of this fact is a simple exhaustion of cases.

A property of interest in its own right and also useful for further discussion of extremal cases is that of *completeness*.

DEFINITION 4. A disjoint system (covering system) is called complete if no k-tuple may be added (deleted) leaving the system disjoint (covering).

Clearly, an extremal system is complete but a complete system need not be extremal. For example $m(3, 2, 10)=13$ but Desargues' configuration is a complete $D(3, 2, 10, 10)$.

An obvious problem is to find the minimum number of k-tuples in a complete $D(k, l, n)$ and the maximum number in a complete $C(k, l, n)$. The answers are known only for special cases.

Many methods are known for composing two tactical systems to obtain a third (larger) system. When these methods are applicable to disjoint or covering systems they yield disjoint or covering systems of larger order but, in general, neither extremality nor completeness is preserved.

We will consider two known compositions of tactical systems which extend immediately to the general situation.

(1) Suppose R and S are sets of triples of objects chosen from sets V and W respectively. Form triples from $V\times W$ as follows, where ax stands for the ordered pair (a, x):

(a) (ax, ay, az) for $a \in V$, $(x, y, z) \in S$.
(b) (ax, bx, cx) for $(a, b, c) \in R$, $x \in W$.
(c) (ax, by, cz) for $(a, b, c) \in R$, $(x, y, z) \in S$.

(2) Suppose R and S are sets of quadruples of objects chosen from sets V and W respectively. Form quadruples from $V\times W$ as follows where ax stands for the ordered pair (a, x):

(a) (ax, ay, az, aw) for $a \in V$, $(x, y, z, w) \in S$.
(b) (ax, bx, cx, dx) for $(a, b, c, d) \in R$, $x \in W$.
(c) (ax, by, cz, dw) for $(a, b, c, d) \in R$, $(x, y, z, w) \in S$.
(d) (ax, ay, bz, bw) for $a, b \in V$, $(x, y, z, w) \in S$.
(e) (ax, bx, cy, dy) for $(a, b, c, d) \in R$, $w, y \in W$.
(f) (ax, ay, bx, by) for $a, b \in V$, $x, y \in W$.

In (d), (e), (f) we assume $x\neq y$, $a\neq b$.

If R and S are both (3, 2)-disjoint or (3, 2)-covers, the resulting system from (1) will have the same property. The equivalent statement applies to the (4, 3)-case for (2).

Completeness is, in general, not preserved. For example if the pair (1, 2) is not covered by R or S as disjoint systems, the triple (12, 11, 21) may be added to the result. Again if the triple $(1, 2, 3)$ is not covered by R or S, the quadruple $(12, 13, 21, 31)$ may be adjoined.

References

1. M. K. Fort, Jr., and G. A. Hedlund, *Minimal coverings of pairs by triples*, Pacific J. Math. **8** (1958), 709–719. MR **21** #2595.

2. H. Hanani, *The existence and construction of balanced incomplete block designs*, Ann. Math. Statist. **32** (1961), 361–386. MR **29** #4161.

3. J. Schönheim, *On coverings*, Pacific J. Math. **14** (1964), 1405–1411. MR **30** #1954.

4. ———, *On maximal systems of k-tuples*, Studia Sci. Math. Hungar. **1** (1966), 363–368. MR **34** #2485.

University of California, Los Angeles

(1, 2, 4, 8)—SUMS OF SQUARES AND HADAMARD MATRICES[1]

OLGA TAUSSKY

1. There are a number of classical facts and problems in algebra and topology in which a dimension n enters and which only holds for $n=1, 2, 4$ or $n=1, 2, 4, 8$. Some of these facts are reviewed briefly here. They are then applied to show that a certain generalization of the concept of Hadamard matrix only occurs for dimensions $n=1, 2, 4, 8$. Further, some other generalizations of Hadamard matrices are discussed.

2. I. Hurwitz **[9]** showed that *identities of the form*

$$\sum_1^n x_i^2 \sum_1^n y_i^2 = \sum_1^n z_i^2$$

when z_i are bilinear forms in the x's, y's are only possible for $n=1, 2, 4, 8$. For these values they can be derived from the norm relations of real numbers, complex numbers, quaternions and Cayley numbers. The case $n=8$ is usually denoted as the 8-square identity. Various proofs have been given for this over the years. The result is connected with the study of anticommuting skew matrices.

Results of Albert, Jacobson and Kaplansky linked the subject to composition algebras and included the infinite case. Composition for sums of squares for all n's which are powers of 2 via rational functions was obtained by Pfister **[12]**, for $n=8$ independently by Taussky **[19]**, for $n=16$ by Eichhorn and Zassenhaus **[22]**.

II. A deeper fact is that *the only algebras over the reals without divisors of zero have as numbers of base elements* 1, 2, 4, 8. No algebraic proof exists for this fact so far. The topological theorem from which this follows was obtained by Bott and Milnor, see **[3]**. A topological proof for Frobenius' theorem concerning associative division algebras was given by Taussky **[18]**.

III. Among the many applications of I and II a special one is now mentioned because of a slight connection with Hadamard matrices.

The only pairs of $n \times n$ matrices $L=(l_{ik})$, $M=(m_{ik})$ with l_{ik}, m_{ik} real linear forms

[1] This work was carried out (in part) under NSF contract 3909. Helpful discussions with D. Estes are gratefully acknowledged.

in n indeterminates and

$$ML = (x_1^2 + \cdots + x_n^2)I$$

where I is the $n \times n$ identity matrix, occur for $n = 1, 2, 4, 8$.

This result can be derived from a study by Taussky [17] of generalized Cauchy-Riemann equations. Generalizing the well-known link between the two classical Cauchy-Riemann equations and the Laplace differential equation the following was obtained:

Let $u_i(x_1, \ldots, x_n)$ be n functions of n variables and let

$$L_j = \sum_{i,k} a_{ikj} \frac{\partial}{\partial x_i} u_k, \qquad j = 1, \ldots, n$$

where a_{ikj} are real constants. Let real constants c_{jhi} exist such that

$$M_i = \sum c_{jhi} \frac{\partial}{\partial x_h} L_j = \frac{\partial^2 u_i}{\partial x_1^2} + \cdots + \frac{\partial^2 u_i}{\partial x_n^2}, \qquad i = 1, \ldots, n.$$

Then there exists an algebra over the reals with n base elements without divisors of zero.

Replacing $\partial/\partial x_i$ by x_i we may write

$$\begin{pmatrix} L_1 \\ \vdots \\ L_n \end{pmatrix} = L \begin{pmatrix} u_1 \\ \vdots \\ u_n \end{pmatrix}$$

where L is a matrix whose elements are linear forms in the x_i. We then have another matrix M of linear forms such that

$$M \begin{pmatrix} L_1 \\ \vdots \\ L_n \end{pmatrix} = ML \begin{pmatrix} u_1 \\ \vdots \\ u_n \end{pmatrix} = (x_1^2 + \cdots + x_n^2) \begin{pmatrix} u_1 \\ \vdots \\ u_n \end{pmatrix}$$

hence

$$ML = (x_1^2 + \cdots + x_n^2)I.$$

The pairs of matrices L, M which actually occur are of a very simple nature, M is the transpose of L and the linear forms are essentially simply the indeterminates x_i in suitable permutation, with suitable factors ± 1. They are the matrices obtained from the products of two real numbers, complex numbers, quaternions or Cayley numbers:

$$A_0 = (x_1),$$

$$A_1 = \begin{pmatrix} x_1 & -x_2 \\ x_2 & x_1 \end{pmatrix},$$

$$A_2 = \begin{pmatrix} x_1 & -x_2 & -x_3 & -x_4 \\ x_2 & x_1 & -x_4 & x_3 \\ x_3 & x_4 & x_1 & -x_2 \\ x_4 & -x_3 & x_2 & x_1 \end{pmatrix},$$

$$A_3 = \begin{pmatrix} x_1 & -x_2 & -x_3 & -x_4 & -x_5 & -x_6 & -x_7 & -x_8 \\ x_2 & x_1 & -x_4 & x_3 & -x_6 & x_5 & -x_8 & x_7 \\ x_3 & x_4 & x_1 & -x_2 & -x_7 & x_8 & x_5 & -x_6 \\ x_4 & -x_3 & x_2 & x_1 & x_8 & x_7 & -x_6 & -x_5 \\ x_5 & x_6 & x_7 & -x_8 & x_1 & -x_2 & -x_3 & x_4 \\ x_6 & -x_5 & -x_8 & -x_7 & x_2 & x_1 & x_4 & x_3 \\ x_7 & x_8 & -x_5 & x_6 & x_3 & -x_4 & x_1 & -x_2 \\ x_8 & -x_7 & x_6 & x_5 & -x_4 & -x_3 & x_2 & x_1 \end{pmatrix}.$$

Returning to the language of partial differential equations, the four-dimensional system is known as the Fueter or Dirac equations, see [7]. The result concerning the generalized Cauchy-Riemann equations was reproved by Stiefel [**16**] who uses algebraic tools, representation theory of Clifford type algebras and the study of anticommuting matrices, as is done in the proof of I. He extends the problem to complex constants which the earlier proof did not permit.

3. Recently Ryser defined a generalized Hadamard matrix to be an $n \times n$ matrix of just the special type described in 2. For $x_1 = x_2 = \cdots = x_n = 1$ an ordinary Hadamard matrix results. The theorem described in III shows that $n = 1, 2, 4, 8$ are the only possible n's. Folkman too had recently studied this question and found an independent proof for this fact. He uses a special case of a theorem of Adams, Lax and Phillips [1]: *Let $A_1, \ldots, A_k$ be a set of real $n \times n$ matrices such that $\sum \lambda_i A_i$ is nonsingular for all real λ_i, except $\lambda_1 = \cdots = \lambda_n = 0$. Let $n = 2^{b+4c}(2a+1)$ with $0 \leq b \leq 3$. Then $k \leq 2^b + 8c$.* The more general application of this theorem is mentioned in the next section.

4. At the suggestion of Szekeres the following generalization of Hadamard matrices had been studied by Spencer [**15**] prior to Folkman's work: Let H be an $n \times n$ matrix whose elements are $\pm x_i$, $1 \leq i \leq k$, when x_i are indeterminates and

$$HH' = (n/k)(x_1^2 + \cdots + x_k^2)I_n.$$

Then Folkman obtains: $k \leq 2^b + 8c$.

The "Cauchy-Riemann equation" aspect for the rectangular case had been touched briefly in Taussky [**17**] at the suggestion of van Dantzig. It is further treated in a recent paper of Eichhorn [**5**] where the whole problem is generalized from several other new angles. This is mentioned in the next section.

5. Eichhorn generalized the whole problem by talking about elements x of a vector space X over a field F of characteristic $\neq 2$ instead of indeterminates $x_1, \ldots, x_n$. He then studies linear mappings $L(x)$ of X into Hom (X, X) for which another such mapping $M(x)$ exists such that $M(x)L(x) = \mu(x)I$ when I is the identity mapping and $\mu(x)$ (not $\equiv 0$) is a mapping of X into F. The function $\mu(x)$ can be interpreted as a quadratic form and if this form is positive definite and of rank n and X an n-dimensional vector space we are back in the problems already discussed. However, this situation is greatly generalized. First of all it is shown

that for any form whether positive definite or not, as long as it has full rank n, the only possible values of n are 1, 2, 4, 8. But also the case of lower rank $r<n$ is considered. For $n=p2^q$, p odd, he obtains $r\leq 2q+2$. This agrees with Folkman's result and indeed Folkman's matrices can be considered as a special case of Eichhorn's in several ways: Firstly, the problem can be generalized to matrices whose elements are linear functions instead of simply indeterminates multiplied by ± 1. Secondly, products of a pair L, M instead of HH' can be studied. Next Eichhorn considers arbitrary fields instead of the reals, and the quadratic form can be arbitrary of rank r, not just a sum of r squares. Although this is not mentioned explicitly, the case of rectangular matrices is covered by Eichhorn's work. However this has nothing to do with the cases where the quadratic form does not possess full rank. E.g. consider

$$L = \begin{pmatrix} x_1 & x_1 \\ x_1 & -x_1 \end{pmatrix}.$$

This gives

$$LL' = 2x_1^2 I.$$

Eichhorn's results are based on the following lemma (which has some links with the result of [1]): Let $n=p2^q$ with p odd. Let F be an arbitrary field of characteristic $\neq 2$. Let A_i, $i=1,\ldots,r-1$, be a set of $n\times n$ matrices with elements in F which have the properties: $A_i^2=\alpha_i I$, $\alpha_i\in F$, $\alpha_i\neq 0$ and $A_iA_k+A_kA_i=0$, $i\neq k$. Then $r<2q+2$.

6. As a consequence of the results mentioned various suggestions concerning the study of ordinary Hadamard matrices evolve: First of all several generalizations are possible.

(1) Rectangular matrices.

(2) The study of matrices H with $HH'=rI$ instead of nI, for $r\leq n$.

(3) The study of pairs L, M with $ML=rI$.

Another generalization could come about if the concept of "transpose" is used in generalized form.

A final suggestion: An $n\times n$ Hadamard matrix H with

$$HH' = rI$$

is a generalization of an orthogonal matrix.[2] Such matrices can be connected with characteristic value problems (see Taussky [20]) in the following way: write the above equation as

$$HIH' = rI.$$

Thus I appears as a characteristic vector of an $n(n+1)/2\times n(n+1)/2$ matrix defined by the operator HXH' on the space of symmetric matrices X. The corresponding characteristic root is r. It is then of interest to study the other characteristic values and vectors.

[2] Such matrices with integral coefficients have been studied recently by G. Pall and pupils [11].

References

1. J. F. Adams, P. D. Lax and R. S. Phillips, *On matrices whose real linear combinations are non-singular*, Proc. Amer. Math. Soc. **16** (1965), 318–322. Correction **17** (1966), 945–947. MR **31** #3432.

2. M. F. Atiyah, *The role of algebraic topology in mathematics*, J. London Math. Soc. **41** (1966), 63–69. MR **32** #4684.

3. R. Bott and J. Milnor, *On the parallelizability of the spheres*, Bull. Amer. Math. Soc. **64** (1958), 87–89. MR **21** #1590.

4. J. Dieudonné, *A problem of Hurwitz and Newman*, Duke Math. J. **20** (1953), 381–389. MR **15**, 5.

5. W. Eichhorn, *Funktionalgleichungen in Vektorräumen, Kompositionsalgebren und systeme partieller Differentialgleichungen*, Aequationes Math. **2** (1969), 287–303.

6. J. Folkman, *A nonexistence theorem for Hadamard matrices in many variables*, Unpublished manuscript.

7. R. Fueter, *Die Theorie der regulären Funktionen einer Quaternionenvariablen*, Comptes Rendus du Congrès Int. des Mathématiciens, Oslo, 1936, pp 75–91.

8. M. Gerstenhaber, *On semicommuting matrices*, Math. Z. **83** (1964), 250–260. MR **28** #3047.

9. A. Hurwitz, *Über die Komposition der quadratischen Formen*, Math. Ann. **88** (1923), 1–25.

10. M. H. A. Newman, *Note on an algebraic theorem of Eddington*, J. London Math. Soc. **7** (1932), 93–99.

11. G. Pall, Unpublished communication.

12. A. Pfister, *Zur Darstellung von −1 als Summe von Quadraten in einem Körper*, J. London Math. Soc. **40** (1965), 159–165. MR **31** #169.

13. J. Putter, *Maximal sets of anti-commuting skew-symmetric matrices*, J. London Math. Soc. **42** (1967), 303–308. MR **35** #4247.

14. H. Ryser, Oral communications.

15. J. Spencer, *Hadamard matrices in many variables*, Unpublished manuscript.

16. E. Stiefel, *On Cauchy-Riemann equations in higher dimensions*, J. Res. Nat. Bur. Standards **48** (1952), 395–398. MR **14**, 38.

17. O. Taussky, *An algebraic property of Laplace's differential equation*, Quart. J. Math. Oxford Ser. **10** (1939), 99–103. MR **1**, 15.

18. ———, *Analytical methods in hypercomplex systems*, Compositio Math. **3** (1936), 399–407.

19. ———, *A determinantal identity for quaternions and a new eight square identity*, J. Math. Anal. Appl. **15** (1966), 162–164. MR **34** #183.

20. ———, *Automorphs and generalized automorphs of quadratic forms treated as characteristic value relations*, Linear Algebra and Appl. **1** (1968), 349–356.

21. F. van der Blij, *History of the octaves*, Simon Stevin **34** (1961), 106–125. [This paper contains many references to relevant work.] MR **24** #A149.

22. H. Zassenhaus and W. Eichhorn, *Herleitung von Acht- und Sechzehn-Quadrate-Identitäten mit Hilfe von Eigenschaften der verallgemeinerten Quaternionen und der Cayley-Dicksonschen Zahlen*, Arch. Mat. **17** (1966), 492–496.

California Institute of Technology

DICHROMATIC SUMS FOR ROOTED PLANAR MAPS

W. T. TUTTE

Several papers on the enumeration of rooted planar maps have appeared in the literature. We take our terminology from the first three sections of one entitled "A census of planar maps," published in the Canadian Journal of Mathematics in 1963 [3].

We recall that a planar map M is determined by a finite connected nonnull graph G embedded in the 2-sphere or closed plane. The vertices and edges of G are called the *vertices* and *edges* of M respectively. The rest of the surface consists of a finite number of disjoint simply connected domains called the *faces* of M. It is permissible for G to have loops and multiple edges.

A planar map M is *rooted* when one edge A is chosen as the *root* or *root-edge*, a direction of description of A is specified, and the two sides of A are distinguished as "right" and "left". It is convenient to refer to the negative end of A as the "root-vertex" and to the face on the left of A as the "root-face".

Let M_1 and M_2 be two rooted or unrooted planar maps on the 2-sphere or closed plane S. A *homeomorphism* of M_1 onto M_2 is a homeomorphism of S onto itself which maps vertices, edges and faces of M_1 onto vertices, edges and faces of M_2 respectively, and which preserves the root, if any, together with its direction and right and left sides. In this paper we do not distinguish between homeomorphic rooted planar maps. We say that two rooted or unrooted planar maps are *combinatorially distinct* if and only if there is no homeomorphism of one onto the other.

The *valency* of a vertex in a graph or planar map is the number of incident edges, loops being counted twice. The *valency* of a face of a planar map is the number of incident edges, that is the number of edges contained in its boundary, isthmuses being counted twice.

In [3] we determine the number of combinatorially distinct rooted planar maps with n edges. In [7] we investigate the number of combinatorially distinct rooted planar maps of $i+1$ faces and $j+1$ vertices. We may generalize such problems of enumeration by asking for the sum of a specified integral or polynomial function $f(M)$ of a rooted planar map M over a specified class of rooted planar maps. Enumeration corresponds to the case $f(M)=1$. Thus R. C. Mullin [2] has investigated the case in which $f(M)$ is the number of spanning trees in the graph defining

M. Here we discuss the more general case in which $f(M)$ is the *dichromatic polynomial* $\chi(M; x, y)$ of M.

The polynomial $\chi(M; x, y)$, in two independent indeterminates x and y, is actually a function of the graph G determining M. In order to explain its definition and properties we introduce the following notation.

For any graph G we denote the number of vertices by $\alpha_0(G)$ and the number of components by $p_0(G)$. We write $p_1(G)$ for the cycle-rank of G, that is the least number of edges whose deletion destroys every circuit. If S is any set of edges of G we write $G:S$ for the graph obtained from G by deleting all the edges not in S, and retaining all the vertices.

If A is any edge of G which is not a loop or an isthmus we write G'_A for the graph obtained from G by deleting the edge A. We also write G''_A to denote the graph obtained from G by identifying the two ends of A to form a single vertex, and then deleting A.

We define the *dichromatic polynomial* $\chi(G; x, y)$ of a finite graph G as follows.

$$\chi(G; x, y) = \sum_S \{(x-1)^{p_0(G:S)-p_0(G)}(y-1)^{p_1(G:S)}\}, \tag{1}$$

where S runs through all possible sets of edges of G. If G is embedded in the plane to form a planar map M, rooted or unrooted, we write

$$\chi(M; x, y) = \chi(G; x, y).$$

If M is a *vertex-map*, consisting of a single vertex, a single face, and no edges, we have

$$\chi(M; x, y) = 1,$$

by (1). By a special convention we count the vertex-map as a rooted map.

A *loop-map* has one vertex, one edge and two faces. The edge is necessarily a loop. For such a map we find from (1) that $\chi(M; x, y)=y$. A *link-map* has two vertices, one edge and one face. The edge must be a link, and an isthmus of the defining graph. For a link-map (1) gives $\chi(M; x, y)=x$.

In addition to these trivial evaluations the following recursion formulae can be deduced from (1).

THEOREM 1. *If G is the union of two subgraphs H and K having at most one vertex in common, then*

$$\chi(G; x, y) = \chi(H; x, y)\chi(K; x, y).$$

THEOREM 2. *If A is an edge of G that is not a loop or an isthmus, then*

$$\chi(G; x, y) = \chi(G'_A; x, y)+\chi(G''_A; x, y).$$

THEOREM 3. *If G has n loops, m isthmuses and no other edges, then*

$$\chi(G; x, y) = x^m y^n,$$

by Theorem 1 *and our results for the graphs of the vertex-map, link-map and loop-map.*

The dichromatic polynomial of any given graph can be calculated by repeated application of these theorems. Evidently two isomorphic graphs must have the same dichromatic polynomial. Hence any two homeomorphic rooted planar maps have the same dichromatic polynomial.

The dichromatic polynomial $\chi(G; x, y)$ is studied in [**4**]. It differs slightly from the polynomial $Q(G; x, y)$ called "dichromatic" in [**6**], being related to it by equation (21) of that paper. But the index -1 in the equation should be replaced by $-p_0(G)$ to allow for the case in which G is not connected.

If G is connected the number $\chi(G; 1, 1)$ is the number of spanning trees of G. More generally it is the number of spanning forests of G having just one component in each component of G. It can be shown that the number of n-colourings of G on the vertices (in which not all the n colours need be used) is

$$(-1)^{\alpha_0(G)}(-n)^{p_0(G)}\chi(G; 1-n, 0).$$

These results can be obtained by simple inductive arguments based on the above three theorems [**1**], [**5**].

Given a rooted planar map M we write $m(M)$, $n(M)$, $i(M)$ and $j(M)$ for the valency of the root-face, the valency of the root-vertex, the number of faces other than the root-face, and the number of vertices other than the root-vertex respectively. We say that M is of Type (m, n, i, j) if $m(M)=m$, $n(M)=n$, $i(M)=i$ and $j(M)=j$.

We introduce the generating function

$$\begin{aligned}\Phi &= \Phi(u_1, u_2, v_1, v_2, x, y) \\ &= \sum_M u_1^{m(M)}u_2^{n(M)}v_1^{i(M)}v_2^{j(M)}\chi(M; x, y).\end{aligned} \tag{2}$$

Here the sum is over all combinatorially distinct rooted planar maps M, including the vertex-map (which contributes a constant term 1). The coefficient of

$$u_1^m u_2^n v_1^i v_2^j$$

in the series for Φ is a polynomial in x and y. It is the sum of the dichromatic polynomials of the combinatorially distinct rooted planar maps of Type (m, n, i, j). We denote it by $N(m, n, i, j)$.

Our object in this paper is to exhibit a functional equation for Φ which permits the recursive calculation of the polynomials $N(m, n, i, j)$. We do not find an explicit formula for $N(m, n, i, j)$, but we do exhibit one for $N(m, n, 0, j)$.

The notion of duality is helpful. We recall that each unrooted planar map M has a *dual* unrooted planar map M^* related to it by a 1-1 correspondence. The correspondence maps vertices, edges and faces of M onto faces, edges and vertices of M^* respectively, and both it and its inverse preserve incidence relations. It can be shown that M^* is uniquely determined to within a homeomorphism, and hence that the dual of M^* is M.

It is well known that two dual unrooted planar maps can be exhibited in the closed plane so that each vertex of each map lies in the corresponding face of the other, so that an edge of one map meets an edge of the other only if the two edges correspond under the duality, and so that two corresponding edges meet at exactly one point, at which they cross.

The vertex-map is its own dual (to within a homeomorphism), and the link-map is the dual of the loop-map.

If two dual unrooted planar maps M and M^* are exhibited in the closed plane in the manner described above, then any rooting of M determines one of M^* by the following rule: the roots of M and M^* correspond under the duality and each crosses the other from the left side to the right. We recall that in rooting a map the choice of a "left" side is arbitrary. The notion of duality can thus be extended to rooted planar maps. Each such rooted map M has a combinatorially unique rooted planar map M^* as its dual, and is itself the dual of M^*.

Now consider an unrooted planar map M. Let N denote its dual map. Let A be an edge of M and let B be the corresponding edge of N. We suppose that A is neither a loop nor an isthmus of M. Then, by duality, B is neither an isthmus nor a loop of N. We form an unrooted planar map M'_A from M by deleting the edge A, except for its endpoints, and then uniting the two faces incident with A with the part of A previously deleted to form a new face F. We form an unrooted planar map N''_B from N by identifying all the points of the closed edge B as a single new vertex F. Thus if M and N are determined by graphs G and H respectively the maps M'_A and N''_B are determined by the graphs G'_A and H''_B respectively. It is easy to verify that M'_A and N''_B are dual maps.

By Theorem 2 we have

$$\chi(M; x, y) = \chi(M'_A; x, y)+\chi(M''_A; x, y),$$
$$\chi(N; y, x) = \chi(N''_B; y, x)+\chi(N'_B; y, x).$$

These relations can be used to prove inductively that

$$\chi(M; x, y) = \chi(M^*; y, x), \tag{3}$$

where M and M^* are any two dual planar maps, rooted or unrooted.

It is clear that the sum Φ does not alter when each rooted planar map M occurring in it is replaced by its dual. Hence

$$\Phi(u_1, u_2, v_1, v_2, x, y) = \Phi(u_2, u_1, v_2, v_1, y, x). \tag{4}$$

We write Φ_1 for the series in u_2, v_1, v_2, x and y obtained from Φ by making the substitution $u_1=1$. Thus the coefficient of

$$u_2^n v_1^i v_2^j$$

in Φ_1 is the sum of the dichromatic polynomials of all combinatorially distinct rooted planar maps M for which $n(M)=n$, $i(M)=i$ and $j(M)=j$. For such maps only a finite number of values of $m(M)$ are possible. For $m(M)$ cannot exceed twice the number of edges of M, and the number of edges is $i(M)+j(M)$ by the Euler polyhedron formula. Because of this fact the series Φ_1 is well defined.

Similarly there is a series Φ_2 in u_1, v_1, v_2, x and y obtained from Φ by making the substitution $u_2=1$. The coefficient of

$$u_1^m v_1^i v_2^j$$

in it is the sum of the dichromatic polynomials of all combinatorially distinct rooted planar maps M for which $m(M)=m$, $i(M)=i$ and $j(M)=j$.

Let W denote the constant term of Φ, X the sum of the terms corresponding to the maps having isthmuses as roots, Y the sum of the terms corresponding to maps having loops as roots, and Z the sum of the remaining terms.

The only map contributing to W is the vertex-map. Hence

$$W = 1.$$

Consider a map M whose root A is an isthmus. Let the root-vertex and root-face be s and F respectively, and let the second vertex incident with A be t. Such a map is shown in Figure 1. We suppose it drawn so that the terms "left" and "right" have their usual meaning with respect to the directed arc A.

When A is deleted, except for its end points s and t, the graph G of M is transformed into a disconnected graph with just two components G_s and G_t. These include the vertices s and t respectively. The graphs G_s and G_t determine planar maps M_s and M_t respectively. Either or both of these may be vertex-maps.

If M_t is not a vertex-map we root it as follows. Its root B is the edge of M that is next to A at t and on the left of A. It is directed from t. If it is a loop we may imagine it subdivided by a point p into two half-edges. It is then directed from the half-edge next on the left to A at t, and into the other half-edge. If B is not an isthmus we take its left side in M_t to be that on which F lies. If B is an isthmus its left side is taken to include the angle α of F (in M) at t lying immediately to the left of A.

The rooting of M_s, if this is not a vertex-map, is defined similarly, but we first reverse the direction of A and interchange its right and left sides. We must then use s in place of t. In Figure 1 the roots of M_t and M_s are the directed edges B and C. They are crossed by arrows directed from left to right sides.

We note the following relations.

$$\begin{aligned} m(M) &= m(M_s)+m(M_t)+2, \\ n(M) &= n(M_s)+1, \\ i(M) &= i(M_s)+i(M_t), \\ j(M) &= j(M_s)+j(M_t)+1. \end{aligned}$$

By Theorem 1 we have also

$$\chi(M; x, y) = x\chi(M_s; x, y)\chi(M_t; x, y).$$

We note also that we can reverse the decomposition and obtain a unique rooted map M having any desired rooted maps as M_s and M_t. We may therefore deduce from the foregoing results that

$$X = u_1^2 u_2 v_2 x \Phi \Phi_2.$$

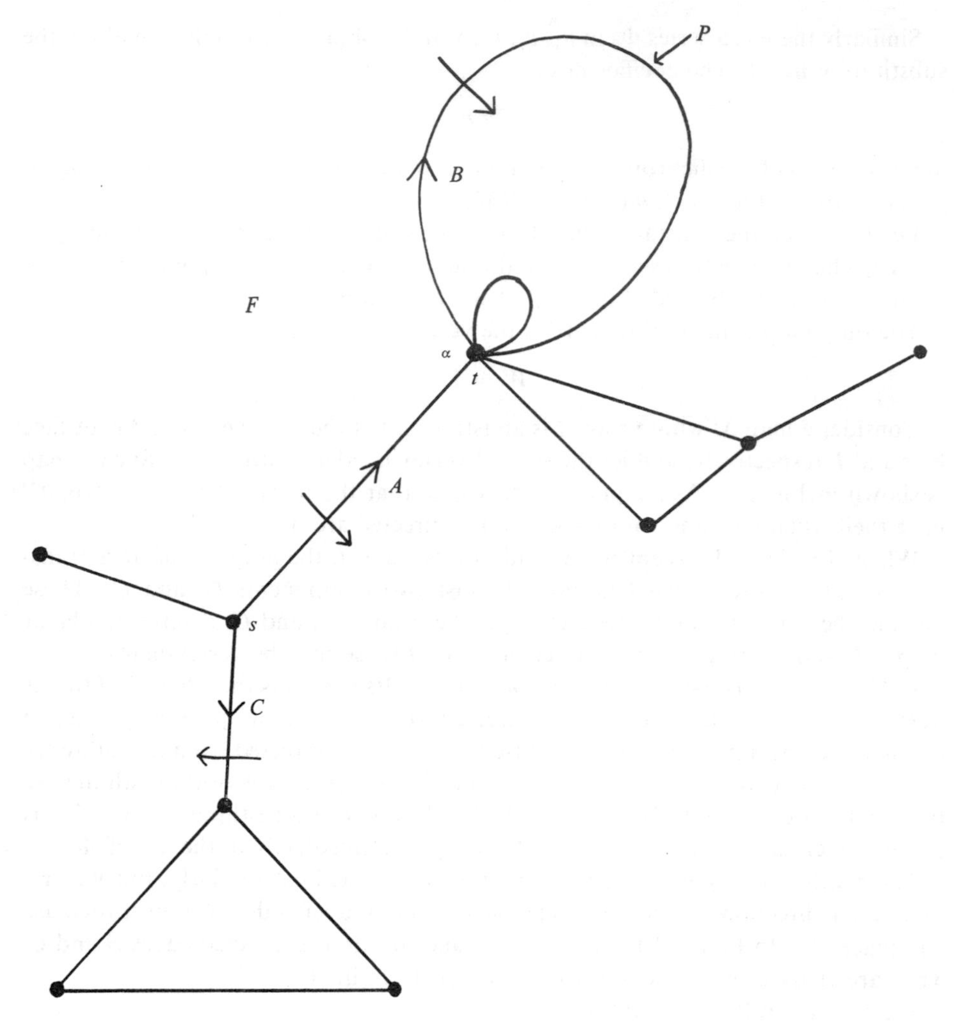

FIGURE 1

The factor Φ in this formula takes account of all the possible maps M_s. The factor Φ_2 deals with the maps M_t, for which the value of $n(M_t)$ need not be given explicitly.

The rooted maps having loops as roots are the duals of those having isthmuses as roots. We may therefore write

$$Y = u_1 u_2^2 v_1 y \Phi \Phi_1.$$

We next investigate the sum Z. Consider any rooted planar map M with a root A that is neither a loop nor an isthmus. We derive a rooted map M'_A from it by deleting the edge A and rooting the residual map as follows. Let the negative end of A in M be s and the positive end t. Let the root-face of M be F and let the other face of M

incident with A be K. Such a face K exists since A is not an isthmus. Consider the angle α of K at s and immediately to the right of A (see Figure 2). We denote the other edge bounding α by B. We take B as the root of M'_A. We direct it from s and take it to have the face F' of M'_A containing F and K on its left. If B is a loop we subdivide it into two half-edges and direct it from the one bounding α to the other. If B is an isthmus we take it to have α on its left.

Let us write $p(M)$ for the valency of K in M. We observe that

$$m(M'_A) = m(M)+p(M)-2,$$
$$u(M'_A) = n(M)-1,$$
$$i(M_A) = i(M)-1.$$
$$j(M'_A) = j(M).$$

Let Q be the class of all rooted planar maps having root-edges that are neither loops nor isthmuses. Using Theorem 2 we can write

$$Z = Z_1+Z_2,$$

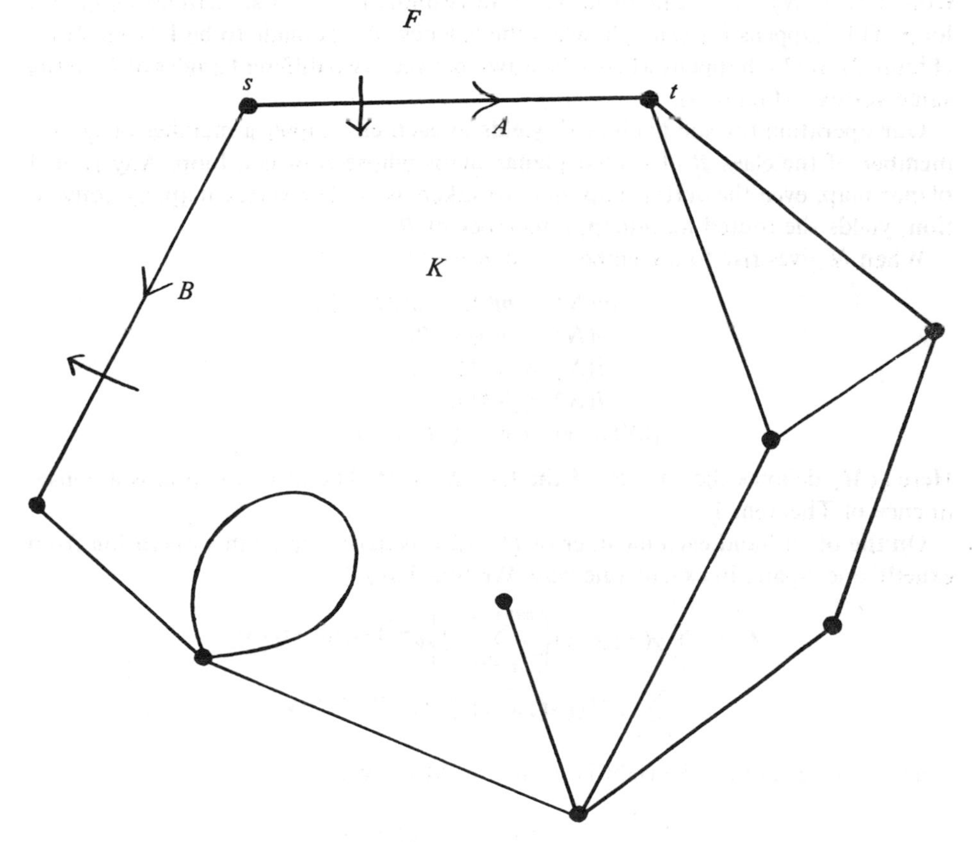

FIGURE 2

where

$$Z_1 = \sum_{M \in Q} \chi(M'_A; x, y) u_1^{m(M)} u_2^{n(M)} v_2^{i(M)} v_2^{j(M)},$$

$$Z_2 = \sum_{M \in Q} \chi(M''_A; x, y) u_1^{m(M)} u_2^{n(M)} v_1^{i(M)} v_2^{j(M)}.$$

To express Z_1 in terms of Φ we consider a general rooted planar map N with root B, root-vertex s and root-face F'. We ask in how many ways can N appear as M'_A in the sum Z_1. For each such way the corresponding map M is formed from N by drawing a new edge A across the face F' from s so as to subdivide F' into two new faces F and K, as in Figure 2. We take A to be the new root. It must be directed from s, and the part of its first half-edge near s must lie in the angle θ of F' immediately to the left of B in N. We take into consideration the case in which A is a loop, though this does not give rise to a member of Q. We adjust the notation so that the angle ϕ of K bounded by the first half-edge of A lies between that half-edge and B in θ. We take F as the root-face of M, so that ϕ is on the right of A in M.

The construction can be carried out in $m(N)+1$ ways, one for each valency of F from 1 to $m(N)+1$. It cannot make A an isthmus but does sometimes make it a loop. This happens for example when the valency of F is made to be 1 or $m(N)+1$ (Figure 3). It also happens when A is drawn between two different angles of F' at the same vertex s (Figure 4).

Our operation for subdividing F' yields in each case either a member of Q or a member of the class R of rooted planar maps whose root is a loop. Any rooted planar map, even the vertex-map, may be taken as N. The vertex-map, by convention, yields the rooted loop-map, a member of R.

When N gives rise to a member M of R we have

$$\begin{aligned} m(N) &= m(M)+q(M)-2, \\ n(N) &= n(M)-2, \\ i(N) &= i(M)-1, \\ j(N) &= j(M), \\ \chi(N; x, y) &= y^{-1}\chi(M; x, y). \end{aligned}$$

Here $q(M)$ denotes the valency of the face K of M. The fifth equation is a consequence of Theorem 1.

On the other hand each member of Q and R is derivable by this operation from exactly one N, and in exactly one way. We thus have

$$Z_1 = \sum_N \chi(N; x, y) \left\{ \sum_{p=1}^{m(N)+1} u_1^p \right\} u_2^{n(N)+1} v_1^{i(N)+1} v_2^{j(N)}$$
$$- \sum_{M \in R} y^{-1} \chi(M; x, y) u_1^{m(M)} u_2^{n(M)-1} v_1^{i(M)} v_2^{j(M)}.$$

The second sum on the right is $y^{-1}u_2^{-1}Y$. Moreover

$$\sum_{p=1}^{m(N)+1} u_1^p = u_1(u_1-1)^{-1}\{u_1^{m(N)+1}-1\}.$$

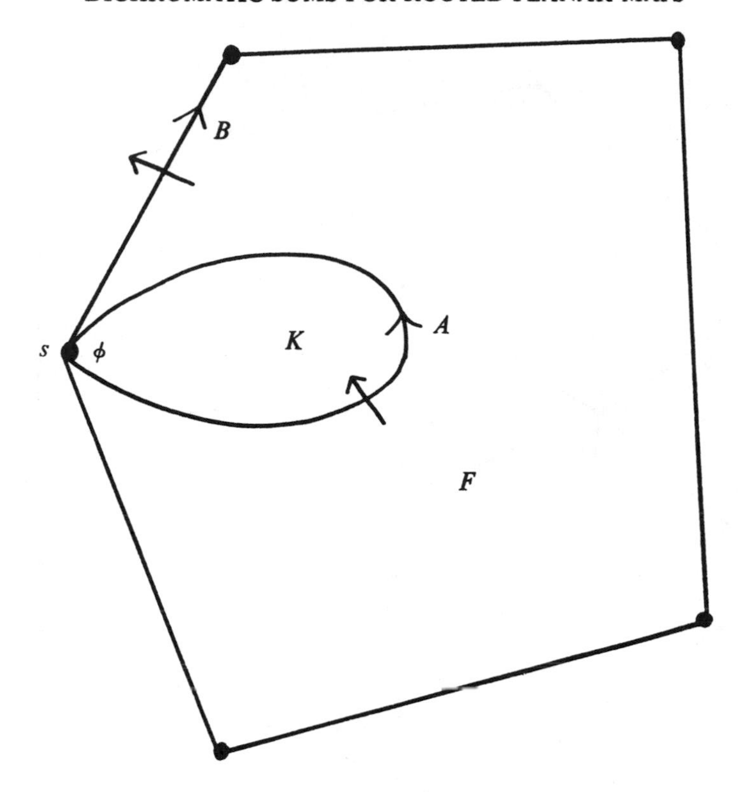

FIGURE 3

Thus

$$Z_1 = u_1u_2v_1\left\{\frac{u_1\Phi-\Phi_1}{u_1-1}\right\}-u_1u_2v_1\Phi\Phi_1.$$

By duality, and with the help of (3), we have also

$$Z_2 = u_1u_2v_2\left\{\frac{u_2\Phi-\Phi_2}{u_2-1}\right\}-u_1u_2v_2\Phi\Phi_2.$$

But $\Phi = W+X+Y+Z_1+Z_2$. Hence

$$\begin{aligned}\Phi = {} & 1+u_1u_2v_2(u_1x-1)\Phi\Phi_2 \\ & +u_1u_2v_1(u_2y-1)\Phi\Phi_1 \\ & +u_1u_2v_1\left\{\frac{u_1\Phi-\Phi_1}{u_1-1}\right\} \\ & +u_1u_2v_2\left\{\frac{u_2\Phi-\Phi_2}{u_2-1}\right\}.\end{aligned} \tag{5}$$

This is our functional equation for Φ. If Φ is known as far as the terms of degree $<r$ in v_1 and v_2 we can derive the terms of degree r in these two variables by substitution in the right-hand side of (5). We can start with $r=1$, for the only term of

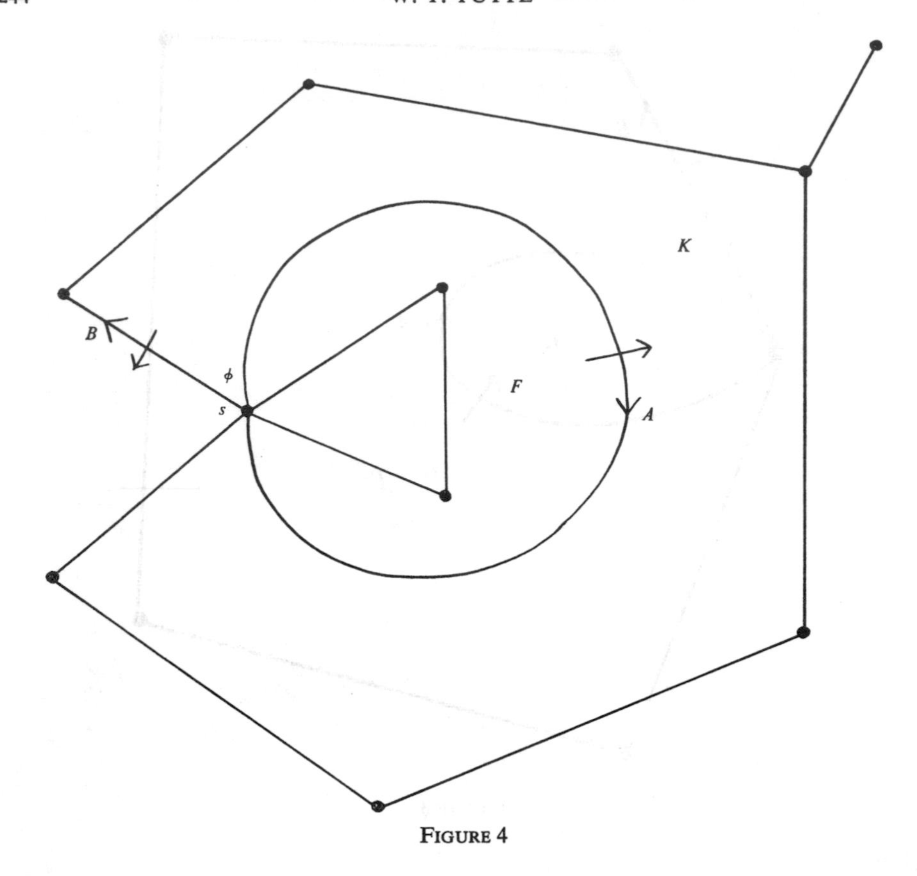

FIGURE 4

degree 0 in v_1 and v_2 is the constant term 1. Thus in principle it is possible to calculate each of the polynomials $N(m, n, i, j)$ from (5). By the same argument (5) has a unique solution for Φ as a power series in u_1, u_2, v_1, v_2, x and y without negative indices. (We know that one such solution exists.)

As a simple special case let us consider the sum Ψ of those terms of Φ in which the index of v_1 is zero. Thus the coefficient of $u_1^m u_2^n v_2^j$ in Ψ is $N(m, n, 0, j)$.

The terms of Ψ correspond to rooted planar maps with no face other than the root-face. These are the "rooted plane trees", the maps whose graphs are trees. If T is such a rooted map we have

$$m(T) = 2j(T).$$

Thus

$$N(m, n, 0, j) = 0 \quad \text{if } m \neq 2j.$$

Putting $v_1 = 0$ in (5) we obtain the following equation for Ψ.

$$\Psi = 1 + u_1 u_2 v_2 (u_1 x - 1) \Psi \Psi_2 + u_1 u_2 v_2 \left\{ \frac{u_2 \Psi - \Psi_2}{u_2 - 1} \right\},$$

where Ψ_2 is obtained from Ψ by means of the substitution $u_2=1$. As with Φ in (5) we find that the equation has a unique solution for Ψ, and it can be verified by substitution that this solution is

$$\Psi = (1-u_1^2u_2v_2x\gamma(u_1^2v_2x))^{-1}, \tag{6}$$

where

$$\gamma(z) = (1-(1-4z)^{1/2})/2z.$$

Now $\gamma(z)$ satisfies

$$\gamma(z) = 1+z\gamma^2(z).$$

Applying Lagrange's Theorem [**8**, pp. 132–133] to this we obtain

$$\gamma^k(z) = \sum_{r=0}^{\infty}\left\{\frac{(2r+k-1)!\,kz^r}{r!\,(r+k)!}\right\},$$

for each positive integer k. By (6) the coefficient of u_2^n in Ψ is

$$u_1^{2n}v_2^nx^n\gamma^n(u_1^2v_2x) = u_1^{2n}v_2^nx^n\sum_{r=0}^{\infty}\left\{\frac{(2r+n-1)!\,nu_1^{2r}v_2^rx^r}{r!\,(r+n)!}\right\}.$$

To find $N(2j, n, 0, j)$ we must pick out the term in this sum corresponding to $r=j-n$. Thus

$$N(2j, n, 0, j) = \frac{(2j-n-1)!\,nx^j}{j!\,(j-n)!}. \tag{7}$$

References

1. R. L. Brooks, C. A. B. Smith, A. H. Stone and W. T. Tutte, *The dissection of rectangles into squares*, Duke Math. J. **7** (1940), 312–340. MR **2**, 153.

2. R. C. Mullin, *On the enumeration of tree-rooted maps*, Canad. J. Math. **19** (1967), 174–183. MR **34** #5708.

3. W. T. Tutte, *A census of planar maps*, Canad. J. Math. **15** (1963), 249–271. MR **26** #4343.

4. ———, *A contribution to the theory of chromatic polynomials*, Canad. J. Math. **6** (1953), 80–91. MR **15**, 814.

5. ———, *A ring in graph theory*, Proc. Cambridge Philos. Soc. **43** (1947), 26–40. MR **8**, 284.

6. ———, *On dichromatic polynomials*, J. Combinatorial Theory **2** (1967), 301–320. MR **36** #6320.

7. ———, *On the enumeration of planar maps*, Bull. Amer. Math. Soc. **74** (1968), 64–74. MR **36** #1363.

8. E. T. Whittaker and G. N. Watson, *A course of modern analysis*, Cambridge Univ. Press, New York, 1940.

University of Waterloo

AUTHOR INDEX

Roman numbers refer to pages on which a reference is made to an author or a work of an author.

Italic numbers refer to pages on which a complete reference to a work by an author is given.

Boldface numbers indicate the first page of the articles in the book.

SUBJECT INDEX